**For
Penny and Jack
Lai-Yet, Jessica, and Michael**

stitute of Tech

Lotus Notes
and Domino
Network Design

Other McGraw-Hill Books of Interest

GILLMOR • *Lotus Notes Domino Toolkit*

THOMAS & HOYT • *Lotus Notes and Domino 4.5 Architecture, Administration, and Security*

THOMAS & PEASLEY • *Lotus Notes Certification*

YEE • *Lotus Notes Fat FAQ*

Lotus Notes and Domino Network Design

John P. Lamb
Peter W. Lew

McGraw-Hill

New York San Francisco Washington, D.C. Auckland Bogotá
Caracas Lisbon London Madrid Mexico City Milan
Montreal New Delhi San Juan Singapore
Sydney Tokyo Toronto

Library of Congress Cataloging-in-Publication Data

Lamb, John P.
      Lotus Notes and Domino network design / John P. Lamb, Peter W.
      Lew.
            p.   cm.
      Includes bibliographical references and index.
      ISBN 0-07-913241-3 (acid-free paper)
      1. Lotus Notes.   2. Groupware (Computer software)   3. Lotus
Domino.   4. Web servers—Software.   5. Computer network.
architecture. 6. Business—Computer programs.   I. Lew, Peter W.
II. Title.
HF5548.4.L692L357   1997
004.6'16—dc21   97-30082
                                                                         CIP

McGraw-Hill

A Division of The McGraw·Hill Companies

1 2 3 4 5 6 7 8 9 0   FGR/FGR   9 0 2 1 0 9 8 7

P/N 036727-2
Part of 0-07-913241-3

The sponsoring editor for this book was John Wyzalek, and the production
supervisor was Pamela Pelton. It was set in Vendome ICG by Wanda Ditch
through the services of Barry E. Brown (Broker—Editing, Design and Production).

Printed and bound by Quebecor/Fairfield.

McGraw-Hill books are available at special quantity discounts to use as premiums
and sales promotions, or for use in corporate training programs. For more
information, please write to the Director of Special Sales, McGraw-Hill, 11 West
19th Street, New York, NY 10011. Or contact your local bookstore.

 This book is printed on recycled, acid-free paper containing a minimum
of 50% total recycled fiber with 10% postconsumer de-inked fiber.

CONTENTS

Contents

Contents

xiv

Contents

TRADEMARKS

ADSTAR is a registered trademark of the International Business Machines Corporation.

Advantis is a registered trademark of the Advantis Company, jointly owned by IBM and the Sears Roebuck Corporation. *(Note:* Advantis will become part of IBM Global Services by the end of 1997.)

Advanced Peer-to-Peer Networking (APPN) is a registered trademark of the International Business Machines Corporation.

AIX is a registered trademark of the International Business Machines Corporation.

Ami Pro is a trademark of the Lotus Development Corporation.

Andrew File System (AFS) is a trademark of Transarc Corporation.

APPC is a registered trademark of the International Business Machines Corporation.

AppleTalk is a registered trademark of Apple Computer, Inc.

Adaptive Replication Engine (ARE) is a trademark of Technology Investments, Inc.

AT&T is a registered trademark of the American Telegraph and Telephone Corporation.

AT&T Network Notes is a trademark of the Lotus Development Corporation and licensed to AT&T.

Aviion is a trademark of the Data General Corporation.

British Telecom is a registered trademark of the British Telecom, Ltd.

Callup is a trademark of the International Business Machines Corporation.

cc:Mail is a registered trademark of cc:Mail Inc.

CERFnet is a trademark of the California Education and Research Federation.

CompuServe is a trademark of CompuServe Incorporated.

DB2 is a registered trademark of the International Business Machines Corporation.

DB2/2 is a trademark of the International Business Machines Corporation.

DCE is a trademark of the Open Software Foundation, Inc.

Delphi is a trademark of the Delphi Internet Service Corporation.

Domino is a trademark of the Lotus Development Corporation.

ES/9000 is a trademark of the International Business Machines Corporation.

Ethernet is a trademark of the Xerox Corporation.

Freelance Graphics is a trademark of the Lotus Development Corporation.

GE Information Services (GEIS) is a trademark of the General Electric Corporation.

HP is a registered trademark of the Hewlett-Packard Corporation.

HP-UX is a trademark of the Hewlett-Packard Corporation.

IBM is a registered trademark of the International Business Machines Corporation.

IBM Global Network is a trademark of the International Business Machines Corporation.

Intel is a registered trademark of the Intel Corporation.

InterfloX is a registered trademark of the International Business Machines Corporation.

Interliant is a trademark of Interliant, Inc. (formerly known as World-Com)

InterNotes is a trademark of the Lotus Development Corporation.

Iris is a trademark of Iris Associates, a wholly owned subsidiary of the Lotus Development Corporation.

ISSC is a registered trademark of the International Business Machines Corporation.

Java is a trademark of Sun Microsystems, Inc.

LAN Distance is a trademark of the International Business Machines Corporation.

LAN Hop is a trademark of the International Business Machines Corporation.

Lotus is a registered trademark of the Lotus Development Corporation.

Lotus Notes is a registered trademark of the Lotus Development Corporation.

LSX is a trademark of the Oracle Corporation.

Microsoft is a registered trademark of the Microsoft Corporation.

Microsoft Exchange is a registered trademark of the Microsoft Corporation.

Motif is a trademark of the Open Software Foundation, Inc.

NETCOM is a trademark of NETCOM On-Line Communications Services, Inc.

NetFinity is a trademark of the International Business Machines Corporation.

Netscape is a trademark of Netscape, Inc.

NetView is a registered trademark of the International Business Machines Corporation.

NetWare is a registered trademark of the Novell Corporation.

Notes is a trademark of the Lotus Development Corporation.

Notes for UNIX is a trademark of the Lotus Development Corporation.

NotesPump is a trademark of the Lotus Development Corporation.

NotesSQL is a trademark of the Lotus Development Corporation.

Novell is a registered trademark of the Novell Corporation.

Novell Groupwise is a registered trademark of the Novell Corporation.

NTT is a registered trademark of the Nippon Telegraph & Telephone Company.

Office Vision is a registered trademark of the International Business Machines Corporation.

Office Vision/VM is a trademark of the International Business Machines Corporation.

OS/2 is a registered trademark of the International Business Machines Corporation.

OSF is a trademark of the Open Software Foundation, Inc.

Pentium is a registered trademark of the Intel Corporation.

Person to Person is a trademark of the International Business Machines Corporation.

Personal System/2 is a registered trademark of the International Business Machines Corporation.

PowerPC is a trademark of the International Business Machines Corporation.

PROFS is a registered trademark of the International Business Machines Corporation.

PS/2 is a registered trademark of the International Business Machines Corporation.

PSINet is a registered trademark of Performance Systems International, Inc.

REXX is a trademark of the International Business Machines Corporation.

RISC System/6000 is a registered trademark of the International Business Machines Corporation.

RPC is a trademark of Sun Microsystems, Inc.

RS/6000 is a trademark of the International Business Machines Corporation.

SAP is a trademark of SAP, AG.

SCO is a trademark of the Santa Cruz Corporation.

Sears is a registered trademark of the Sears Roebuck Corporation.

SNA is a registered trademark of the International Business Machines Corporation.

Smarticons is a registered trademark of the Lotus Development Corporation.

Sniffer is a registered trademark of the Network General Corporation.

Solaris is a trademark of SUN Microsystems, Inc.

SOM is a trademark of the International Business Machines Corporation.

SP2 is a registered trademark of the International Business Machines Corporation.

SPARC is a registered trademark of SPARC International, Inc.

SUN is a trademark of Sun Microsystems, Inc.

SUN Solaris is a trademark of Sun Microsystems, Inc.

SupportPacs is a trademark of the International Business Machines Corporation.

SURAnet (Southeastern Universities Research Association Network) is a service mark of BBN Planet.

System/370 is a registered trademark of the International Business Machines Corporation.

ThinkPad is a registered trademark of the International Business Machines Corporation.

Time and Place is a trademark of the International Business Machines Corporation.

UNIX is a registered trademark of Unix System Laboratories, Inc.

Visual Basic is a trademark of the Microsoft Corporation.

VM/ESA is a registered trademark of the International Business Machines Corporation.

VMS is a trademark of the Digital Equipment Corporation.

WebExplorer is a trademark of the International Business Machines Corporation.

Windows is a trademark of the Microsoft Corporation.

Windows NT is a trademark of the Microsoft Corporation.

Windows 95 is a trademark of the Microsoft Corporation.

WIN-OS/2 is a trademark of the International Business Machines Corporation.

WordPro is a trademark of the Lotus Development Corporation.

Workplace Shell is a registered trademark of the International Business Machines Corporation.

X/Open is a trademark of X/Open Company, Ltd.

FOREWORD

Although host computers have been communicating with each other for more than 30 years to share data between Information Technology (IT) professionals, it has only been in the last 10 years that Local Area Networks (LANs) of personal computers at enterprise sites have become widespread for sharing data between business professionals. In the last 5 years, a subset of that trend that was pioneered by Lotus and called *groupware*—the real-time sharing of information across physical and virtual networks among groups of users—has emerged as a significant productivity tool and versatile application development platform.

In the next 10 years, we can not only expect that Wide Area Networks (connections between enterprises and their customers or suppliers) will become more common as the cost and availability of communications technology improves, but that network-centric computing will become the dominant model as the telecommunications, information technology, and media/entertainment industries converge. The collaboration feature of Lotus Notes is frequently the application that drives companies to extend their LANs beyond a single site and outside the enterprise to create virtual teams.

These trends have been dramatically accelerated over the last 3 years by the advent of the World Wide Web, the multimedia segment of the Internet. Point-and-click user interfaces providing access to hyperlinked applications and databases have brought networking into the popular vocabulary as millions of individuals "surf the 'Net" and thousands of enterprises and institutions create electronic addresses via home pages and begin to conduct business online. Already, we are seeing the dawn of a new method of carrying out many daily activities—from banking and shopping to medicine and education—in the form of electronic commerce over private and public networks, including intranets, extranets, and the Internet.

Lotus Notes can be a powerful lever for companies and institutions to extract more value from the Internet. Using Lotus Notes and Domino technology, users can convert and exchange information between Web sites and Notes databases, easily manage connections to other Web sites and extend powerful Notes search, sort, filter, redistribution, and broadcast functions to the Web.

With its replication functions providing remote users with current versions of documents, up-to-the-minute e-mail, or Internet access just a phone call away, Notes and Domino can maximize productivity for the growing population of mobile workers. Thanks to Notes' multilevel security features—such as cryptography, encryption, and digital signature—users can work in confidence both onsite and on the road.

The emergence of global networking nevertheless brings with it technically challenging tasks, especially when it involves multiprotocol connections and hardware or software from a variety of vendors. This book collects in one place an extensive range of vital information to provide a common foundation of knowledge for CIOs and their staffs who design, implement, or administer Notes and Domino, including such critical issues as:

- Hub-and-spoke network design for efficient bandwidth utilization.
- Corporate naming standards to assure interoperability.
- Backup and restore strategies to provide high availability.

Topics cover design methods to improve such areas as replication, mail routing, dial access, connection to outside domains, and bandwidth management. Methods for leveraging the Internet for many of the capabilities of Notes and Domino are highlighted throughout this book.

With information applicable to Notes Release 4 and 5, subjects also include links to the mainframe, tips for Windows NT, uses of Java technology with Notes, and comparisons with competitive products and the Web's cache methods. In addition, there are chapters on network and application management, useful network tools for Notes remote administration, and examples of enterprise implementations of Notes. The authors draw on experiences from IBM's own creation and use of Notes and Domino applications and functions.

To derive the most value from a Notes system in an intra-enterprise or inter-enterprise environment today, the authors recommend careful consideration of these factors:

- Robust network design, which allows high-performance replication of Notes databases, using TCP/IP protocols for effective use of Notes over Wide Area Networks.
- Consistent gateway design, which allows easy routing of multiple E-mail systems.
- Pervasive dial access for mobile users anywhere in the world, which can most easily be obtained through a network service provider.
- Strong network security features, especially when Notes interfaces with the Internet and other intranets or extranets via dial connections.
- Additional Notes and Domino services to fully leverage Notes core capabilities, such as fax gateways, organization-wide database repository, and phone and video options.

For the future, the rapid evolution of network technology and capability will enhance the value of successive versions of Lotus Notes and Domino. The spread of Internet Protocol networks, provided by IBM Global Services and others, will make electronic services widely available on both the public Internet and on private intranets and extranets. Finally, the emergence of intelligent software agents (which rapidly locate, filter, and route information geared to an individual's needs) and data mining techniques (which enable firms to exploit traffic data and survey information to customize services) will allow users and providers to fully exploit the power of digital technology.

IBM and Lotus are fully engaged in all of these activities and constantly developing new services that will make network-centric computing easy, accessible, and affordable for individuals, groups, and enterprises to benefit from the rapidly evolving electronic marketplace where time and distance are no longer barriers to choice and value.

Richard B. Anderson
General Manager
IBM Global Services-Network Services

PREFACE

This book is a substantial update to its predecessor, *Lotus Notes Network Design*, which was published by McGraw-Hill in May of 1996 and mostly referenced Release 3 of the Lotus Notes product. This book, as the name implies, covers both Notes and the Lotus Domino server. Release 4 and Release 5 of Notes and Domino are discussed in detail.

With Release 4.5 of the code, the Lotus Notes Server became known as the Lotus Domino 4.5 Server to emphasize the significant Web technology that was included with the server code. The client software kept its name of Lotus Notes Client 4.5. The Domino 5.0 servers and Notes 5.0 clients, which will not be formally available until late in 1997, will have full support of Java in Lotus APIs and Java Beans and, in general, fully embrace Web technology. Because of the rapid rate of enhancements to Lotus Notes and "add on" products that relate to Notes, over one-half of this book consists of new material added to the predecessor book. This makes the book far more than a second edition or a simple revision. All of the material carried over from the predecessor book has received extensive editing, and many changes were made based on the latest release changes to Notes and the latest changes to the case studies, such as the IBM roll-out of Lotus Notes and Domino.

Lotus Notes is a wonderful tool. It's quite easy to use, yet it has very extensive capabilities, and it's getting better. Both Lotus Notes Release 4 and 5 have many improvements. Some of these improvements are geared for large Notes installations. Lotus Notes was originally designed to be used on a Local Area Network (LAN), with most users confined to a single site. When large companies started using Notes across the corporation with thousands of users accessing the same information, the basic Notes design was sometimes "stretched" beyond its capabilities. There were questions about how "scalable" Notes was for the large enterprise. Many of the improvements in Release 4 and 5 address the issue of Notes being scalable for the large enterprise. However, there are many design issues for the enterprise beyond the functions included in Release 4 or 5 of the product. These design issues are discussed in this book and are relevant to all versions of Lotus Notes and Domino.

The areas covered in this book are physical and logical network design, Notes domain design, replication strategies, robust mail routing, mail gateway implementation, dial-access options, Lotus Notes security concepts, incorporating the Internet with Notes, X.500 directory concepts, examples of corporate standards for network topology and naming, network and application management, and migration to Lotus Notes from existing mail systems. In addition, there is a chapter on useful network tools for Lotus Notes that covers remote administration and NotesView. Other chapters discuss Notes services over the network and give examples of the enterprise-wide use of Notes. The final chapter discusses the future of Lotus Notes, which includes the evolution from the client/server model to the "Network Computing" model.

Currently, Lotus Notes is implemented by most organizations using the client/server model. An end-user (the client) shouldn't have to know (or care) on which servers his data is located. However, to make this client/server concept work well, the Lotus Notes network design must be carefully considered. That's where you come in.

This book was written primarily for Notes Administrators and Notes Network Designers. It should also be of interest to Chief Information Officers (CIOs) and their staffs, Notes Application Developers, and Notes end-users who want to know how it all works.

The book assumes that readers have some familiarity with computer networks but vary in their understanding of Notes. Therefore the book starts with an overview of Notes, concentrating on its network features and requirements. Also, the book draws extensively on IBM's experiences with Lotus Notes; therefore, an ingredient of most chapters is "how IBM does it" and "how you can do it".

What has made Lotus Notes such a popular product? After all, in a sense, it's just a fancy bulletin board system. The secret to this groupware product is that it makes it very easy for everyone on a project to read documents from a common server and to share ideas and comments using text and graphics. In this way, it's like an electronic discussion or a virtual conference room. Notes is very good at organizing all the documents under its control.

Another reason for the popularity of Notes is its remote access facility. Once you've tried the workstation-based mail feature of Notes, you'll be addicted. You start off by replicating your encrypted mail and important proposal database to the hard-disk on your laptop. Then you spend the next few hours on a train or airplane working on your mail and the proposal database. Then, when you arrive at your destination, you dial your Notes network, and magically you've sent the mail you worked on, you've received your new mail, you've approved network changes that get automatically routed to Network Operations as part of a workflow application, and you've synched up your work on the proposal database with the work done by your colleagues who were perhaps in different parts of the globe. These examples demonstrate the strengths of Lotus Notes that make it so popular. These strengths include replication, security, and workflow.

Notes has always been easy to work with once your desktop has been set up with the database icons you need. However, if you didn't know the names of the databases you needed, "navigating" to the right place could be a problem. That was made easier under Release 4 with the new window structure. Release 5 contains further improvements.

These releases of Lotus Notes contains many improvements that affect all of us. For example, Notes Release 4 and 5 allow databases to be opened quicker, and replication of databases is more efficient, which helps assure that we're looking at the latest information. Also, for those of us who use dial access for Notes, we now have direct dial access using the TCP/IP protocol, and, using XPC or TCP/IP for direct dial, we can access more than just our home Notes server. A summary of Release 4 and 5 features are listed in appendix B of this book.

As you probably know, it's easy to design new application databases in Lotus Notes, especially if you start with the "example templates" provided with the Notes product. More complicated Notes applications can be left for the Notes Application Developers to write. One of the strengths of Notes is the concept of *views* and *forms*. Views allow the Notes user to present or sort the data in the way he or she wants. Forms are for the Notes Application developer to worry about. Data is entered into a Notes database via forms.

Lotus Notes is *groupware* and is designed to allow groups to share information. For this concept to work best, your group should have a culture that shares readily and gives people credit for their intellectual contributions. This sharing concept is basic to the "teamwork" ideas that most companies have been fostering over the past few years. The use of groupware such as Lotus Notes will help enhance the teamwork concept at your company. Groupware will help to transform a hierarchical organization into more of a flat, information-sharing enterprise.

Popular books—such as Dr. Steven Covey's *The Seven Habits of Highly Effective People*—relate to this concept of information sharing. For example, the seven habits listed in Covey's book can be summarized and the habits relating to Lotus Notes listed here:

- Individual goals and accomplishments
  1. *Be Proactive*—Decide to help make things happen.
  2. *Start with "the end in mind"*—Envision the project completely.
  3. *"First things first"*—Prioritize tasks to complete the project.

- Teamwork (e.g., Lotus Notes, "working together")
  4. *Win/Win*—Don't think in terms of winner and loser.
  5. *Understand and listen to other points of view.*

6. *Synergize.*

■ Re-engineer and revamp your process. "Don't use the same old dull saw."

7. *"Sharpen the saw"*—Work on the first six habits!

Lotus Notes is a very "hot" product. There are over 15 million users of this application, and the numbers are growing rapidly. The press has reported that a Lotus Development Corporation goal is to have 20 million Notes users by the end of 1997 and 30 million users by the end of 1998. We will continue to see more and more Notes users!

With the rapid growth in the number of Notes users, we will see significant growth in the number of large scale Notes and Domino networks. One of the things this book does is share the IBM experience of developing a working, large-scale Notes network as a case study. However, sometimes the level of detail, scope of work, nuances of functionality, and level of integration with the existing environment pose an overwhelming challenge for companies looking at the design of large-scale Notes and Domino networks. That is why it is often a good idea to involve people who have actually lived through the planning and/or implementation of large-scale Notes and Domino systems. These Lotus Notes and Domino network service providers or consulting groups (such as the Lotus Consulting Group and IBM Global Services) bring the experience needed for such challenges.

In addition to network design, this book focuses on the security aspects of both the Lotus Notes and Domino application and network. Another focal point is the use of the Internet with Lotus Notes. Thus three of the most intriguing and exciting technologies of today are combined in one book:

■ Lotus Notes and Domino

■ The Internet

■ Computer/network security

ACKNOWLEDGMENTS

The ideas presented in this book are based on the Lotus Notes networks used within the IBM Corporation. Therefore, we would like to thank all those involved with the building of IBM's Notes networks and especially the Lotus Notes commercial offering set up by the IBM Global Network. In May of 1997, IBM announced that they were buying the Sears investment in the Advantis Company. When that buyout is complete (by the end of 1997), Advantis will be part of IBM Global Services, Network Services and the name Advantis will be completely phased out. However, for these acknowledgements and within the book itself, the Advantis Company is mentioned many times because it was still very much in existence when the book was written.

Special thanks goes to our management team for their support in this project. The management team consists of our direct managers Dann Schlegel and Rich Tamborski, our Vice President Eurael Bell, and our Group Executives Pete Hicks and Denise Custer. Special thanks also go to Richard Wood of IBM/Advantis, who was our most dedicated reviewer. "Woody" looked over almost all of the revisions of the manuscript with nary a complaint!

Paul Singer, Frank D'Apice, Gary Norton, Tony Cusato, David Frank, Paul Bouwmeester, Mike McGuire, Thorne Ventura, Mary Keough, Chip Carlson, and Ames Nelson of Advantis and IBM read the different versions of the manuscript and contributed many valuable comments. Also, David Carno and Peter Lantry of the Lotus Consulting Group and Barbara Mathers and Dave Newbold of the Iris Associates subsidiary of the Lotus Development Corporation provided valuable insight from the Lotus perspective. Barbara and Dave graciously allowed us to include much of their "Lotus Notes Internet Cookbook" in appendix C of this book. Peggy Bovaird and John Falkl of IBM Global Services contributed to the description of IBM's Notes architecture through their white paper on "Enterprise Deployment of Lotus Notes."

To get an international perspective, the manuscript was also reviewed by Andrew Wainwright of IBM UK, by Philippe Bondono of IBM France, Axel Tanner of IBM Research in Switzerland, Takanori Seki of IBM Japan, and Owen Price and Darryl Miles of IBM Australia.

Most of the information on Lotus Notes products was taken with permission from Lotus Development Corporation's "Lotus Notes Knowledge Base." The Lotus Notes Knowledge Base is available on the Lotus Notes Network (LNN) for Lotus Business Partners. The Knowledge Base was also the source of much of the information in chapter 11 on "Migration to Lotus Notes from Existing Mail Systems." Most of the information on IBM products came from IBM Product Information Press Releases.

We would also like to thank Pete Stair, the author of several books on networking and the Internet, who gave us very useful advice on publishing a book of this type. Fred Kauber of the DigiVentures Company, Rich Jenkins of IBM, Bill Conklin of Lotus, and Jim O'Donnell and Kim Rankin of IBM/Advantis reviewed the manuscript and contributed to the case studies in chapter 15.

Finally, we would like to thank all the people at McGraw-Hill who worked with us on this project. Special thanks go to John Wyzalek and Judy Brief, our editors, who showed great patience and gave the needed encouragement to help us complete this project in a relatively short time.

While the authors appreciate the assistance provided by IBM, Advantis, and Lotus, we would like to state that the opinions expressed in this book (and any inaccuracies within it) are those of the authors and in no way reflect the opinions of IBM, Advantis, or the Lotus Development Corporation.

ABSTRACT

Lotus Notes, the LAN-based groupware product for sharing information, has grown extremely popular over the past few years. At the end of 1996, the Web-enabled Lotus Domino server replaced the Lotus Notes server. This change is beginning to revolutionize the design of intranets.

The Lotus Notes client and Domino server are sometimes used within a single office, but more often they are employed over an entire enterprise. They are frequently the applications that drive companies to extend their LANs beyond a single site.

To be effective within an enterprise, the network design concepts for Lotus Notes and Domino are very important. This book describes the different ways to approach network design for Lotus Notes and Domino. IBM's implementation of Lotus Notes over its corporate enterprise is used as a reference throughout the book. This includes design methods to provide effective replication, mail routing, dial access, network and application management, and security for Lotus Notes and Domino over the enterprise. Methods for leveraging the Internet for many of the network requirements of Lotus Notes and Domino are emphasized throughout the book.

This book was written primarily for Notes Administrators and Notes and Domino Network Designers. It should also be of interest to Chief Information Officers (CIOs) and their staffs, Notes Application Developers, and Notes end-users who want to know how it all works. The book assumes that readers have some familiarity with computer networks but vary in their understanding of Notes.

ABOUT THE AUTHORS

JOHN LAMB has been involved in the design aspects of Lotus Notes networks for over four years. His main focus has been on the interconnection of Notes domains across IBM's worldwide internal and commercial wide area networks. Overall, he has more than 10 years of experience in network architecture and design for IBM's internal networks, including management of IBM's Network Performance Design and Analysis Department. He holds a B.A. from the University of Notre Dame, M.S. degrees from the University of Utah and Pace University, and a Ph.D. in Engineering Science from the University of California at Berkeley. John has published more than a dozen technical papers and co-authored the book "*Lotus Notes Network Design*" (published by McGraw-Hill, 1996) with Peter Lew.

PETER LEW has held a wide variety of technical and managerial positions at IBM since 1983. He is currently focused on Internet mail and security, with emphasis on Lotus Notes. His work applies to the IBM global network used by IBM internally and to commercial networks as well. He holds a B.S. from the California Institute of Technology (Caltech) and an M.S. and a Ph.D. in Engineering Science from Stanford University. Peter has published a number of papers in technical journals and conference proceedings and co-authored the book "*Lotus Notes Network Design*" (published by McGraw-Hill, 1996) with John Lamb.

CHAPTER 1

Introduction

Groupware handles three kinds of user interaction, the "three Cs":

- *Communication*—Generally means e-mail and related functions.

- *Collaboration*—Online discussion groups and common access to folders of shared data.

- *Coordination*—The most complex part, which allows workers to jointly accomplish specific procedures and tasks.

Lotus Notes combines all three aspects of groupware into one product and has been elegantly designed to do so from day one. It includes e-mail, a database, a directory of users, security, tools to program custom applications, and support for remote access from portable computers.

Fortune magazine, July 8, 1996

Lotus Notes has become the standard software product used by many companies for sharing information across the enterprise. This book draws extensively on the experiences of the IBM Corporation in the use of Lotus Notes across its enterprise. Lotus Notes mail is sent and database information shared by the IBM Corporation throughout the United States, Canada, Mexico, Japan, Australia, Ireland, the UK, Spain, France, Sweden, and many other countries.

The use of Lotus Notes over the IBM enterprise has surfaced requirements for the efficient distribution of Notes databases, effective replication and mail routing, connections to Lotus Notes domains outside of IBM, naming standards, a company-wide Lotus Notes address book, an enterprise-wide IBMNOTES Database Repository, and remote management and monitoring. The Internet has been leveraged as much as possible, both for IBM's internal use and for its Lotus Notes commercial offering.

You might want to design your own Lotus Notes network (or perhaps you already have), you might want to have an outsourcing provider do the design and perhaps even provide the Notes servers, or you might want to do some of the network yourself and outsource the wide area network part. In any case, this book will be helpful in providing you with design information for both small and large scale Notes implementation.

Designing Your Own Notes Network

Much of what this book describes is what the IBM Corporation has found in its implementation of Lotus Notes across the enterprise. IBM has over 150,000 Notes licenses, making its Notes installation one of the largest in the world. Much of what IBM has discovered in its use of Lotus Notes will help you in your design decisions. Some of these areas of importance for the effective use of Lotus Notes over the enterprise are:

- The need for robust network design for the effective replication of databases among the major site LANs in the enterprise.
- Strong network and database security, especially when the Notes network connects to the Internet and/or directly to other companies. Security and the concept of replication are the greatest strengths of the Lotus Notes product.
- Easy dial access for mobile users anywhere in the world.
- Many other Notes services, such as fax gateways, SMTP mail gateway to the Internet, dial access via the Internet, dial access to a company's LAN, calendar function, pager, and phone and video options.
- Enterprise directory services, which include one large name and address book for all users of LAN and host-based office, is a requirement for smooth migration from a mainframe-based system to Lotus Notes.
- LAN naming/addressing standards on the enterprise level are required. A centralized registration process for LAN names and addresses is an effective way to manage these standards.

All of these areas are discussed in this book. So read on and learn what we've discovered working with Notes on an enterprise level and what you can do to improve your Notes network.

Lotus Notes as Enterprise Groupware

The idea behind Lotus Notes is to make it easy to share information among a group of people, thus the term *groupware*. This could mean sharing information among the people in your immediate department. In this case, you might be all right down the hall from each other. It could also mean that you need to share information with people in the US, the UK, Sweden, Japan, and Australia. In the first case, network design for Lotus Notes might just be Local Area Network design. In the second case, network design needs to include careful consideration for the "wide area" network. In both cases, Lotus Notes is being used as *enterprise groupware.*

Figure 1.1 shows an example of the familiar Lotus Notes desktop with icons for different databases. This desktop, with its access to the Lotus Notes Name and Address Book, Lotus Notes logs, and Lotus Notes servers via Wide Area Networks (WANs), provides the way to implement many of the network design changes and to monitor much of the network usage that you might want to incorporate into your Lotus Notes network. The real work is done with the Notes Name and Address (N&A) Book and Notes logs by using the Notes views, documents, and forms associated with these databases. Implementing the changes is often quite easy. Deciding what changes should be made is the difficult part. The next few sections serve to review some of the basic types of Lotus Notes applications. These are the applications that will tend to drive your Notes network design requirements.

Figure 1.1

Example of a Lotus Notes user's desktop.

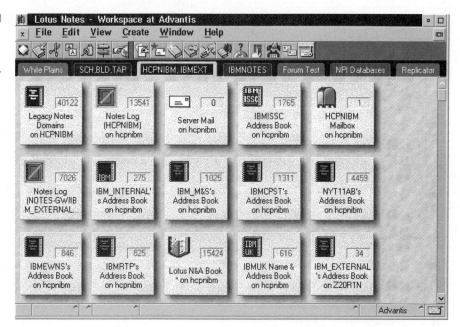

Database Repository

One way to share information over the enterprise is to have a Lotus Notes *database repository*. This repository might contain databases from the Lotus Corporation, such as the "Notes Knowledge Base" and the "Lotus Partner Forums." It could also contain Notes applications that developers in your company are willing to share with others in your company.

Discussion Databases

A *discussion database* is usually the first database established when a new Notes domain is set up. You probably have several discussion databases on your Notes servers. A discussion database is the Lotus Corporation's name for a database that others call *bulletin boards, newsgroups,* or *forums.* Discussion databases or bulletin boards all have these features in common:

- Users can, of course, read any of the information in the database.
- They can also "respond" to questions posted by other users.
- They can enter new items.

The Lotus Notes Access Control List (ACL) can be used to control which users are permitted to respond to questions, enter new questions, or even permitted to read a database. Chapter 7 on the security aspects of Lotus Notes will give you much more information on this type of database access control.

Shadowing Databases from Mainframe

If your company is migrating applications from the mainframe to Notes servers, then one way to help with the migration is to *shadow* some of the databases from the mainframe to the Notes server. Shadowing is equivalent to making a copy of a database and then keeping that copy up to date. It's the same as a one-way replication in Lotus Notes, except that, in the case of going from the mainframe database to the Notes server, database conversion is required.

METHODS. IBM routinely converts its VM Forums and Customer Forums to Notes databases. The conversion program uses InterfloX (IBM's REXX API to Notes) and a REXX program. InterfloX and REXX can be thought of as a development alternative to using the Notes C API and the C language. In chapter 14, the section "Enterprise Conferencing with VM and Lotus Notes" discusses the conversion methods in detail.

IBMNOTES DATABASE REPOSITORY EXAMPLE.   The IBM-NOTES database repository is an example of a Notes repository that is used throughout the whole IBM enterprise. IBMNOTES is accessed by IBM Lotus Notes users all over the world and contains information from many different sources, including Notes databases from the Lotus Development Corporation, IBM VM and Customer Forums downloaded from the mainframe, and shareware application packages written by Notes application developers within IBM. Additional details on this repository are given in chapter 14 in the section "IBMNOTES Database Repository."

Process Automation/Workflow

One of the most important applications for Lotus Notes is *process automation*, or *workflow*. Any workflow that is currently based on distributing paper from person to person, department to department, and so on, according to some sort of timetable or schedule, can be streamlined and automated in Notes. If this type of information is currently decentralized among many people in many different formats, Notes can be used to centralize it into a single application. A Notes application enables people to see information in its entirety, thus giving them the big picture to aid them in making decisions. Notes e-mail, where documents are automatically routed to one or more recipients (a user and/or a database), is often a part of this process. Corporate project tracking applications, network problem and change tracking systems, and Notes administration are some examples of processes and workflows that are often automated with Notes applications.

Lotus Notes Network Service Providers

There are several companies that will provide help in setting up your Notes network. In fact, most of them will do everything for you, if that's what you want. The Lotus Notes Public Network (LNPN) members are a worldwide group of telecommunication carriers and information service providers offering Lotus Domino World Wide Web hosting services and Lotus Notes outsourcing services for customers seeking faster deployment, easy administration, cost management, and value-added services. You can look at descriptions of the LNPN members at http://www.lotus.com/lnpn. This Web page contains information needed to help you choose a LNPN. Geographic coverage is one important aspect covered.

The following sections discuss some of the major players.

Interliant (formerly WorldCom)

This service offering is from a "little guy" that is a relatively new, very aggressive company. Interliant (formerly called WorldCom), which is based in Houston, Texas, has all of its Notes servers at its headquarters. It has done a good job leveraging the Internet. In addition to offering Notes access via the Internet, Interliant offers XPC access (the dial access built into the Notes product) through nationwide 800 numbers and local numbers. (This WorldCom changed its name because the long-distance carrier LDDS also chose the name WorldCom, Inc. in May 1995.)

IBM Global Network

This service offering, by the world's largest computer company, is based around the global data networks that IBM has provided for customers over the last 10 years. Pilots with several companies began in mid-1995. General availability, with service via the Internet or IBM's private IP network, began in May of 1996. The IBM Notes service interoperates with other Notes service offerings.

CompuServe

This company has been providing network service since 1980 and is one of the "pioneers" in the field. Their Notes access is offered through CompuServe's commercial network.

Nippon Telegraph & Telephone (NTT)

This service, first available in Japan in August of 1996, is part of the Lotus Notes Public Network (LNPN) services. The name of the service is

NNAS (NTT Network Application Services by Notes). NTT provides ISDN and multimedia services. NTT's stated reasons for the service were to:

- Provide a secure infrastructure to share information and applications economically and easily.
- Allow client/server messaging, global access, and distribution of the World Wide Web, together with a platform for rapidly developing and deploying strategic groupware.

US West

The US West offering for Lotus Notes is via INTERACT, the company's Internet-based computer networking service. The INTERACT offering includes complete implementation services including Lotus Notes application development, consulting, systems integration, and training. The high-speed data communications offered with INTERACT includes Frame Relay Service, Asynchronous Transfer Mode (ATM) Service, Private Line LAN Interconnect, Switched Multi-megabit Data Service (SMDS), Integrated Services Digital Network (ISDN), a variety of video related services, and related Internet services, such as Internet access, Web hosting, collaborative computing, and electronic commerce.

British Telecom (BT)

This Lotus Notes service is offered in Europe and Australia. ISDN is available for dial access. A company's global data centers can be connected to the BT offering using Frame Relay. In the UK, high data bandwidth demands can be supported by the service at multi-megabit speeds using SMDS.

Telstra: Multimedia Global Network for Lotus Notes

This is the first Australian-based Notes public network. This service, first offered in September of 1996, is aimed at Australian businesses. It provides interoperability with the other LNPNs in the US and Europe. Users access the Telstra Notes Public Network (TNPN) via X.25, ISDN, and the Internet.

The Competition: Other Groupware Products

The groupware competition for Lotus Notes is increasing. Some of the more popular competitors are described in the following sections.

Microsoft Exchange

The initial release of Microsoft Exchange positions the product between Lotus Notes and its Web-based groupware competitors. The Microsoft Exchange Client software builds on the Exchange Inbox included with Windows 95. With Exchange, everyone in an organization can access public folders in addition to their personal folders. You can assign several predefined levels of permission to users and groups, ranging from Contributor to Reviewer to several types of Author.

You can configure the Client to sort folder messages by author, date, keywords, or content type. Items can be grouped up to four levels deep by subject, importance, and other MAPI or OLE properties, and they can be filtered by the same, or different, criteria. Connecting to an Exchange Server allows the use of AutoAssistants that can process your Inbox and public folders for expected documents and then move or forward them based on rules.

The Exchange Client incorporates a client/server-enabled version of Microsoft Schedule+ 7.0 with which you can maintain appointments, tasks, events, and contacts, as well as schedule meetings, while sharing data with authorized users across the enterprise. Exchange Server supplies OLE Automation interfaces to both MAPI (i.e., OLE messaging) and Schedule+ (i.e., OLE scheduling), giving visual Basic programmers powerful tools for integration with Microsoft Office and other OLE-enabled products.

Exchange's tight integration with Windows NT as the base for the Exchange server and NT or Windows 95 as the base for the Exchange client provides both positive and negative feelings. If you have a pure Windows shop, then the tight integration is a plus. If you run an IS shop with a mixture of vendor equipment and software, then the Lotus Notes cross-platform capabilities provide a clear advantage.

Novell Groupwise

Groupwise, originally developed by WordPerfect and now under the stewardship of Novell, has evolved into a mature, rock-solid calendaring, scheduling, and e-mail platform. The product is an expanded client/server e-mail system that is the first to include complete document- and image-management capabilities as part of the Universal Mail Box concept. It seamlessly integrates calendaring/scheduling, task management, shared folders, threaded conferencing, workflow, remote access and Internet all into the Universal Mail Box.

The product is administered through NetWare Directory Services, which reduces administration costs by greatly simplifying the job of managing individual users, groups of users, network devices, and applications.

Novell has improved Groupwise to make it the way a messaging system should be. It is not merely an e-mail system, but a cooperative workflow effort of information objects among users. Groupwise allows users from different platforms—Unix, Mac, Windows 3.x, Windows NT, and Windows 95—to share messages, schedules, calendars, documents, and Web information in a click of a button.

Users can determine that documents, tasks, notes, or calendar they want to share and to what extent by setting different access levels. Users also are given the choice of providing their own workflow management rules. A team can work on a document through a check-in/check-out process (Document Management System) that can create a log entry for every step of a document's progress. This feature includes the actions taken by other outside applications, such as those in Microsoft Office or Corel WordPerfect Suite.

Built-in features allow the administrator to devote more time to planning and executing strategic moves with a higher degree of automation than in the past.

Internet fever also has hit Groupwise. Users now can extract information from the Internet or an intranet using their browser, can view meetings, can check the progress of a project, and can send e-mail. It is now far easier to find that unknown file you worked on a week ago with a colleague that had a graphic of your client. Search criteria nested to almost any level can be stored for future searches.

Despite strong advances in the Groupwise product, it still has a ways to go to catch the groupware leader, Lotus Notes. Nevertheless for Novell shops, it should be given serious consideration.

Netscape SuiteSpot

Netscape SuiteSpot is a flexible suite of integrated servers that enables businesses and workgroups to communicate and collaborate using open Internet technology. Netscape SuiteSpot provides a platform for information, applications, and collaboration. It can include Netscape Enterprise Server, Netscape Proxy Server, Netscape Catalog Server, Netscape

News Server, and Netscape Mail Server, plus the LiveWire Pro development environment.

Companies use SuiteSpot to let project teams work collaboratively on a document and discuss issues with their partners, automate business processes, and then extend these interactions to their customers. Netscape SuiteSpot provides these functions using an architecture that can scale easily from the LAN to the Internet.

The SuiteSpot server suite offers:

- Live content publishing from any desktop
- Rich, standards-based messaging
- A platform for applications based on Java and JavaScript

Integrated management services include:

- SNMP version 1 and version 2
- SSL 3.0-based security
- Web-based secure remote administration

Cross-platform and cross-database support offers:

- Unix and Windows NT on all major architectures
- Native support for Informix, Oracle, Sybase, and Illustra
- ODBC connectivity to others

The flexible licensing scheme includes:

- Any combination of servers, plus LiveWire Pro
- A license that is not restricted to a single machine

Oracle Interoffice

Oracle Corp., Digital Equipment Corp., and Digex provide the collaborative services of Oracle InterOffice—e-mail, directory services, and calendaring and scheduling—to anyone with Web access. Benefiting from www.interoffice.net, users can subscribe to and access Oracle InterOffice without having to absorb the costs of software ownership, installation, and maintenance. This new online service is provided through Digex, a leading Internet service provider running InterOffice on a single Digital Alpha Server.

Oracle has been a big proponent of Network Computing and has often taken the lead in developing technology for Network Computing. Oracle's InterOffice enables any Web user to access their own personal e-mail account and to work on workgroup documents from any Web browser, anywhere in the world. It enables even small companies to operate as "virtual corporations" by allowing business users to send and receive e-mail, schedule group appointments, work on shared documents, and access valuable information regardless of the location of the user or the information.

Consumers will find www.interoffice.net a useful collaboration tool. For example, a parent could use the e-mail functionality to provide medical records to a student studying abroad. The student could access the records via any Web browser from anywhere in the world. The family could also use InterOffice's calendaring and scheduling functionality to arrange for the student's trip home.

Leveraging Oracle's recently announced Network Computing Architecture, Oracle InterOffice, which is built on Oracle's industry-leading Oracle7 database, separates applications into a *thin-client* component while relegating the complex application logic to the application server. This enables users to easily run all applications from within a Web browser and lowers the cost of ownership due to reduced software maintenance on the client.

Product Information InterOffice, the world's first Web-based collaboration software, can scale from workgroup to enterprise, connecting users across corporate intranets and public networks. While competitors can support a few hundred simultaneous users on a single server, InterOffice can support thousands. Built from the ground up using open Internet standards and Web interfaces, Oracle InterOffice provides corporate intranets with e-mail, scheduling, workflow, and document management services that improve collaboration across the enterprise and increase user productivity.

Some Comparisons

From the previous descriptions of some of the competitors to Lotus Notes, it's clear that the competition for Notes is continuing to build. Notes, with its early lead in groupware, is still out in front of the pack and is continuing to innovate at a furious pace. That pace is "furious" because of the competition. This section offers some comparisons between Lotus Notes and two of its biggest competitors: Microsoft Exchange and groupware-based completely on Web Technology.

Notes vs. Exchange

As mentioned in the description on Microsoft Exchange earlier in this chapter, Exchange's tight integration with Windows NT as the base for the Exchange server, and NT or Windows 95 as the base for the Exchange client, provides both positive and negative aspects in a comparison with Lotus Notes. If you have a pure Windows shop, then the tight integration is a plus. If you run an IS shop with a mixture of vendor equipment and software, then the Lotus Notes cross-platform capabilities provide a clear advantage.

For further information on Exchange versus Lotus Notes, please see appendix E. This appendix is in the form of a "debate" on the merits of Microsoft Exchange and Lotus Notes. The debate is in the form of Q&A, or more correctly comments and responses.

Notes vs. "Pure" Web Technology

One appeal of groupware via Web Technology is the dream of choosing groupware products based on open Web standards. However, although scattered pieces of Web-based groupware are beginning to appear, we're a long way from widely adopted standards that allow a mix-and-match system that works.

Notes, the most widely used groupware, already does that. Yes, it used to be proprietary. No, the source code is not in the public domain; you must license the software to get at the APIs to the platform. However, Notes is interoperable: It's available on Windows, OS/2, the Mac OS, and Unix. It supports SPX, NetBIOS, AppleTalk, and Banyan Vines. Release 4 and 5 support Internet protocols, including TCP/IP, SMTP, HTTP, and Hypertext Markup Language (HTML).

The standards for groupware based on Web Technology still have to be worked out. On the other hand, Lotus Notes, with its Domino server, is continuing to embrace Web Technology as a major thrust. So Notes and Domino currently give you a way to use the powerful functions of Notes and yet include the benefits of Web Technology. With its POP3 client, Lotus is filling in the gaps between a Web browser and a full Notes mail client.

Lotus Notes as Part of "Network Computing"

The trend for use of tools like Lotus Notes will be towards the concept of Network-centric computing or more simply *Network Computing*. This is where the network becomes the computer. All you'll have to be concerned about is the "plug in the wall" for all your services.

A few years ago, most of corporate computing was on the "mainframe-centric" model. This is where a large computer, the mainframe, served all the computer needs of users who communicated with the mainframe through "dumb terminals" or PCs emulating dumb terminals. There were some important advantages to this model. Because almost all the computer smarts were centered on the mainframe, the computer professionals could concentrate on this machine and the applications it ran. Not too much could go wrong with the dumb terminals.

To gain more flexibility and effectively utilize the power of today's PCs, corporate computing has been migrating towards the client/server model. In this model, many servers take the place of the mainframe, although often the mainframe takes on a new role as one of the servers, a type of "super server." Truly "dumb" terminals don't play a role in this model. The PCs that had emulated dumb terminals now run client code that allows them to participate in the computing role.

Because the client role is more demanding than the terminal-emulation role, some of the PCs might have needed a boost in computer power (more memory, more hard disk space, and maybe even a faster processor).

In the client/server model, the PC on your desktop (the client) takes on a role of doing a significant amount of the computing and uses the servers as sources of data and, sometimes, the place where heavy computation is done (which is perhaps beyond the ability of a simple PC). Lotus Notes fits this client/server model very well. The Notes client code you install on your PC is quite sophisticated and, in fact, will allow you to do many of the server functions right on your PC before it is connected to a network. This is one of the powerful aspects of Lotus Notes. You can replicate your mail and other important databases to the hard disk on your laptop computer and do all your Lotus Notes work independently of the Notes server. Then, you can connect to the network and replicate the changes you made to your mail and other databases back to the servers. This reduces your costs by reducing the bandwidth requirements on leased lines and the connect time on dial lines.

The current client/server computing environment is now seen as shifting to a foundation of public wide-area networks, which follow the Network Computing model. Network-enabled application services (which would include Lotus Notes) are believed to represent a large business opportunity in the years to come. Because of this, you'll probably be reading a lot more about Lotus Notes Network service providers in the future. These "service providers" will include the current Lotus Notes Public Network (LNPN) providers, but the services will be greatly expanded when compared to the current LNPN offerings.

The Network Computing model also brings up the role of the *Network Computer*. A Network Computer, in its most basic concept, is a computer with no hard drive. That makes it inexpensive to produce and also makes it only useful when connected to the network. Because it has no hard drive, the client part of a client/server application (such as Lotus Notes) must be downloaded to the computer's memory before the application can be run.

This sounds like it might be a big step backward for users who have customized their PCs over the years. However, there are benefits to this model from IT Management's viewpoint. Whenever an application, such as Lotus Notes, is updated on the server, all of the Network Computers will immediately start using that new release of client software. Costs are therefore reduced for workstation software upgrades. Problem determination is also easier. Because all Network Computers use the Lotus Notes client software from the server, the help desk doesn't have to be concerned about Notes databases or files that might have gotten "clobbered" on the user's workstation. However, of course, the concept of Network Computing isn't dependent on the use of Network Computers.

At the publication of this book, there was a lot in the press about the battle between the Network Computer (NC) and the Network PC (NetPC). The Network Computer was being pushed by Oracle, Sun, and IBM. The NetPC, essentially an inexpensive simpler PC, was being pushed as an alternative to the NC by Microsoft and Intel. The basic idea behind the NC and NetPC is the same: to reduce the life cycle costs of workstations for large companies. The end result is probably clear. Both NCs and NetPCs will play a significant role as workstations for corporations. Most corporations will have both Network Computers and PCs (or NetPCs).

Reading this Book

You can read the following chapters sequentially or skip to an area of immediate interest. This book has been written so that each section can be read separately. Thus the book can be read effectively from "cover to cover," or you can skip to a section of interest and find that the information stands by itself. The book makes the assumption that you have familiarity with the basics of Notes and computer networks, but every attempt is made to review the basic concepts before getting into more complex details.

Network Design Aspects for Lotus Notes

The Network is the Computer.
Sun Microsystems Advertisements for the last several years

Network design for Lotus Notes over the enterprise can be looked at from both a physical and a logical aspect. The physical design has to do with how the local area networks, wide area networks, etc., are laid out. Because the physical design is usually the design used for all of your communications network needs, it might have been completely installed before you even thought about deploying Lotus Notes.

Physical Network Design

Figure 2.1 shows an overall "generic" Lotus Notes network design for a company. The physical design is indicated by how the components, such as Notes servers and Notes clients, are physically connected. The physical connections for the Notes components include LANs, WANs, leased lines, and dial connections. Other physical network design components—such as bridges, routers, and intelligent switches—are part of the physical design but are not explicitly shown.

Figure 2.1 includes the concept of hub and spoke (e.g., the spoke domain IBMUK connecting to the hub domain IBM_INTERNAL), connection of the private network to the Internet via a firewall, dial access to the Internet, and dial access to the company's private network. Dial connections can include a wide variety of options. XPC dial, LAN dial, and Internet dial are depicted. Leased line access is shown with the private network, but could have also been shown for the Internet. Dial access via ISDN, although not explicitly depicted, will soon become one of the most popular access methods for Lotus Notes. A variety of network protocols could be used with this configuration, although TCP/IP is recommended. TCP/IP is required for the Internet and can easily be used as the protocol for LAN dial or ISDN connections.

Logical Network Design

The logical aspect of Lotus Notes network design for the enterprise is based on a hierarchical structure around the Lotus Notes domain concept. The Lotus Notes domain is a group of Notes servers with the same public Name and Address Book. Domains are used to define the scope

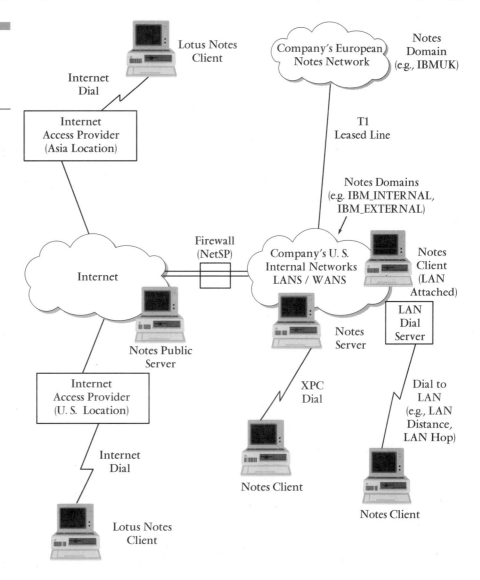

Figure 2.1
Example of Lotus Notes "physical" network design for the enterprise.

of a Notes mail environment, and they are the basis for the design of a Lotus Notes hierarchical structure.

NOTE: *This concept of hierarchical structure is independent of the use of hierarchical certifiers. Hierarchical certifiers are discussed in this chapter in the section "Using a Hierarchical Hub and Spoke Domain Structure." The term* hierarchical structure *in this book refers only to the hub-and-spoke logical design concept. Also, Notes domains are not the same as TCP/IP domains or Windows NT domains.*

At the top of the hierarchical structure are hub domains. The Notes servers in these domains act as hub servers for the spokes that go to different sites within a company. Figure 2.2 shows the basic schematic for the Lotus Notes hub-and-spoke concept. This concept is used extensively by Lotus Notes designers and is recommended by the Lotus Development Corporation for large-scale implementations, both for domains and for the servers within a domain. For example, the words "hub domain" in Figure 2.2 could be replaced by "hub server," and the words "spoke domain" replaced by "spoke server." Then, the hub-and-spoke concept can be applied to the servers in a single domain. As an IBM example, the hub domain would be IBM_INTERNAL, and the spoke domains could include IBMUK, IBMNORDIC.

The major advantages of the "hub and spoke" domain architecture are: minimized administration, replication efficiency, minimized name and address resolution, centralized routing problem determination, and centralized performance monitoring.

Minimized Administration

It is only necessary to swap one certificate (i.e., cross certify) with the hub domain instead of swapping certificates with each site with which you need to communicate. The certificate concept is basic to the security structure of Lotus Notes. Every Notes domain you set up has a Certifier that is used to stamp each User ID and Server ID that needs access to Notes resources in the domain.

A User ID or Server ID can hold many certificates. Each certificate is unique, with its own private and public key information as well as an

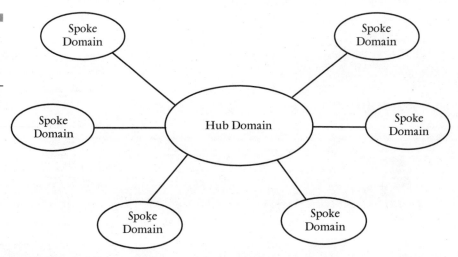

Figure 2.2
Lotus Notes
hub-and-spoke
domain architecture.

expiration date. Of course, it's best to keep the number of certificates you have on a User ID or Server ID to the minimum required. That's where a hub domain helps. In the case of IBM_INTERNAL, which has over 100 spoke domains already cross certified with it, a new spoke domain needs only to cross-certify with IBM_INTERNAL rather than 100 individual spoke domains.

Replication Efficiency

With the hub-and-spoke architecture, a single spoke server needs only to replicate once with the hub server. Then the hub server will replicate to all other spoke servers. This is in contrast to having each spoke server replicate with all other spoke servers. This frees up the spoke servers for other tasks. In the case of IBMNOTES, which has connections to over 100 spoke servers, an individual spoke server needs to replicate only once with the IBMNOTES hub server instead of replicating separately with over 100 servers.

Minimized Name and Address Resolution

With the hub architecture, addressing is simpler. It is only necessary to use the name of the person being addressed along with his domain (e.g., Tom Smith@ADVANTIS, instead of Tom Smith@ADVANTIS@XYZ@etc.). Non-adjacent domain entries pointing to the hub domain (e.g., IBM_INTERNAL) as the next domain resolve the complete address when using the hub concept.

The hub domain knows how to route to hundreds of Notes domains with either direct connections or indirect connections (using non-adjacent domain entries). Let the hub domain administrator be concerned about changes in the routing. When a routing configuration does occur, one change in the hub domain fixes the problem for all domains using the hub domain.

Centralized Routing Problem Determination

The manager of the hub domain takes care of routing problems, when they occur. It's the manager's problem to make sure alternate paths are available when a serious network problem occurs. He or she also works with the managers of the physical network when that network is the problem.

Centralized Performance Monitoring

Again, it's better to let the hub domain administrator/manager worry about whether there is enough bandwidth to handle replication requirements or mail routing. Chapter 12 discusses all of the things to be concerned about for Notes network management and monitoring.

Different Methods of Interconnecting Notes Servers and Domains

There are several methods that can be used to connect your Lotus Notes servers and/or domains. The method you choose will depend on where your servers are located (e.g., all in the same building) and what you currently use for wide area network connectivity. The following sections cover some of the options.

XPC Dial

XPC is the dial protocol that comes with every copy of the Lotus Notes software (client or server). No additional software is required for the customer to use this dial access to a Notes server. With Release 3 of the Notes software, XPC dial will only provide access to one Notes server. Releases 4 and 5 give much greater flexibility, including "XPC pass-through" capability to other Notes servers.

XPC dial service will be an option for IBM Global Network Lotus Notes customers; however, dial access that allows use of the TCP/IP protocol provides greater flexibility and will be the preferred option. TCP/IP's flexibility is based on the fact that it is a routable protocol (as opposed to XPC or NetBIOS), it's the protocol of the Internet, and it's the protocol of choice for most corporate wide area networks.

XPC only allows Lotus Notes access, while TCP/IP allows access to Lotus Notes, Web servers, mainframes, and many other servers with one dial connection. Also, both NetBIOS and IPX generate more broadcasts than TCP/IP, which gives another plus to TCP/IP as the protocol for Notes. Of course if TCP/IP isn't already installed on your Notes servers and end-user machines, this could add several hundred dollars to the cost of each platform. However, recent platform operating systems, such as OS/2 WARP Connect and Windows 95, ship with a TCP/IP stack that eliminates the cost for licenses, although the cost for support and deployment of an additional protocol must still be considered, and that could be significant.

The XPC dial protocol used by Lotus Notes in Release 3 is inherently very secure because it provides only Notes access, and the access is limited to only the two servers participating in the connection. This type of Notes dial access is being used at several IBM locations for access to outside domains (e.g., the Lotus Corporation's "Lotus" domain) and has passed all the IBM Network security requirements. Release 4 or 5 of Notes, with the passthrough function, actually is more difficult to use for this purpose.

One concern with using dial for replication and mail routing between Notes servers is the limited bandwidth available. Dial access might not provide sufficient bandwidth to complete replication of large databases during the required time interval. Also, because Notes mail can contain any amount of embedded graphics and attached databases, dial speeds might be inadequate for mail. However, as a matter of policy, sending large databases as mail attachments should be discouraged, and dial should handle most mail requirements. Large databases should be distributed via replication. For example, replication between Lotus Notes servers on the IBM Global Network is ordinarily over a LAN with 16 Mbps or a WAN with T1 (1.544 Mbps) bandwidth.

Nevertheless, XPC dial between servers provides a way to almost immediately set up a connection between any two servers no matter where they are. All you need is a modem at each server and an available analog telephone line. There is never a question of protocol incompatibility (e.g., I only use TCP/IP on my Notes server, and you only use IPX) because *all* Notes servers have the built-in XPC protocol available. In addition, dial technologies are continually improving. 56-Kbps modems and switched ISDN are becoming quite popular, and very-high-speed telephone line access such as ADSL, with speeds in the megabits per second, are starting to be deployed. Chapter 6, on dial access, gives more information on these new technologies.

Leased Lines

IBM Global Network offers leased line access to both the OpenNet (the IBM Global Network part of the Internet) and to the IBM Global Network commercial multi-protocol network, Internetworking 1.1. Access speeds for these leased lines ordinarily range from 56 Kbps to T1. With the roll-out of the IBM Global Network ATM network, bandwidth options will increase significantly. Customer Frame Relay connections to the IBM Global Network has been offered since 1995. Those connections will be up to T1.

Customer leased lines to the IBM Global Network part of the Internet will be limited to the TCP/IP protocol, but customer leased lines to the IBM Global Network commercial MPN will allow use of many protocols including TCP/IP, NetBIOS, SDLC/SNA, and IPX.

Customers will be encouraged to use TCP/IP to access IBM Global Network Lotus Notes servers over leased lines to the commercial MPN, but NetBIOS and IPX will also be supported. If a customer only has the IPX internally, for example, he could add the TCP/IP protocol to only the one server that would connect to the IBM Global Network.

Hints on Using More than One Network Protocol on a Notes Server

Lotus Notes will support several different network protocols on the same LAN adapter on a Notes server. For example, Notes servers in the IBM domains generally have one LAN adapter with both NetBIOS and TCP/IP ports enabled. In addition, they might have COM ports enabled to allow XPC dial.

Most Notes clients on the same LAN at IBM access Notes servers via the NetBIOS protocol. That's simply because most client workstations are already set up to access an IBM LAN server, and this means that the NetBIOS protocol is already available.

Using TCP/IP requires that the TCP/IP protocol suite be loaded on the client workstation. If TCP/IP is available on a client workstation, then that workstation can also have both the NetBIOS and TCP/IP protocols enabled for Lotus Notes. There are a few important things to understand when either a Notes server or Notes client has multiple ports enabled.

Notes (either server or client) will use the port that appears first in the port list (under File|Tools|User Preferences|Ports) when it tries to access another Notes server. For example, let's say you have two ports defined and enabled on your workstation (or server) called NETBIOS and TCPIP, and they are listed in that order. If you then do a File|Database|Open, type in a server name, and click on Open, Notes will try to access that server using the NETBIOS port because NETBIOS appears first in the port list (under R3 the port list would default to be in alphabetical order). If the server is at a different site that is only accessible via TCP/IP, your workstation will tell you "Server not responding."

One way to override the default listing is to edit the NOTES.INI file in the NOTES directory on your workstation or server. Look for the

PORTS= line, and rearrange the order of the ports listed. A better way to handle the situation is to do a FileIToolsIUser PreferencesIPorts and use the Reorder button to get the ports listed in the way you want them.

It's a good idea to choose port names that are meaningful. LAN0 is the default name of the first port you define. That's not a very meaningful name. I usually use the name "NETBIOS" for my NetBIOS port and "IP" for my TCP/IP port. Because "IP" is alphabetically before "NETBIOS", and I always want to use TCP/IP as my primary protocol, these names will also give me the correct default order for my Release 3 Notes client.

You can "force" the port to be used by employing what Lotus refers to as a "bang, bang" approach ("bang" is a nickname for an exclamation point). The "bang, bang" is three exclamation points (i.e., !!!). To try this, go to your server, which we'll assume has ports NETBIOS and TCPIP defined, and do a FileIDatabaseIOpen. Then, to force the use of the TCPIP port type in the server field TCPIP!!!SERVER1 (where SERVER1 is the name of the server you want to access). Likewise to "force" the use of the NETBIOS port, you could enter NETBIOS!!!SERVER1.

It's important to also understand that Notes pays no attention to connection records when routing mail to servers for which there is a network connection (in the same domain and Notes network or having a path via common networks). In this situation also, Notes will use the first port listed in the PORTS= line in NOTES.INI to do the routing. If that connection fails, it does not try another.

If you always want to make a server route mail using TCP/IP, you could make this happen by defining only one port on your server (e.g., called TCPIP), you could order the ports in the priority you want using the Reorder button under FileIToolsIUserPreferencesIPorts, or you could edit the PORTS= line in the NOTES.INI file.

To change the port names on a server so that they appear in the order you want them, be sure to change the server connection records and use FileIToolsIPreferencesIPorts to change the port order while the server is down. Changes to ports can be made when the server is up; however, for the change to take place, the server will have to be brought down and back up again.

In Release 3 of Notes, Apple users experienced the port usage problem because AppleTalk's port is typically named ATALK. If the Apple user with Release 3 of Notes wants to use TCP/IP for the initial connection when replicating, then it's tough to find a name for the TCP/IP port that appears alphabetically before ATALK! One solution that a lot of people use is to let the first port name default to LAN0 (let's say for the NetBIOS or ATALK port). Then, if you use LAN1 as the next port name (for TCP/IP), you'll have TCP/IP as the first port in the list (because LAN1 gets listed before LAN0).

Recommendation: If you want your TCP/IP port to be primary, use the default LAN0 name for your NetBIOS or AppleTalk port, and use IP as the name for your TCP/IP port. This way you use the default LAN0 for the protocol that is generally used when you first bring a client or server up on Notes, and you have a recognizable name (IP) for the port that is often added later and becomes your primary port. This works well for all Releases of Notes.

Use of the Internet

There are many commercial services that offer access to the Internet via dialing to local numbers or through leased line access. Go to any computer store, and you'll find Internet access kits. Often the kits offer easy access to any of several Internet access providers. Dial access speeds of up to 56 Kbps are usually supported with the kits. Leased line access from a service provider usually is up to T1 speed (1.544 Mbps).

Internet access providers offer corporate Internet registration and billing services. The corporate service includes dial access diskettes for mobile users, so the trip to your local computer store won't be necessary. If security on the Internet is a major concern, several network providers offer access to private TCP/IP networks. Each customer is usually on a "virtual private network" so that only authorized users have physical access to the TCP/IP subnet. IBM Global Network (IGN) provides this type of TCP/IP service on their private commercial IP network called Internetworking 1.1.

SNA Connect Option

The Lotus SNA Connect companion product allows Notes server-to-server (or workstation-to-server) communication over an SNA link. This product will be one option for providing Notes service over an SNA link. The IBM ANYNET product could also be used as a way to transport NetBIOS or TCP/IP protocols over SNA lines and thus provide Notes services. Which SNA option (SNA Connect or ANYNET) is chosen depends on the unique environment of the company. They might already be using one of the products. If so, that product should be used.

Because the IBM Global Network operates the largest SNA network in the world, the SNA Connect option is of considerable interest. IBM Global Network customers can use their current SNA dial or SNA leased lines as the way to connect to Lotus Notes servers on the IBM Global Network.

X.25 Connect Option

This companion product is similar to the SNA connect option and is sold separately from Notes Release 4. With this option, you can implement Notes server-to-server communication over an X.25 link. Networks using X.25 transmit data faster than those using the XPC protocol driver and handle long delays that often occur when transmitting data over long distances. Because X.25 is a very popular protocol in Europe, this option is frequently used there.

Domain Design Considerations

How many Lotus Notes domains should your company have? Is one large domain better than several smaller domains? This section gives some recommendations on those areas. Remember, though, that Lotus Notes network design depends on many factors, some technical and some nontechnical (e.g., company policies and politics), so there are few absolute answers in this area.

Generally, one large Lotus Notes domain is better than several smaller domains. One large domain has these advantages:

■ Anyone outside your company can send to anyone inside your company using the same Lotus Notes domain. Let's say your company's name is AJAX and you choose to call your company's Lotus Notes domain AJAX. Then, to send to John Doe at the AJAX company, the Notes address would be John Doe@AJAX. That's quite simple and easy to remember. If the AJAX company decided to have three Notes domains called AJAX1, AJAX2, and AJAX3, then I'd have to know that of these domains has the user John Doe.

■ Administration is simpler when there is only one N&A Book as opposed to several. Because the Name and Address book contains connection records, domain definitions, etc., that show how your domain interfaces with other domains and Notes networks, it's much easier to administer a single Notes domain (and hence single N&A Book).

On the other hand, there are often good reasons (and sometimes bad reasons!) to have multiple domains within your company. The IBM Corporation has over 100 domains defined. The following sections cover some of the reasons for this.

Size of a Single Domain

Based on IBM's experience, with R3 of Notes, it was a good idea to keep the number of people defined in a single N&A book (and hence domain) below 10,000. The IBM PC Company had stretched this to well over 10,000 users to a single domain; however, because of performance and administration concerns, their domain was broken into multiple domains.

Release 4 of Lotus Notes has improvements that make it easier to have large domains, but remember that Notes was originally designed with the concept of a group of users on a single LAN. There will be some growing pains associated with the easy use of Notes for the global enterprise for some time to come.

Even with R4 of Notes, a very large N&A Book could be a concern. The IBMUS domain (an R4 domain used for IBM employees in the US) has a single Name and Address Book that has close to 100,000 names. This NAMES.NSF file is about one gigabyte in size, so it's not a Name and Address book that mobile users are likely to replicate to their laptops! Release 4 of Notes appears to have alleviated the performance concerns with such a large N&A Book, but administration will always be a concern for this database because it is so large and contains so much vital Notes information.

Corporate Structure

Many of the domains in the IBM Corporation are based on IBM's corporate structure. For example, a Notes domain might be based around a single large site, on a division level, or on a country level. The Advantis company (jointly owned by IBM and Sears) has an ADVANTIS domain. Because there are about 5000 employees in Advantis, a single domain creates no problems with size. Likewise, the IBM PC Company has a BCRNOTES domain for their Boca Raton, Florida, site.

Also there are domains, such as IBMUK (for IBM United Kingdom), IBMJAPAN (for IBM Japan), IBMNORDIC (IBM Nordic countries), and IBMCONSULT (IBM Consulting). These domains were created, not by an overall corporate strategy, but by the separate IBM divisions or countries obtaining the funding and other resources to build a Notes domain (or domains). In mid-1995, IBM started to put a strong corporate focus on Notes for the whole enterprise. This changed the direction from a bottom-up approach to a top-down approach for Lotus Notes implementation.

Administration

IBM Japan will remain as a separate Notes domain with separate administration because of language differences in addition to corporate structure reasons.

Special Domains

In any corporation, there will be a need for specialty domains, such as those used for test and development. Also, if your company needs to communicate with Notes to other companies (e.g., via the Internet or through XPC dial), it makes sense to have a special domain for that purpose, which can be thought of as an "isolation domain." Network security is one of the main reasons for having a special domain of this sort. IBM has a domain called IBM_EXTERNAL, which is used for this purpose.

Mergers and Acquisitions

The "Lotus" domain will remain separate from IBM Notes domains, because the "Lotus" domain was well-established before the IBM acquisition of Lotus.

The Use of Domain Statements for Routing

The domain section of your Lotus Notes Name and Address book can play an important part in mail routing. There are two ways to route information from one domain to another.

The first way is to have a direct connection to the other domain. In that case, you need to "cross certify" your server with a server in the other domain. Usually you place a connection record in your server's Name and Address book to indicate the type of connection, and a schedule for replication and mail routing.

The second way to route information to another domain is to route through another domain. Let's say your company has a domain called DOMAINA, that has a direct connection to another company's domain called DOMAINB. Now you want to route mail to another company's domain, called DOMAINC. You do not have a direct connection to DOMAINC, but you know that DOMAINB has a direct connection to DOMAINC. Then you can route from your DOMAINA to DOMAINC by routing through DOMAINB. To do this, you use the concept of "non-adjacent" domain entries in your N&A book. To look at the domains in your N&A book, click on the icon representing your N&A book, and then do a ViewlDomains.

Figure 2.3 shows the domain view from the IBM_INTERNAL N&A book. IBM_INTERNAL is a "mail hub" for the IBM Corporation, and it has direct connections to over 100 other IBM domains. It also has many non-adjacent domain entries. In the figure, all those entries with a domain name in the "Next Domain" column are non-adjacent domains. For example, if you look for the IBMMTP name in the first column (the column called "domain"), you'll see that the Next Domain is given as MTPNHUB. This means that IBM_INTERNAL doesn't have a direct connection to IBMMTP, but it can route mail to IBMMTP by routing the mail to the MTPNHUB domain for that it has a direct connection. MTPNHUB is also a hub domain that has many connections. See Figure 2.16 later in this chapter, which shows a diagram of a routing hub design that includes for the primary hub domain, IBM_INTERNAL, and a secondary hub domain, MTPNHUB.

Figure 2.3
Example of domain statements for routing.

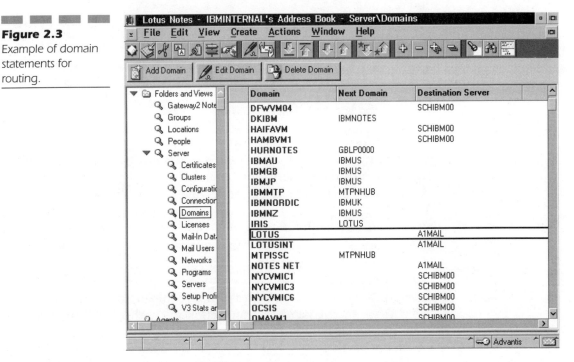

The next sections tell you more about setting up direct connections to your Notes domain and entering non-adjacent domain entries for those domains for that you don't have direct connections.

Direct Connections and Gateway Definitions

Direct connections from your domain to another domain require you to put *connection records* in your Name and Address book to schedule replication and define the protocol to use for replication and mail routing. With a direct connection, generally, entries are not made in the domain section of your Name and Address book, but they can be. For example, if there are several servers in a mail hub domain, you might want to show how to get directly to a server that connects to an outside domain. The IBM_INTERNAL N&A book has many of these entries (see Figure 2.3). Without a domain statement (called a *Foreign Domain*) the mail router will go down the connection records in order and route to the server in the first connection record it finds to the desired domain. Entering a Foreign Domain entry allows you to designate the routing that should be used first.

To enter this type of domain statement, click on your N&A book and do a Create|Domain|Foreign. Figure 2.4 shows an example of the type of display you'll see. There are only three items to enter: Foreign domain name,

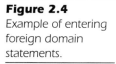

Figure 2.4

Example of entering foreign domain statements.

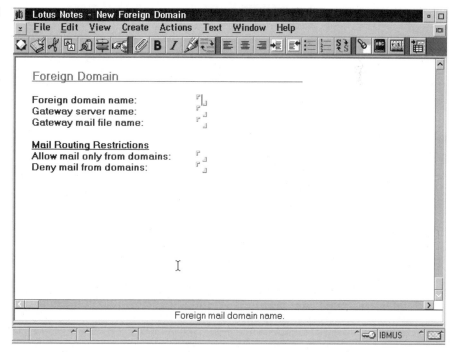

Gateway server name, and Gateway mail file name. The Foreign domain name for the direct connection is just the domain name. The Gateway server is the name of the connecting server in the domain. The Gateway mail file name will always be MAIL.BOX. An example of this in Figure 2.3 is the domain LOTUS with server name A1MAIL (with the R4 or R5 N&A Templates, the Gateway Mail filename does not appear on this screen).

The Foreign Domain entry is also the way to route mail to mail gateways, and that's why the entries refer to "Gateway server name," etc. Mail gateways, such as the IMLG/2 from IBM and the SMTP gateway from Lotus, require these entries. An example of an entry for the IMLG/2 entry is shown in Figure 2.3. The "Foreign Domain" is NYCVMIC1, the IMLG/2 gateway name is SCHIBM00, and the "Gateway Mailbox" is DGATEWAY.NSF (with the R4 or R5 N&A Templates, the Gateway mailbox file name does not appear on this screen.)

Non-Adjacent Domains

If you have a friend or colleague in a Lotus Notes domain that doesn't directly connect to your Lotus Notes domain, then non-adjacent domain entries in your domain's N&A book can be used to make the mail routing easy. With the correct setup, the addressing would be *FriendsName@HisDomain* (i.e., just the friend's name at his domain name). Otherwise the routing might need to be of the form *FriendsName@ HisDomain@Domain1@Domain2@*etc., where *Domain1*, *Domain2*, etc., are the domain names that are between your domain and his domain.

To enter a non-adjacent domain statement, click on your N&A Book, and do a Create|Domain|Non-Adjacent. Figure 2.5 shows the screen that will appear. All you have to do is enter two items:

- The destination domain (the non-adjacent domain)
- The adjacent domain you'll route through to get to the non-adjacent domain

Figure 2.3 shows examples of non-adjacent domain entries in the IBM_INTERNAL N&A book. In this figure, domain IBMNORDIC is a non-adjacent domain that is reachable via the adjacent domain IBMUK.

Using a Hierarchical Hub-and-Spoke Domain Structure

The hub-and-spoke domain design should have a hierarchical structure when used for the enterprise. For example, a large site within a company

Figure 2.5

Example of entering
non-adjacent domain
statements.

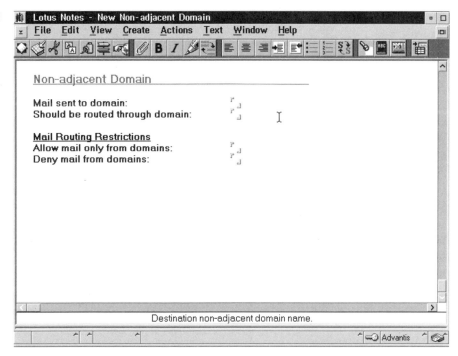

might have several different domains, each relating to a different division or function within the company. Those site domains could all connect to a site hub domain, which would provide inter-domain routing for the domains at the site. In addition, this site hub domain could provide connectivity to the company-wide hub domain. This type of hierarchical hub-and-spoke domain structure is shown in Figure 2.6. The secondary hub domain shown in the diagram could be the site hub domain that we just discussed, while the primary hub domain shown in the diagram could be the company-wide hub domain. An example showing the way the IBM Corporation has implemented this concept of hierarchical hub and spoke domain structure is shown in this chapter in the section "A Worldwide Example."

The main reason for the hub-and-spoke architecture is for more effective and efficient connections, which means more effective and efficient replication and mail routing for the enterprise. Of course, there could be many levels of hierarchical hubs. However, if you keep in mind the rule of limiting the number of domains within a company (by not having a lot of small domains) and the fact that the more levels the more complicated the routing and management, then you should find that two levels is sufficient. Let's keep it simple.

It should be pointed out that this hierarchical domain structure is independent of whether or not hierarchical certificates are used (that's the next topic).

Figure 2.6
Hierarchical
hub-and-spoke
domain structure.

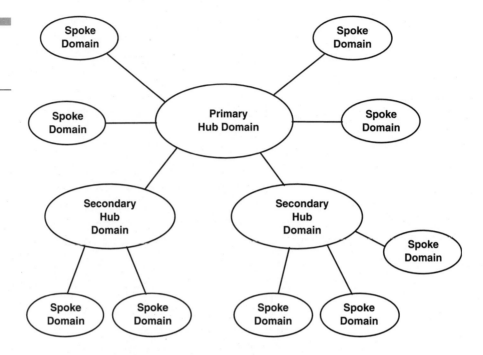

Figure 2.6
Hierarchical
hub-and-spoke
domain structure.

Domain Certificate Strategies

We just discussed the hierarchical domain structure. That had to do with how the domains were linked together in a hierarchy. Each certificate for a domain can be what Lotus calls hierarchical or non-hierarchical. This is a separate concept from the hub-and-spoke hierarchical structure we just discussed. Lotus Notes first used the term *hierarchical certificate* with Release 3 of Notes (1993). Up until then, all versions of Notes had used non-hierarchical or flat certificates.

The hierarchical Lotus Notes certificate structure was designed to follow the X.500 standard. X.500 is the OSI directory standard that works with the X.400 message handling standard. With Release 3 of Notes, the Name and Address Book was enhanced to allow the use of distinguished user and server names compliant with X.500 addressing and hierarchical structure. The elements of a distinguished name are:

- *Common name*—The user's or server's name (e.g., first name, middle initial, and last name).

- *Organization unit*—There can be from zero to four organizational units indicating, for example, the user's or server's department and location.

■ *Organization*—This is often a company name. There can be only one organization. For example: IBM or Lotus.

■ *Country name*—This is the CCITT-defined, optional, two-letter country code.

In the ADVANTIS domain, the hierarchy used is:

```
John Q Public/White Plains/Advantis/US
```

This follows the X.500 direction, where the highest level (US) is the country. For Advantis Notes Servers, the "Who can access servers" section in the Server records in the ADVANTIS Name and Address Book has the entry "*/ADVANTIS." This is an easy way of giving everyone in the ADVANTIS domain access, while denying access to everyone else. The ability to use the "*/ADVANTIS" entry is based on the hierarchical certificate structure used for the ADVANTIS domain. It should be pointed out, however, that groups will always need to be defined in the N&A Book in order to control database and server access for groups of people other than very generalized group consisting of such categories as everyone in a domain.

NON-HIERARCHICAL (OR FLAT) CERTIFICATES.   Most domains established with non-hierarchical certifiers are those that were first established before Release 3 of Lotus Notes was made available. Once you have an established domain with a non-hierarchical certifier, it's difficult to change to hierarchical. Updating the release of Notes on your servers keeps the structure non-hierarchical. The IBM_INTERNAL hub mail domain and the IBMNOTES database repository are both based on non- hierarchical certificates because they were established during the days of Notes Release 2.0.

Non-hierarchical Notes certificates are simpler to administer, but the structure in a hierarchical domain provides for more control over the authority given to users, and hence provides greater security.

HIERARCHICAL CERTIFICATES.   If you establish a new Lotus Notes domain, the default will be a hierarchical certifier structure. The hierarchical certificate structure has been recommended by Lotus since Release 3 of Notes, and all new function and support will be built around the hierarchical structure. Therefore the great majority of new Lotus Notes domains will be based on hierarchical certificates.

To cross-certify with a flat system, a hierarchical system cannot use its hierarchical certifier. A non-hierarchical certifier ID must first be created.

The Notes Administrator for the hierarchical system selects File/ Administration/Register certifier/Non-Hierarchical... With this ID, the administrator for the hierarchical must certify the safe.id of the non-hierarchical server and his own server's ID file. He then sends to the flat domain's administrator the non-hierarchically certified safe.id of the non-hierarchical server along with a new safe copy of his server's ID file (created after it was certified with the newly created non-hierarchical certifier). This will be stamped and returned to be merged into his domain server's ID.

It is always a good idea to make a backup of your server.id before merging in a certified safe.id. This is because an improperly stamped safe.id can corrupt a server.id and render it unusable. That's a disaster, unless you have a backup! Using a hierarchical certifier to stamp the safe.id of a server in a flat domain is one way to have the disaster happen.

Finally, the Notes Administrator uses the non-hierarchical certifier to recertify the IDs of all users in the domain who want to access the server in the non-hierarchical domain.

Recommendations for the Enterprise

The recommendation is to use a hierarchical certifier structure and only one domain (if possible) for your company. If you do have more than one domain (not counting "test" or "development" domains), then you can cascade all the domains together to get some of the same effect as having one domain.

If you're using Notes R4.5 or later, then the "Directory Assistance" feature is the best way. You set up directory assistance by creating a replica of a Master Address Book on all servers in all domains. The Master Address Book tells each server:

■ The naming rules associated with a domain so that it can search secondary Public Address books efficiently.

■ The location of one or more replicas of Public Address books that should be searched.

You can replicate a Public Address Book on strategic servers outside of its domain if that will help performance. In general, directory assistance is a more efficient method of managing multiple Public Address Books than cascading. However, directory assistance requires Notes R4.5 or later, while the function of cascading N&A Books is available in all releases of Notes.

The procedure for cascading Name and Address Books on your Notes servers is described in chapter 12 in the section "Management of Both

Hierarchical and Non-hierarchical Domains." Remember that there is a limit of 10 cascaded Name and Address Books, including the Name and Address Book for your own domain (NAMES.NSF).

Example of Hierarchical Notes Design used Internally by IBM

The IBM Corporation uses a hierarchical architecture for its internal Notes domains. The two Lotus Notes peer domains at the top of this hierarchy for mail routing are the IBM_INTERNAL and IBM_EXTERNAL domains. The IBM_INTERNAL domain is used for communications between IBM domains, while the IBM-EXTERNAL domain is used for Notes communications between IBM and other companies.

Figure 2.7 shows the IBM_INTERNAL and IBM_EXTERNAL domains along with the IBMNOTES Database Repository domain. These three domains are used by the IBM Corporation to send Notes mail within the corporation; to other companies, organizations, or individuals; and to share Notes databases and other Notes information within the corporation. Each of these domains is described in detail within this section.

IBM_INTERNAL Domain

This domain is used for the efficient connectivity of IBM Lotus Notes sites to other IBM and IBM subsidiary sites through the IBM internal network.

IBM Lotus Notes domains are connected to the IBM_INTERNAL domain either with the TCP/IP or NetBIOS protocols.

To easily send mail to users in other domains connected to IBM_INTERNAL, those domain names should be entered as non-adjacent domains in the site domain's Name and Address book. This allows a user in that domain to address a user (Tom Smith) in another domain (ADVANTIS) by typing Tom Smith@ADVANTIS instead of typing Tom Smith@ADVANTIS@IBM_INTERNAL.

Some examples of non-adjacent domains added to the ADVANTIS domain's N&A book are shown in Figure 2.8. Note that, because the ADVANTIS domain uses the IBM_INTERNAL domain as a hub domain for mail routing, almost all of these non-adjacent domains point to IBM_INTERNAL as the way to get to the desired domain.

Figure 2.7
IBM_INTERNAL,
IBM_EXTERNAL, and
IBMNOTES domains.

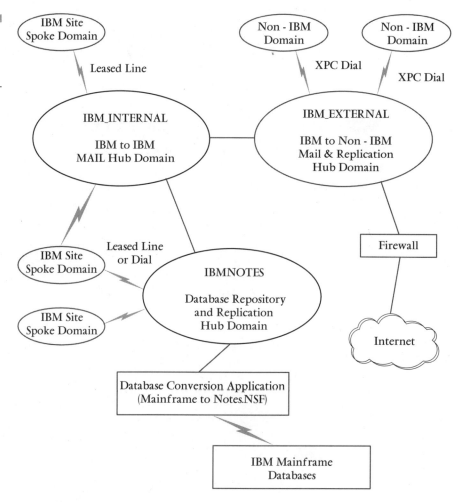

IBM_EXTERNAL Domain

This domain is used for the connectivity of IBM Lotus Notes sites to
non-IBM sites. These connections follow the Inter-Enterprise System
Connection (IESC) standards for IBM (a security set of regulations). With
the use of the NetSP product (a fire-wall between IBM's private network
and the Internet), Lotus Notes access to non-IBM users connected to the
Internet is available. NetSP helps provide the required security.

IBMNOTES Domain

This domain contains a server with Notes databases that are shared by
all areas of IBM. The databases in this IBM Corporate Notes Database

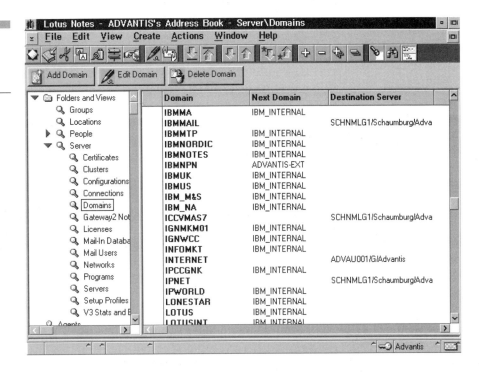

Figure 2.8
Example of non-adjacent domains.

Repository include information from Lotus (e.g., Lotus Knowledge Base, Lotus Partner Forums), Notes "shareware" applications written by IBMcrs world wide, discussion databases, and hundreds of IBM Forums downloaded from the mainframe and converted to Notes databases.

Servers in IBM spoke domains and individuals directly access the databases in IBMNOTES, although it's preferred to have a spoke domain server replicate popular databases so that all of the users in that spoke domain can access the database locally.

Of course individual Notes users could also replicate databases directly to the hard disk on their workstation. However, generally that is not advised. For one thing, some of the databases are very large (e.g., the Lotus Notes Knowledge Base and IBM Announcements databases are each over 60MB).

IBM's "New" Notes Architecture

In June of 1995, the IBM Corporation, looking to create a single unified Lotus Notes architecture capable of supporting its entire organization, chartered an effort to investigate the requirements and resources needed to design, build, and maintain such an environment. This effort was primarily aimed at defining a scalable, maintainable, and economical infrastructure capable of supporting IBM's anticipated business requirements well into the future. This new Notes architecture is replacing the "legacy" Notes architecture within IBM that had been established mostly at a grassroots level.

To create the new Notes architecture, a multi-skilled international team was formed, composed of members from all parts of the IBM family. Experienced Notes technologists, administrators, and users from Lotus Consulting and the different IT groups from within IBM collaborated using Notes to define both the technical, organizational, and procedural infrastructure required to support a large-scale Notes deployment and a manageable migration path from the many existing Notes infrastructures to the newly defined environment.

The "new" Notes domain structure within IBM is based on the concept of dedicated domains by functional process. The two primary functions to be supported are Notes electronic mail and database services.

The concept of a domain dedicated for a functional purpose is relatively new. In the past, this distinction might have been made at the server level, where a single domain would contain some servers dedicated to mail and other servers dedicated to database functions.

In large deployments, structuring the domains by dedicated functional purpose has several benefits:

- Makes sizing of computing hardware easier because functions are isolated. Allows for maximizing capital investment.
- Simplifies the ACLs required for replication to and from the database domain. There is a single point of entry and exit for database replication by Site.
- Provides for easier segregation of administrative duties by domain function.
- Limits clients to one domain location by site to peruse for database access.
- Offers economies of scale.

IBM "Server Farm" Concept

Mail and database services are provided to IBM personnel on a site basis. IBM computing sites provide a broad set of computing services to surrounding geographic areas. There is a defined number of 24 strategic IBM computing sites worldwide. This is the "Server Farm" or Geoplex concept. Figure 2.9 shows the locations of the 24 IBM Server Farms throughout the world. (EMEA is Europe, Middle East, Africa.)

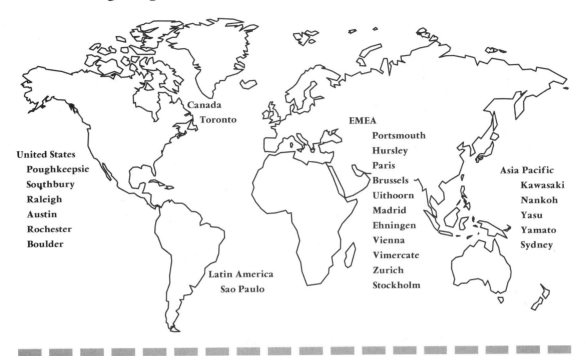

Figure 2.9
IBM worldwide server farm locations.

IBM Lotus Notes Global Architecture Components

The IBM worldwide Notes architecture has several component levels. At the highest level is the concept of Geoplex. A Geoplex (also called a Server Farm or sometimes a "Site") is defined as a physical data center location containing a number of Notes domains. There is a defined limit of 24 Sites worldwide.

In a way, the concept of centralizing all of the Notes services at a Geoplex or Server Farm is a throwback to the old mainframe days. There are many advantages to this centralized concept from an administration and management aspect. Just as in the "mainframe days," the computer professionals can concentrate on a relatively few Server Farm locations rather than hundreds of distributed server locations. Maintaining service 24 hours a day and 7 days a week is much easier when those support personnel can concentrate on a few server farms locations rather than hundreds of distributed server locations. It is a throwback to the mainframe days!

A Notes domain is a collection of Notes servers that utilize the same Public Address book. Each Notes domain has a defined functional purpose. The two principal types of domains are:

- *Mail domain*—A collection of servers used to provide electronic mail services.

- *Database (DB) domain*—A collection of servers used to provide for nonmail services. The DB server type category has two further subclassifications:

  - Production applications requiring high availability, reliability, and standard support services.

  - Nonproduction includes development or test environments where the primary purpose is development, functionality, or integration testing prior to the application being released to the development environment.

PRODUCTION DATABASE SERVERS. Each Site requires production database services. Although a database domain might service clients at numerous Sites, the domain itself is housed within a given Site.

Servers within the production database domain are segmented by the type of information they house. One or more of the following server constructs are used for facilitating information management within a database domain:

- Enterprise-wide server
- Business-unit server
- QuickStart server/RAD or Rapid Applications Development server

The enterprise-wide server contains production database information of general interest to the entire company. Specific enterprise servers might be identified by IGS as mission-critical. Machine configurations for these mission-critical systems will vary depending upon service level agreements. As a general rule, these systems will receive the highest level of protection (e.g., hardware redundancy, backup power, etc.) in the production environment.

The business-unit server contains production database information of interest to a given business unit or work group within a business unit. Specific business-unit servers might be identified by IGS as mission-critical. Machine configuration for these mission-critical systems will vary depending upon service level agreements. As a general rule, these systems will receive the highest level of protection (e.g., hardware redundancy, backup power, etc.) in the production environment.

The QuickStart Server/RAD or Rapid Applications Development server contains databases that:

- Might have a limited life.
- Do not necessarily have a business process owner.
- Have local-only or no IT support.
- Will not have mission-critical status.
- Are never replicated to other servers or domains.

The previous segmentation provides additional benefits as IBM's Notes installation grows in terms of servers and databases. As the installation matures, the need for a clear information management strategy increases. This is driven by the client's requirement to easily and quickly access different types of information. Providing such an organized information structure helps meet this need. It also benefits the navigational tools that will be used to facilitate access to information on the Notes network.

Two additional types of servers reside in the database domain. These servers provide control points for production processing and template delivery services:

- *Production-control server*—Used for running production jobs (API or otherwise) for applications within the site.
- *Template server*—Contains all approved application templates (e.g., mail template).

To support mail-enabled workflow applications, routing services are enabled in each database domain to the nearest mail domain. Subsequent routing (e.g., to and from Office Vison/WM) is accomplished using the established mail topology. User mail addresses are provided to applications via redirection to the mail domain's dedicated Public Address Book server or via a custom directory application. Database spoke servers are equally distributed among the three database domain hubs (high, medium, and standard) to avoid overburdening one hub. This design avoids having to replicate secondary N&A book information to each and every server, thereby saving disk space.

PRODUCTION MAIL SERVERS. Each Site's electronic mail requirements are met through the implementation of one or more Notes mail domains. A given mail domain might service clients across several sites. Unlike database domains, the servers that comprise the domain itself might also be housed cross-site. This capability is provided to support distributed work groups that have high-end communication needs.

The mail server architecture provides mail routing services to:

- IBM Notes installations worldwide. This includes mail routing:
  - Within a Notes domain
  - Across Notes domains but within the same Site
  - Across Sites
- IBM partners/interenterprise via IBM Global Network services:
  - To/from key IBM computing platforms using gateways
  - Secure connections

Clients are allowed to elect one of three mail routing service levels at the point where electronic mail is composed (Table 2.1). (Although these three routing service levels are in the architecture, the general design calls for immediate routing of all mail, so routing priorities do not buy you anything!)

TABLE 2.1

The Three Possible
Mail Routing
Service Level

Service Level	Service Description
High	Routes mail at once to next hop
Medium*	Routes mail at next connection interval
Low	Routes mail overnight (12am - 6am)

*Indicates template default

Each mail server in the domain has its primary N&A book replicated on that server. Secondary N&A book look-up services are available through the Notes R4.5 Master Address Book function. Within each mail domain, all of the mail server spokes share a single Notes named network so that messages addressed to other users in the same domain are routed instantaneously. We have found that this structure is very effective in the IBM environment where the majority of mail is sent to people in the same country, and therefore the same mail domain.

SINGLE MESSAGE STORE. Notes R4 has the enhanced capability of single message store. This feature allows administrators to conserve server disk space by storing a single copy of a message sent to multiple individuals on the same server. Clients then access the message using their pointer file. It is estimated that this architecture will result in a 30% savings of disk space on mail servers.

It is not a requirement to implement single message store on Notes R4 mail servers. However, it has significant potential benefits given the IBM mail culture, where courtesy copying and/or multiple addressees is quite common. It is, therefore, the default message store option.

Figure 2.10 shows the IBM Global Architecture for Lotus Notes. The top section of the diagram shows a LAN that would be located at a user site and would contain all user workstations, local file servers (including print servers), and a hub machine that provides Notes connection to the IBM Geoplex site. The Geoplex houses the Lotus Notes Mail and Database servers (designated as the "Notes Strategic Application Center" in the diagram) and the "Notes Global Network Hub" that provides routing to other domains and gateways to the Internet, to fax machines, pagers, etc.

Figure 2.11 goes beyond the architecture for Lotus Notes and shows the computer site structure for housing all client-server applications (including Lotus Notes). Although the current wide area connection between IBM user sites and Geoplexes is mostly T1 router based using the TCP/IP protocol, we are migrating to an ATM-based wide area network as depicted in the diagram. T3 circuits (45 Mbps) provide the bandwidth for the ATM implementation.

Figure 2.12 goes another step further and shows all of the components that will be in IBM's server-based network computing architecture. This is the basis for all of IBM's future computing needs, which is built around the Network Computing model. The architecture includes both IBM's Global Notes Architecture (GNA) and our current client/server architecture. ATM provides the wide area network for multi-protocol transport (including TCP/IP and SNA).

Figure 2.10
IBM's Global Notes
Architecture (GNA).

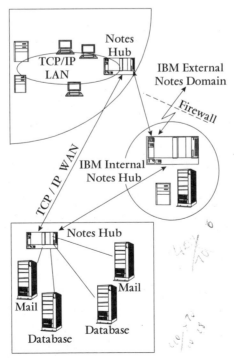

- **NOTES CAMPUS SERVICES**

  - ALIGNED WITH WORKGROUPS
    AND TEAMS
  - COMMON DATABASE STANDARDS
    (INCLUDING MAIL)
  - CENTRALLY ADMINISTERED
  - TCP/IP LAN COMMUNICATION

- **NOTES GLOBAL NETWORK HUB**

  - SEAMLESS CONNECTION FOR ALL
    IBM DOMAINS
  - PROVIDES ENTERPRISE-WIDE
    SERVICES
    ● MAIL GATEWAY, FAX, ETC.
  - SECURE EXTERNAL NOTES
    CONNECTIONS
  - TCP/IP WAN COMMUNICATION

- **NOTES STRATEGIC APPLICATION
  CENTER**

  - GEOGRAPHIC MEGACENTER
    LOCATIONS
  - AIX SERVER BASED ON SP2s

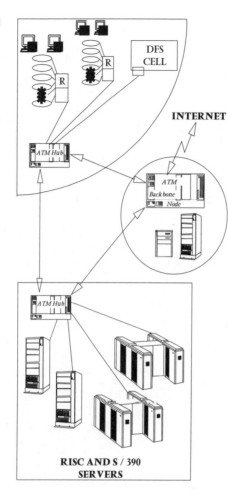

Figure 2.11
IBM collaborative
computing
server-based network
computing
architecture.

Campus Network and Servers
- Structured Layered Networks based on Austin Model and ATM Technology
- DFS Cells Provide Common Libraries of Binary Large Objects including
  —Engineering/Mechanical Designs
  —Software Source and Object Libraries
  —Common Application Software
  —Image, Video, and Voice Media
- Distributed Systems Management Technology

Intelligent Network Platform
- Geographic Network Hubs
- ATM Backbone provides Scalable Bandwidth
- Network Management and Intelligent Services
  —Universal Electronic Mail Exchange
  —Secure Internet Access
  —Remote Access Gateway (LIG)

Strategic Application Centers
- Strategic Application Delivery and Management
- Megacenter/Megaplex Delivery Centers
- S/390 CMOS in Parallel Sysplex Configuration
  —Information Warehouse and Transaction Systems
- Scalable RISC Application Servers
- Lotus Notes Super Servers
- Corporate WorldWide Web Servers

Bandwidth Requirements for Central Geoplexes

At the direction of the IBM CIO, the IBM Global Network worked with application owners and business units to develop a model to predict the impact that IBM's new strategic client server applications will have on the IBM global TCP/IP network.

Two major issues are highlighted:

■ Certain client/server applications that use many transactions to build a single screen will have a perceived response time that is many times higher than the network's latency.

■ Significant capacity upgrades are needed to accommodate additional traffic load on the network caused by Lotus Notes mail

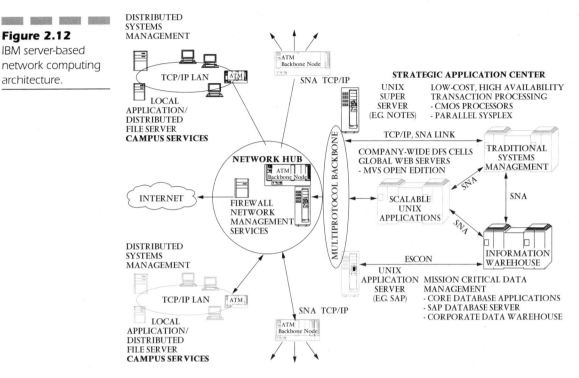

Figure 2.12
IBM server-based network computing architecture.

and applications, Web usage, four manufacturing applications, one finance application, and SAP fulfillment.

Modeling techniques included projections based on current data, LAN sniffer traces, application documentation, user profiles, and interviews with application owners' technical staff. On a personal note, I have found that measuring actual usage is a much better way to model than by "pure simulation." This strong belief started with my doctoral work at U. C. Berkeley where I modeled mechanical failure using approximation techniques and Monte Carlo simulation. This belief was supported when I spent several years doing building energy analysis both by measurement and projections from real data and by using simulation programs funded by the U.S. Department of Energy. If it is available, use "real" data as much as possible. As a second choice, use a combination of "real" and "pure simulation" data. The last choice should be projections based solely on simulation.

NETWORK LATENCY. Client-server applications that were developed for a LAN are now being pushed onto the wide area network. For many of these applications, a single transaction causes many round-trip flows back and forth "under the covers." Packets are sent in half-duplex flip-flop mode, which is the most inefficient session protocol. After a packet is sent, this half-session stops and waits to get a packet from the other side before sending its next packet, rather than sending multiple packets in succession. Although each individual round-trip flow by itself takes well under a second, the sum of many turnaround flows looks like long response time to the user.

> From the network perspective, client-server applications are incredibly inefficient.... Getting anything done...requires hundreds of packets to be passed back and forth between client and server.... Response times can go glacial.... Quick fixes like bumping up the bandwidth won't do the trick.... The solution is...deploying...servers...and getting the software developers involved: They're the one who can optimize application code to speed up communications.... Finally run a health check on the network itself.
>
> A data request generates hundreds of small packets. Each of these is sent individually and acknowledged before the next packet is sent.... On the local area, this inefficiency is masked by...low latency.... And while net managers are likely to worry about bandwidth, latency is the real killer.
>
> Data Communications August 1996, McGraw Hill, page 53

Lotus Notes mail, for example, requires six half duplex flip-flops between the user's client and the server after the user clicks on Send. This causes six times more latency than PROFS Mail. Trace results also indicate that Notes Mail will use about three times as much bandwidth as PROFS Mail (IBM Mainframe mail).

The discussion of Product Manager, Aspect, and EPPS, in the section "Measuring Bandwidth Requirements" later in this chapter, highlights some of the network latency concerns for these applications.

NETWORK CAPACITY.   Significant capacity upgrades are needed to accommodate additional traffic load on the network caused by:

■ Lotus Notes mail and attachments (WordPro, Freelance, 1-2-3, etc.)

■ Lotus Notes applications (new applications and applications migrating from VM)

■ Web browsing—Tools and some applications are expected to migrate to intranets; IBMers will browse the Internet.

■ Manufacturing applications
 ■ Product Manager (engineering records)
 ■ Enterprise Repository Environment (mechanical design)
 ■ Aspect (component selection)
 ■ Enterprise Parts Procurement System (demand and supply planning) (EPPS)

- Financial Information Warehouse (ledger)
- SAP (fulfillment)

Measuring Bandwidth Requirements

Based on the previous requirements for Notes, Web, and other client/server applications, IBM/Advantis devised the following bandwidth measurement methodology for each application.

LOTUS NOTES. For Notes, we assumed that $\frac{1}{3}$ of all Notes users would be active during daily busy hours. This assumption was based on many years of studying usage on IBM's SNA network. With 60,000 to 70,000 users of the SNA network over the past 15 years, the $\frac{1}{3}$ of the users active on the network at any one time has proven to be a very accurate number. We determined that an active Notes user requires 3.6 Kbps between the client and the server. This was determined by profiling a dozen users. While they used Notes, we ran an analysis program on their client that was based on the OS/2 IP Trace function. After using Notes for an hour or so, they stopped the analysis program and recorded the results, which included the Kbps used during the period.

This was done many times per user to get an average. Users were using mail and a few databases. Obviously the numbers can vary widely. If the user sends notes with a large file or spreadsheet attached, it's high; if they are writing a note, it's low. However, as with everything, the value is in the averaging.

Marc Auslander, from IBM Research, also came up with a similar number by analyzing the Notes server traffic at the IBM Watson Research Center. If you conservatively say you want to plan to have a T1 at about 60% utilization, then that makes it 250 active Notes users per T1 or 750 users, $\frac{1}{3}$ of which will be active. We did the same thing for Web browsing and came up with 5.0 Kbps per active user. We got that by profiling and verifying by analyzing the NetSP firewall server logs. We assume that 5% of all WEB capable users will be surfing during the busy hours.

The Notes flows themselves, during typical Notes transactions, are typical of IP client/server applications. Namely they tend to have a number of transmissions back and forth before the user gets the next screen. These transmission "turnarounds" do cause higher response times; however, the response times seem to be reasonable given a high-speed IP network, a reasonable number of hops, and clients supported from servers in the nearest Geoplex.

A Lotus Note is about 4 or 5 times the size of a comparable PROFS note. A trace shows that a simple Lotus Notes message of about two screens with no graphics is about 14,000 bytes. A comparable PROFS note of two screens is about 3000 bytes. The difference is because Notes sends information describing the Note (the form) and sends multiple transmissions between client and server.

WEB BROWSING.　This activity was measured at 5 Kbps per active user. Our assumption was that 5% of the users were active at any one time. The methodology was the same as described in the previous section.

FINANCIAL INFORMATION WAREHOUSE (FIW).　We evaluated FIW by placing a LAN sniffer on the test bed and filtering for client and server traffic while development manager Stu Horn performed expected typical transactions. The application now uses a Web front end. Queries are directed to a single server in Southbury, CT, with DB2 data in Sterling Forest, NY.

ASPECT.　This legacy hardware development application is used on servers at Yamato, Japan; Poughkeepsie, NY; and Portsmouth, UK. TCP/IP traces were used to estimate total capacity required.

ENTERPRISE PARTS PROCUREMENT SYSTEM (EPPS).　Capacity requirements were based on a predicted number of users at each site performing a predicted number of complex, medium, and simple EPPS transactions. TCP/IP traces were used to estimate total capacity required.

ENTERPRISE REPOSITORY ENVIRONMENT (ERE).　The ERE is a centralized "run-once" data repository (Ehningen, Germany) that contains the IBM corporation's design data (mostly mechanical but some electrical card/board data).

The ERE allows customers to interactively enter as well as extract data in support of the IBM mechanical design and release process. ERE technical data is tightly coupled with the administrative release systems in IBM (DPRS/PM). Both client and server were originally mainframe-based, but are now client/server based.

A typical user (client) can:

- Query the relational database in Germany for product/user information.
- Update the data base with user/product information.
- Send data to the repository for inclusion into the database.
- Receive data from the repository for incorporation into the design environment.

PRODUCT MANAGER (PM). The application owner produced an excellent document that rigorously analyzes PM's behavior on a network, itemizes transactions between server and client and between client and terminal, and thoroughly lists expected user boarding by server, by business unit and by time period. In addition, traces and analysis were performed to understand the number of bytes transferred and, perhaps more importantly, the numbers of transactions and acknowledgments back and forth needed to build a single user screen.

SAP ASSET MANAGER. This asset accounting application produces heavy file transfer activity between Sterling Forest and Poughkeepsie. 900MB files will be sent weekly, and 1.8GB files quarterly. In addition, there will be files ranging from 100MB to 700MB on specific workdays. Because bandwidth between these sites is currently 5 x T1 and traffic is projected to be off shift, we do not anticipate significant network impacts.

The typical TCP/IP traffic per user generated by each of these applications is depicted in Figure 2.13.

Figure 2.13
IBM TCP/IP
application modeling
methodology.

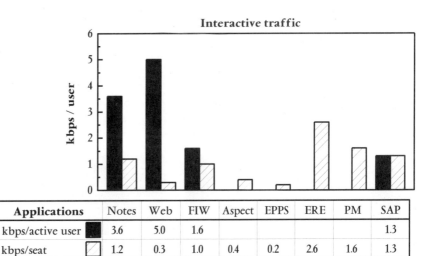

Applications	Notes	Web	FIW	Aspect	EPPS	ERE	PM	SAP
kbps/active user	3.6	5.0	1.6					1.3
kbps/seat	1.2	0.3	1.0	0.4	0.2	2.6	1.6	1.3

Methodology								
	user profile Watson, server analysis	user profile, NetSP analysis, industry trends	trace	trace	user profile	trace, legacy traffic	trace	SAP doc, bus units

Network Design Examples

This section gives network design examples for a single Notes domain (the ADVANTIS domain), multiple domains (the IBM PC Company in the U. S.), and a worldwide example with over 100 domains and a hierarchical domain structure (the IBM Corporation).

Single Notes Domain

The Advantis Company was formed at the end of 1992 by IBM and the Sears Corporation (Note: At the end of 1997 Advantis will be merged into IBM). Advantis designs and manages the wide area networks for IBM and Sears and for commercial customers. The company consists of about 5000 employees at four major sites: White Plains, NY; Schaumburg, IL; Tampa, FL; and Boulder, CO.

The size of this company easily fits the general guidelines for one Notes domain (one group that needs to work together with 10,000 or fewer people). The size of the N&A book is one of the governing factors. The IBM PC Company originally had one large N&A book of well over 10,000 names. That worked, but it was approaching the limits of practicality for that earlier version of Notes.

The IBM Corporation currently has about 220,000 employees worldwide, and with the design in Release 3 of Lotus Notes, it would *not* be a good idea to have 220,000 names in one "IBM" domain Name and Address Book. Actually, IBM does have an IBM N&A book, which is the consolidation of dozens of smaller N&A books within IBM. This N&A book has over 70,000 names, but it is just used as a directory and has no connection records, server records, domain records, etc., and has only basic information for each of the 70,000 people entries. Because of these restrictions in this 70,000 entry N&A book, it "performs" quite well (i.e., it opens very quickly, etc.).

As discussed earlier, Release 4 or 5 of Lotus Notes does make it easier to have very large N&A Books, and future releases will also help solve performance problems associated with large N&A Books. Therefore, in the long term, performance considerations will not be a reason for limiting the size of your Lotus Notes Name and Address Book.

Figure 2.14 shows the ADVANTIS Lotus Notes domain. Because there are about 5000 people in the Advantis Company, one domain was chosen for the whole company for the reasons discussed previously.

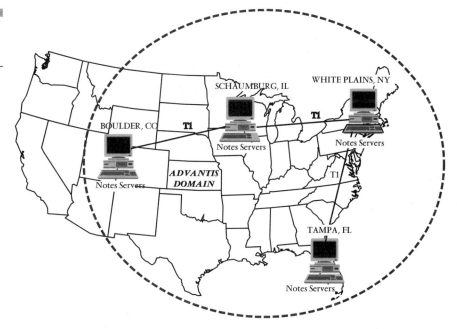

Figure 2.14
Advantis Lotus Notes
domain.

Multiple Domains

One example of the use of multiple domains within a corporation is the IBM PC Company in the U. S. The PC Company has converted all of their employees to Lotus Notes. Originally, all 12,000 employees were in a single Lotus Notes domain called IBMPSLOB (for IBM Personal Systems Line of Business).

With R3 of Notes, 12,000 names in a single N&A Book created some performance problems, especially because the PC Company wanted to have a great deal of information on each employee (e.g., fax number, address, etc.) and in addition have an extensive domain view (non-adjacent domains, foreign domains, etc.). Therefore, in 1995, the PC Company created three major domains for about 10,000 employees.

They still had the original IBMPSLOB domain that covered the Somers, NY and Raleigh, NC sites. In addition, they had the BCRNOTES domain for the Boca Raton, FL site and the AUSNOTES domain for the Austin, TX site.

At each site, all three N&A books were cascaded together, so some of the benefits on one large N&A book were still available. In addition to the performance benefits, there was also the aspect of each site being responsible for managing their own N&A book. This is undoubtedly seen as a benefit by some independent-minded Lotus Notes administrators and can be thought of as following the concept of distributed computing. The benefits of centralized versus decentralized computing and systems management will always be debatable!

The IBM PC company has been migrating all of their Notes users to the IBMUS domain, based on R4 of Notes. When this migration is complete, the separate domains used by the IBMPC company will no longer be needed. The enhancements in R4 of Notes have apparently eliminated any performance problems with one large N&A Book.

Figure 2.15 shows the multiple domains used by the IBM PC company when they were on R3 of Notes.

A Worldwide Example

The IBM Corporation has about 220,000 employees world-wide and had over 50,000 Notes licenses at beginning of 1996. Because 10,000 names was probably a good maximum for a single N&A book under R3 of Notes, the IBM Corporation clearly needed many different Notes domains. There were well over 100 major Notes domains defined within IBM (and many more domains that could be considered "test and devel-

Figure 2.15
IBM PC Company multiple domains.

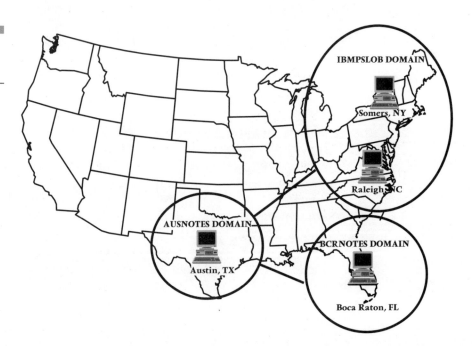

opment" or lab domains). These different "legacy" Lotus Notes domains are combined into a large IBM N&A book for routing purposes (see chapter 9), but the information in that N&A book is "barebones" for efficiency.

Figure 2.16 shows some of the Notes domains within IBM during 1996 and shows an example of the hierarchical hub-and-spoke domain structure. The secondary hub domain shown is at the IBM Mt. Pleasant, NY site that is headquarters for several IBM groups (e.g., IBM Latin America, IBM Asia Pacific, IBM Workforce Solutions, etc.). Each of these major groups at the site needs their own N&A Book, so this lends itself to the "site hub" concept. This site hub domain (called MTPNHUB) connects to the company-wide IBM_INTERNAL primary hub domain. IBM_INTERNAL has connections from IBM Notes domains from all over the world.

This worldwide example is a "snapshot" in time. IBM is migrating toward a new Notes architecture, based around Notes R4 and designed at the corporate level, rather than the "bottom-up" design where each group set up its own Notes domain. In the new "top-down" design, Notes domain names within IBM have the form IBM *xx n*, where *xx* is to be the two-character CCITT country designator, and *n* is a sequence number for those countries where IBM will require multiple domain names. For example, IBMNZ was designated as the Notes domain name for IBM New Zealand (NZ is the CCITT designator for New Zealand).

Figure 2.16
Example IBM
Corporation
worldwide Notes
domains.

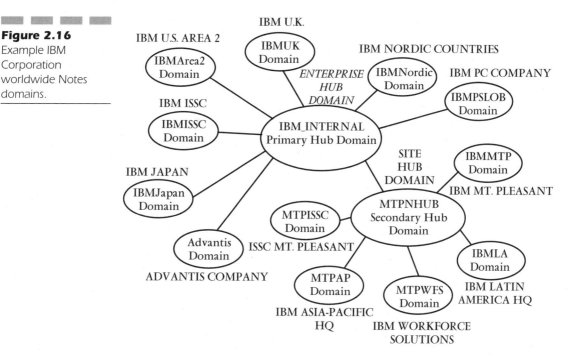

Notes domain names for IBM in the United States could have the names IBMUS1, IBMUS2, etc., although it now appears that the single domain name of IBMUS will be sufficient.

It should be noted that, because IBM does business in 131 different countries, the "new" IBM Notes architecture will still consist of well over 100 domains. However, the consolidation of Notes domains in the larger countries greatly simplifies routing.

Replication
The Heart of Lotus Notes

Replication is really hard. It's the core thing Notes does that
nobody else knows how to do yet.
Eric Schmidt, at the time chief technology officer of Sun Microsystems
(from July 8, 1996 issue of Fortune magazine)

Replication, the method used by Lotus Notes to synchronize databases, is
an essential aspect of the popularity of Lotus Notes. In fact, replication is
the "heart" of the Lotus Notes product. This chapter describes the concept
of replication and the different methods used for effective replication.

Over a small group (e.g., an office environment), the concept of a server for hub replication is unnecessary. However, for a large Notes environment, it is necessary to use the concept of hub replication. In this design,
all spoke severs replicate with the hub server and not with each other.
The hub server can be considered to contain the master copies of each
database. The reduced network overhead with hub replication is significant, because the alternative of allowing each spoke server to replicate
with every other server creates significant additional traffic when there
are a large number of Notes servers.

Replication of databases is a tremendously important feature of Lotus
Notes. However, partly because of the replication concept, Notes is not
suitable for "real time" transactions. There is no record-locking concept,
so there is nothing to prevent two Notes users from editing the same
document at the same time. An often-mentioned application that would
not be applicable for Lotus Notes is an airline reservation system. Two
people could end up with the same airline seat reserved!

Don't be discouraged, though, the vast majority of applications you'll
need work very well with Lotus Notes. Also, in the airline reservation
example, no reservation information would be lost. Both would be in
the database, with a replication conflict indicated. A Notes database
administrator must manually make a decision on which of the conflicting documents to delete (although Notes indicates which request was
submitted first, so you could still give the airline seat to the first person
to get a reservation).

Bandwidth Considerations

To replicate large Notes databases effectively, significant bandwidth is
required. Of course, it's possible to have a very large Notes database that
requires small infrequent updates. Documentation of processes for your
company might be that way, or you might keep all documentation on
Lotus Notes products on Notes databases. Then only the first replication to

a spoke server might be a problem, especially if the new replication is to be over a 9.6-Kbps dial line. 9.6 Kbps is 1200 bytes per second, so it would take 14 hours at 100% efficiency to copy a 60MB Notes databases over that type of line. Notes Release 3 (and below) replicates by shipping an entire document each time there's a change. Release 4 of Notes is smart enough to only send updated fields across the network. Depending on how large the documents are in a database, use of this new feature in Notes could make a significant difference in the efficiency of your Notes replication.

Sometimes it's very difficult to keep a dial line up for a replication that takes many hours. It becomes virtually impossible to create a new replica of a large database of, for example, 60MB. However, once the new replica is established, dial access is often a very economical way to keep a Notes database up-to-date. Some pack rats (like the author) might keep 60MB of Notes mail on a laptop. That mail database could be replicated initially via a LAN attachment. Then replication with dial access is a very reasonable way to keep your mail up-to-date when you're on a trip. Today's 56 Kbps modems make it even more reasonable. Because the replication process keeps track of that updates have been completed, losing a dial connection does not mean you've lost all of the replication work that was completed. Effectively, the Notes replication process has what database people refer to as "checkpoint, restart" (i.e., the process starts again from the interruption point).

For large-scale server-to-server Notes replication, a high bandwidth leased line (e.g., T1 at 1.544 Mbps) is a requirement. IBM has had groups that used dial access between servers as a very quick way to get coverage on a worldwide basis. The dial access was replaced with higher bandwidth links as the network requirements grew.

NOTE: *Lotus Notes with its rich text fields, graphics capabilities, and add-on features such as VideoNotes—will continue to put new stresses on your network. The demand for extra bandwidth will be both on your LANs and WANs.*

Replication is usually the most demanding Notes feature when it comes to bandwidth. Because Release 4 & 5 has made significant changes in replication, the next few paragraphs examine how these changes will impact the need for replication bandwidth.

Release 4 & 5 Changes to Replication

Release 4 & 5 of Lotus Notes include several changes to replication that have an impact on the efficiency of the replication procedure. Here are the important changes with a comparison to Release 3 of Notes and the potential impact on bandwidth requirements.

REPLICATION ON THE FIELD LEVEL. In Release 3 of Notes, any change to a document in a database causes the entire document to be copied to replicas of the database during the next replication. Release 4 or 5 offer more granularity, so only modified fields will be copied. The advantage to this change is faster replication than Release 3 for some databases. The best candidates for improvement are large-document databases in that most changes involve updates to small fields in existing documents. An example would be a training database in that each document has a video portion and a list that adds your name to it once you've viewed the document. Only the list of names will be copied during replication in Release 4 or 5.

A potential disadvantage would be increased overhead and longer replication times for databases that don't fit the previous structure. Release 4 or 5 might take longer to replicate databases made up of documents that contain many small fields. For a server to determine which fields of a document need to be replicated, the bookkeeping of updates must be handled for each individual field rather than just for whole documents. Moreover, even after a server has identified modified fields, updating each of them individually would probably consume more time than simply overwriting the entire document.

SIMULTANEOUS REPLICATORS. In Release 3 of Notes, a server's replicator can pull updates from only one other server at a time. Release 4 and 5 servers are able to run several simultaneous replicators. This change results in less severe consequences of occasional extraordinarily long replications. In a hub-and-spoke replication topology, for example, the hub's replicator can often be the constraining bottleneck. Any unusually long pull (due to a slow connection or a large number of updates at the spoke) ties up the hub's replicator and causes delays in the initiation of subsequent replications by the hub. In severe cases, subsequent replications are simply missed. A Release 4 or 5 server will be able to produce additional replicator processes as necessary and thereby perform simultaneous pulls.

Possible disadvantages to the concept of multiple replicators would be CPU overload, database-engine overload, and database corruption. The replicator is a CPU-intensive process. Several replicators running at once could seriously bog down the server. Several replicators simultaneously updating large numbers of documents could bog down the database engine, and serious problems could arise when several replicators try to simultaneously update the same database or, worse yet, the same document.

IDENTIFYING UPDATES. With Release 4 or 5, servers are able to identify much more rapidly those situations in that there are no new up-dates to be pulled. This results in less overhead associated with initiation of replications. This improvement makes it less risky to schedule frequent replications.

Bandwidth Measurement Methods

What are some of the methods that can be used to estimate the network bandwidth requirements for effective replication? In chapter 2 in the section "Measuring Bandwidth Requirements," some of the methods used by IBM were discussed. These methods can be used to estimate overall bandwidth required and to further break this down into the bandwidth required for Notes replication. In chapter 12, the section "Remote Management of Lotus Notes over the Wide Area Network" gives further details and refers to the measurement programs included with the CD-ROM.

Effective Replication Design

Because replication is the key to providing Lotus Notes users the ability to share information locally or across geographically disperse sites, it must be:

- Efficient
- Timely
- Reliable
- Fault Tolerant

Replication is also a key feature of Lotus Notes that allows mobile users the ability to stay "in synch" with their colleagues while on the road. There are two forms of replication: client-to-server and server-to-server.

Client-to-server Replication

Client-to-server replication is push-pull. This means that the client workstation "pushes" the data to the server and "pulls" the data from the server. In this scenario, the client is doing all the work. Replication performance is more dependent on the client than on the server. The client only has a user session open with the Notes server and is not burdening the server's replicator. This form of replication is independent of schedules and the server-to-server replication scheme. Because client-to-server replication is heavily used by mobile users, there is a strong need for efficiencies. Some of the efficiencies that the mobile user can utilize are:

■ *Selective replication*—This allows you to replicate only a few selected databases, rather than all the databases your workstation has in common with the Notes server. The benefits are a reduction in the amount of disk space needed on your workstation and a reduction in the dial connect time to your Notes server. Dial connect time is often a big cost factor for the mobile user.

■ *Background replication*—This allows you to continue to work with other Notes databases while data updates to selected databases are being replicated in the background. In fact, you can continue to access the database that is being replicated in the background and find out how things are going.

Server-to-server Replication

Server-to-server replication is dependent on a replication scheme. Some of the features of a replication scheme are:

■ Usually the spoke servers initiate replication with the hub server. This provides the greatest efficiency.

■ The interval at that replication occurs is determined by the replication schedules in the connection records in the Name and Address Book.

Setting up Connection Records

The Connection Records in your Notes Name and Address Book give the schedules for replication. It is best to have the spoke server set up the connection record and thus have replication controlled by the hub server's schedule. Figure 3.1 shows an example of a connection record for the spoke server GBLX1822 in the IBMUK domain to the hub server HCPNIBM in the IBM_INTERNAL mail routing domain. In addition to mail routing, servers in the IBM_INTERNAL domain act as replication hubs for the IBM Consolidated N&A Book called IBM.NSF. Therefore IBMUK uses the HCPNIBM server in IBM_INTERNAL both as a mail and replication hub. The connection record indicates that replication should start at 5:06 a.m. every morning with a repeat interval of 60 min-

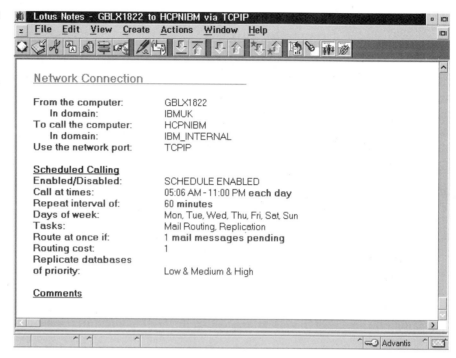

Figure 3.1

Replication schedules
in the Name and
Address book.

utes. TCP/IP is the protocol to be used for replication and mail routing over the network link.

Leveraging the Internet

If you have a Notes server that can access the Internet via dial or leased line, and you have a friend or colleague in any other part of the world with a Notes server and Internet access, then you and your friend can use the Internet to share Notes databases through replication. The protocol you would use is TCP/IP, of course, because that's what the Internet speaks.

This is simple in concept, but there are some warnings. Most dial access to the Internet is via SLIP (Serial Line Internet Protocol) or PPP (Point-to-Point Protocol). With any of these Interface dial methods, your workstation will be assigned a different TCP/IP address every time you connect. Thus you cannot tell your friend the TCP/IP address of your Notes Server on the Internet because it will change every time you connect.

One solution to this is to use a Notes Internet Service Provider and to use that provider's Notes replication server as a hub between your server and your friend's server. If you have a leased line to the Internet, you could have a fixed TCP/IP address for your Notes server, but your company might have a firewall between your TCP/IP domain and the Internet that would prevent complete access to your friend's Notes server (which would also need a fixed TCP/IP address). Some information on Lotus Notes Network Service Providers for the Internet is given in chapter 1.

Replicating a Company-Wide Database Repository

The IBMNOTES Database Repository was described in chapter 1, in the section "Different Methods of Interconnecting Notes Servers and Domains." This database repository replicates databases from many sources, including the databases from the LOTUS domain. Figure 3.2 shows a connection record for replication from the IBMNOTES domain to the IBMLink-External domain. Note that this is a dial connection (using XPC dial). IBMLink-External serves as a dial in Notes repository for other companies and contains the "IBM Announcements" and "IBM

Figure 3.2

Dial replication schedule for the IBMNOTES database repository.

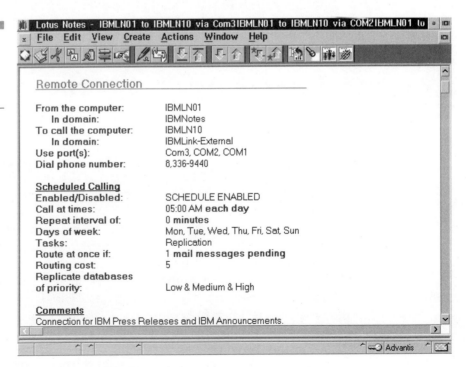

Lotus Notes - IBMLN01 to IBMLN10 via Com3IBMLN01 to IBMLN10 via COM2IBMLN01 to

File   Edit   View   Create   Actions   Window   Help

Remote Connection

From the computer:	IBMLN01
In domain:	IBMNotes
To call the computer:	IBMLN10
In domain:	IBMLink-External
Use port(s):	Com3, COM2, COM1
Dial phone number:	8,336-9440

Scheduled Calling

Enabled/Disabled:	SCHEDULE ENABLED
Call at times:	05:00 AM **each day**
Repeat interval of:	0 minutes
Days of week:	Mon, Tue, Wed, Thu, Fri, Sat, Sun
Tasks:	Replication
Route at once if:	**1 mail messages pending**
Routing cost:	5
Replicate databases of priority:	Low & Medium & High

Comments
Connection for IBM Press Releases and IBM Announcements.

Advantis

Press Releases" databases among others. IBMNOTES dials this database for replication updates once every morning at 5 a.m.

Figure 3.3 shows the connection record to IBMNOTES from a typical spoke server located in the IBMUK domain in the UK. The connection is over a high-speed network link and requests replication every 60 minutes starting at 12:12 a.m. every morning. The Notes log on IBMNOTES will indicate which databases are replicated by IBMUK. To look at the replication events, open the server log, do a View/Replication Events, and open the record for the server in the spoke domain (GBLX1822 for the IBMUK domain from Figure 3.3).

New Strategies for Notes Replication

On the small scale for which it was originally designed, Lotus Notes replication works very well. For example, if you work for a small company and you're on the road, you might dial in to your office's Notes server to synchronize copies of a Notes database. In a matter of minutes, every change to the database since your last replication appears in the database

Figure 3.3
Replication schedule to IBMNOTES from the IBMUK spoke domain.

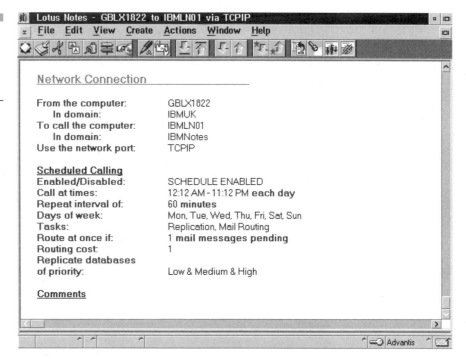

copy on your laptop. Except for the occasional busy signal, the process works very well.

The process, however, doesn't always work so well for large Notes operations. The problems all lie in scale, and a typical large Notes installation might have several thousand active notes databases distributed among hundreds of Notes servers worldwide. By design, any one of these databases can be updated by any of its users on any of the servers, from anywhere and at any time.

In such a dynamic environment, keeping crucial information current is a difficult task. The secret lies in proper scheduling and giving the responsibility of replication to the right people. It also means that there's no "one size fits all" solution. Database replication in your organization often requires a unique solution.

When Lotus designed replication into Lotus Notes, it did so with the model of a small number of servers and a small geographical distribution in mind. Few people foresaw that within a decade, Notes replication worldwide would grow to the point where entire consulting companies would be dedicated to making it work better.

That's because many distributed organizations today choose to have access to a local copy of global information rather than access to centrally stored information. Regardless of the distributed data setting, replication is the underlying process by that multiple copies of the same data are synchronized, creating the illusion that distributed users are all sharing one set of data. Replication clearly serves as a means of overcoming technological and geographical boundaries among distributed members on an organization or workgroup.

The Complexity of Large-Scale Replication

When Notes replication is scaled up for the large installation, the complexity can become overwhelming. For instance, draw 300 points on a piece of paper and consider how you might connect them. Then assign a time zone to each one and think about when you might connect them. Replications in a large organization are initiated according to a fixed schedule. The schedule determines the logical replication topology by stating that pairs of servers should replicate with each other. In this case, the schedule determines which of the two servers should initiate the replication and at what time. Both the complexity and importance of designing a reliable, "custom-fit" replication schedule are hard to overstate. Here are some of the main constraints that you'll face:

- *Duration of replications*—Notes replication takes a long time: minutes or hours, not seconds or milliseconds. The overhead of executing the process often creates more of a bottleneck than the volume of data does. Moreover, the duration of replications can differ at both ends. A hub's replicator can remain idle and not initiate a scheduled replication if the telephone line is tied up while the partner/server from the last replication continues to pull its updates.

- *Single-threaded replicator*—With Lotus Notes Release 3, a server can pull updates from only one other server at a time. Whenever it is pulling, it cannot initiate additional replications and will not respond to replication requests sent by other servers. Release 4 of Lotus Notes allows multiple concurrent replications so that this concern disappears.

- *Time-zone constraints*—For international organizations, scheduling across time zones poses a major challenge. For example, the low-usage lunchtime hour might seem like a good time to schedule replications. However, the difference between 11:00 a.m. and noontime in New York is the difference between getting updates to European users before or after the end of their business day.

Using the Right Replication Topology

In addition to devising the right replication schedules, there are two other factors to consider. First, ongoing monitoring and a periodic, systematic review of the entire replication system are of paramount importance. As Notes matures within an organization, usage and needs grow dramatically. Such growth often renders the existing topology and replication schedule obsolete. This is especially true in the transition from the pilot phase to the substantial roll-out phase. On average, the entire replication strategy should be reviewed annually. It is sometimes very difficult to visualize the replication that's going on in your network. NotesView provides a good tool to map out your replication topology.

Each Notes implementation is unique in size, infrastructure, capabilities, and needs. Some companies use Notes to disseminate a small core of information that is generated in one central location. Others have massive, highly interactive applications in that documents are generated and magnified by users scattered all around the globe. In some organizations, 24 to 48 hours for the propagation of updates is sufficient; others need several cycles per day. Some organizations commit substantial resources to monitoring performance, while others dedicate minimal resources.

On the small scale for that Notes was conceived, replication is a nonissue. As Notes networks have grown, many organizations find that their

Notes servers become completely tied up with the replication process. For example, a Notes server might have to replicate with 20 other servers worldwide and before it completes the 16th server, it is time to start replicating again with the first server.

There are several approaches available to remedy the limitations of Notes replication for the large-scale situation. One approach is to outsource your Notes replication tasks to a Lotus Notes Network Service Provider. In this case, you need only to replicate once to the service provider and that provider has the responsibility to replicate to all your other servers. Now your Notes server resources are free again.

Another solution would be to set up a hub-and-spoke replication design for your organization. This allows the hub server to fan out replication to all the spoke servers rather than having each spoke server replicate with all the other spoke servers.

A third approach would be to use a queuing system that will off-load much of the replication process from Notes and remove the limitation of a single replication at a time. The MQSeries is such a queuing system solution and is discussed in the last section of this chapter. The next section describes the cache method used by Web browsers, which has some of the same function as replication.

Replication vs. Cache

Netscape does not replicate information, but it synchronizes information by preemptively distributing frequently used Web pages to server caches. Preemptive caching limits the impact on network bandwidth.

Netscape's Proxy Server 2.0 will cache frequently accessed Internet or intranet documents and use statistical analysis to determine that ones users are most likely to request. A batch-retrieval process downloads a group of URLs (up to 128GB of data or 70 million URLs) to ensure that the server caches the most popular sites and then makes them available. Netscape claims that, when it deploys Proxy Server 2.0, there will be a 50% to 70% chance that the servers will cache a document locally.

Cache Methods Used by Web Browsers

As discussed in the previous section, the most popular Web browser (Netscape Navigator) synchronizes information by using preemptive caching. All Web browsers use some caching, but it can be assumed that the Netscape model will be technologically in the front of the pack.

Interactive Requirements

For many interactive requirements, on a user's workstation, caching and replication give a user the same benefits. A user is able to work with the latest database information on his workstation. This limits the impact of network bandwidth on his ability to quickly browse through information.

Caching and replication both bring the latest information to the hard drive on the user's workstation and allow the user to work with the information on his local drive. The improved response time makes for much happier users, especially when that user has dial access to his server and bandwidth has a significant impact on interactive response time. Replication, however, shines when that same user wants to work with data while disconnected from the network and then later "hook up" to the network and synchronize his work with the data on his server. This is the *mobile user requirement*.

Mobile User Requirements

Notes fans tout one other important aspect of replication: mobile workers. A traveling sales executive, for example, can connect a laptop to a Notes server, replicate, and then work on databases, documents, Web pages, and even discussion threads while on a plane. When the executive replicates again, that work is updated throughout the system. Intranets have not add-ressed this aspect of replication yet. However, some argue that the universal dial-up capability to the Web and the evolution of wireless modems will lessen the importance of this Notes feature.

Use of MQSeries for Lotus Notes Replication

One way to help solve replication problems when employing Lotus Notes over the enterprise is to off-load the replication process to another system that has an architecture that might solve some of the problems. One such system that has been used within IBM is the MQSeries. In this scenario, MQSeries has a role of a message queuing system. This concept made a lot of sense with Release 3 of Lotus Notes where there was only a single-threaded replicator, and we didn't have field-level replication. With the replication enhancements that came with Release 4 and 5 of the Notes and Domino products, the benefits of off-loading the replication process to another system are not so clear. Nevertheless, it is worthwhile to discuss the concept by using the ARE product as an example.

The Adaptive Replication Engine (ARE) is an application product offered by Technology Investments, Inc. (TI). It is used in conjunction with the Lotus Notes groupware product to provide additional functionality to the inherent Lotus Notes facility for replicating Lotus Notes databases. ARE's functionality addresses the following user considerations when replicating databases on a large scale: connectivity, performance, and database integrity.

Connectivity is a very important consideration for a database and/or network administrator. Each server that participates in a replication must have a dedicated, bidirectional connection and the connections must be active. The connection must be defined to and maintained at both servers. The ARE solution is time-independent, and the sender can continue processing without receiver acknowledgment or operator intervention at a remote site.

Performance plays a significant role in the replication of data. Tight replication windows precipitate delays in daily production start-up or cause inadvertent or unwanted usage of old data. The ARE product is built on top of IBM's MQSeries product, which is considered industrial strength middleware and is quickly becoming an industry standard. ARE's transport layer utilizes the messaging and information backbone provided by IBM MQSeries.

Database integrity is a prerequisite when replicating databases. The ARE product enhances the transfer of data with a high degree of confidence in database integrity. It offers a solution that meets the needs of users with strict requirements for the accessibility to Lotus Notes data and for systems that have high demands on moving Lotus Notes databases. ARE's data delivery layer is implemented using a syncpoint mechanism for the recovery of important data in the event of system failure.

The ARE product, used in conjunction with Lotus Notes, can be used over value-added public networks and/or customer-owned and-managed private backbone networks. The structure of the ARE product's database replication solution makes it easy to add support for new communications and network protocols, thereby adding flexibility in managing and replicating databases in today's complex heterogeneous networking environments.

The ARE solution for database replication is the heart of an adaptable infrastructure that is not only robust and dependable but that can expand with the adoption of new solutions.

Overview

IBM MQSeries contains a set of messaging and queuing services that support data transfer between distributed applications. These services allow applications to communicate without knowledge of the lower levels of the communications network and without specific knowledge of the location of the other applications.

ARE has been implemented with the following features and functions that provide enhancements to the Lotus Notes database replication facility and utilize the strengths of the IBM MQSeries product for distributed application data transfer:

- A stable, high performance, connectionless transport facility that also provides asynchronous processing and assured delivery.
- Persistent resynchronization of databases in the event of a system failure.
- Parallel processing that increases load balancing thereby reducing replication time.
- A single point of administration and control of all servers in the network.
- Administrator manipulation of replications on any database in the network.
- Enterprise-wide replication scheduler.
- Enterprise-wide interrupt on scheduled or manual replications.
- Audit, statistics, security, and exception control.

Connectivity

Remote distribution of Lotus Notes databases is an option depending on the overall business requirements. Databases can be replicated locally without additional communication protocol stacks.

ARE uses IBM's OS/2 version of IBM MQSeries as its transport layer. Distributed database connectivity can be via SNA, TCP/IP, NetBIOS, or Novell IPX using Novell NetBIOS emulation.

Open Standards for Replication

The Lotus replication process is often considered the crown jewels of the Lotus Notes product. No other product handles replication in the sophisticated way that Notes does it. This is a mixed blessing.

During the last couple of years, Lotus has been striving to embrace the open standards of Web technology with its products. This is probably the main idea behind changing the name of the Lotus Notes server to the Domino server. Lotus Notes has long been acknowledged as the leader in groupware technology, but much of that technology was proprietary. Changing the server name to Domino was a way to indicate that this server was based on the open standards of Web technology and that customers should not associate the name with proprietary technology they had associated with the Lotus Notes name.

Okay, so Domino embraces the open standards of Web technology. However, the Lotus replication technology is still proprietary. The press has reported attempts by Lotus to "give away" their replication standards specifications. The reports stated that Ray Ozzie wanted to do this if Microsoft and Netscape would promise not to start a replication war. So far the Lotus replication technology is still proprietary. We'll have to wait to see if this will change in the future.

CHAPTER

Mail Routing

E-mail is a requirement for any groupware application and is the first of groupware's "three Cs," communication, which generally means e-mail and related functions.
paraphrased from Fortune magazine, July 8, 1996

Mail routing, like replication, is also an essential aspect of Lotus Notes over the enterprise. For a large Notes enterprise, the concept of a hub domain is an essential part of Lotus Notes network design. Other domains need only to exchange certificates with the hub domain, and this allows them to send mail to any of the other domains connected to the hub domain. The IBM_INTERNAL domain is used as the hub domain for mail routing within the IBM Corporation. There are over 100 domains connected to IBM_INTERNAL, so the use of this domain prevents a very wasteful alternative of having each of the domains connect to every other domain.

It should be noted that the IBM_INTERNAL hub domain is part of IBM's legacy Lotus Notes architecture. IBM's new Notes architecture (see "IBM's 'New' Notes Architecture" in chapter 2) follows the concept of having one domain for the enterprise. In IBM's case, "the enterprise" is on a country basis. So, for example, there is only one IBM mail domain in the United States (the IBMUS domain). This new architecture eliminates the need to have a hub domain for connecting all of the other IBM mail domains in the United States. In the new architecture, IBM has a common organization certifier (/IBM) that eliminates the need to cross-certify IBM domains for each country. In the new architecture, mail is routed directly from one major country domain to another. However, the concept of a hub domain for mail routing is still very valid, especially for those organizations where each site or major group has their own mail domain (as was the case within IBM with its legacy domains).

Effective Design: Hub and Spoke

The hub-and-spoke design for mail routing follows the same concept as for replication. In fact, often the same hub-and-spoke design is used for both replication and mail routing, but it doesn't have to be. The requirements for replication and mail routing can be quite different. For example, it might be sufficient to replicate large databases only once a day at 5 a.m. However, mail should be routed as soon as possible. Thus many

Notes installations have separate connection records for replication and for mail, and thus potentially different hub-and-spoke designs.

Figure 4.1 shows a hub-and-spoke design for mail routing.

Dial vs. Fixed Bandwidth

Bandwidth for mail routing is not nearly as important as bandwidth for replication or database access. If electronic mail is delayed by a few minutes due to heavily used bandwidth, that's no big deal. Replication of databases is also not seriously impacted by lower bandwidth capabilities, except when the replication cannot be completed in the time requested due to inadequate bandwidth. Interactive access to Notes databases will always be impacted by low bandwidth. If you dial your Notes database at, let's say, 14.4 Kbps, then you'll notice the bandwidth problem, especially if you're used to LAN speeds.

One solution to this last problem is to replicate the database to your workstation's hard disk. Then your response time for this database will

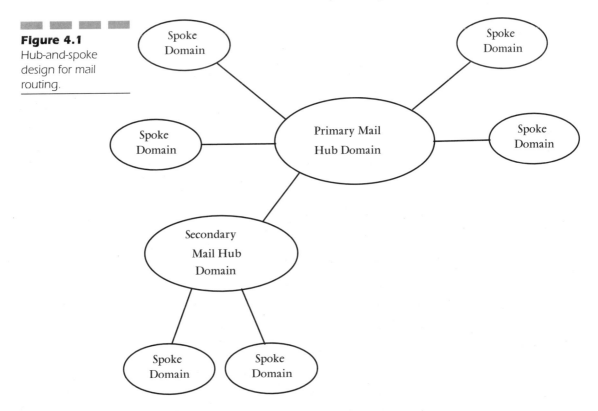

Figure 4.1
Hub-and-spoke design for mail routing.

only be dependent on the speed of your workstation. That sounds fine, but it only works if you have plenty of hard disk space available on your workstation. If the database you want to access is 60MB, and you only have 40MB to spare on your hard disk, you might get away with replicating only part of the database.

If the database is a discussion database that goes back three years, you could just replicate the last three months of the database. Then the storage required on your hard disk might be less than 10% of the entire database, and you might only be interested in the last two or three months of data.

To replicate only the last three months of a database, click on the database icon for the database on the server. Then select File|Replication| New replica. You'll see a screen similar to the screen shown in Figure 4.2. If you click on the button Replication settings, you'll get an option to click on the box indicating "Remove documents not modified in the last 90 days." If you select that option, you'll get what you want. Ninety days happens to be the default setting, but you can change this to anything you want. If this is a discussion database, for example, then you can make appends or responses to appends while your workstation is offline. A two-way replication will then assure that both your appends get back to the server and new appends by others are added to the replica copy on your workstation.

Generally speaking, mail routing should not be nearly as demanding on bandwidth as replication. After all, most mail routing is done on a store-and-forward basis. For Notes, this means that any of the mail servers, mail hub servers, or Notes gateways will store mail until bandwidth is available to send it on its way to the next destination, whether that's the final destination or another intermediate stop along the way.

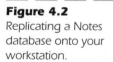

Figure 4.2

Replicating a Notes database onto your workstation.

New Replica "IBM Customer References"		
Server:	Local	OK
Title:	IBM Customer References	Cancel
File name:	IBMREF.NSF	Help

Encryption...   Size Limit...   Replication Settings...

Create: ○ Immediately   ● Next scheduled replication

☑ Copy Access Control List

☐ Create full text index for searching

With Lotus Notes these mail messages can be very large, containing graphics, video, attachments, and whatever. So the bandwidth requirements to get this type of mail delivered can be substantial. Small messages should fly through. Usually, users have a tolerance for waiting for large messages.

Traditionally, for mail systems, very large messages or files have to wait until off hours (e.g., the wee hours of the morning) for transmission. This is a good idea for Notes mail also. Schedule large messages for off-hours. That will free your bandwidth during the day for high priority small messages.

Very large pieces of mail can impact not only bandwidth, but hard disk space on mail hub servers. At times, hub servers in the IBM_INTERNAL domain have had problems routing mail when there was not enough disk storage available for the mail.box when users sent 100MB messages. In this case, the hub server would reject the message stating it didn't have enough storage. The solution was to make more storage available on the hard disk containing the mail.box. In the case I'm referring to, the mail message was so large because a large database was being sent as a mail attachment. With VideoNotes, detailed graphics, and large databases, very large pieces of Notes mail will undoubtedly become more common. A 100MB piece of mail can be transmitted over a dial line; however, at 28.8 Kbps, it will take over eight hours and the chances of failure are very high.

Setting Up Connection Records

There are different options for delivering mail messages, which are controlled by the "Connection Records" in your Lotus Notes Name and Address Book. You can choose to have:

■ *Messages are delivered only during scheduled replication*—This might delay mail delivery by several hours, so this method is not usually recommended. If the connection between servers is by an expensive dial solution, then cost considerations might drive having mail delivery happen only during scheduled replication.

■ *Messages are delivered whenever a threshold of pending messages has been reached*—If you have a leased line between your servers, the "threshold" of pending messages should be set to one (i.e., mail should be delivered immediately). If you don't specify "one," then to avoid having a "poor lonely message" waiting in a queue for other messages to arrive so that a threshold is reached, a maximum wait time is always scheduled. This maximum wait time is usually the same as the replication schedule. For example, if replication between servers occurs at least every two hours, then the maximum wait time for mail to get sent is also two hours, no matter to what value the Notes administrator has set the threshold.

Some sample connection records added to the ADVANTIS domain's N&A book are shown in the following view of that N&A book (Figure 4.3). Notice that some of these connection records are for both replication and mail routing.

Setting Up Non-Adjacent Domains

To send mail to a Lotus Notes domain not directly connected to your domain, you'll have to set up a non-adjacent domain entry in your Name and Address Book. To do this, click on your domain's N&A book, then click on View in the menu bar at the top of your desktop. Then select Server|Domains, and you'll see a list of the non-adjacent and foreign domains, similar to the one shown in Figure 4.4 will. Select Add domain, and you'll be able to add a non-adjacent domain for mail routing (see

Figure 4.3

Example of Lotus Notes connection records for replication scheduling and mail routing.

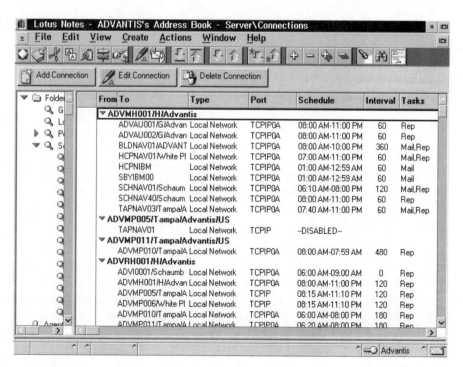

Figure 4.4
Example of a screen for selecting non-adjacent domain records for mail routing.

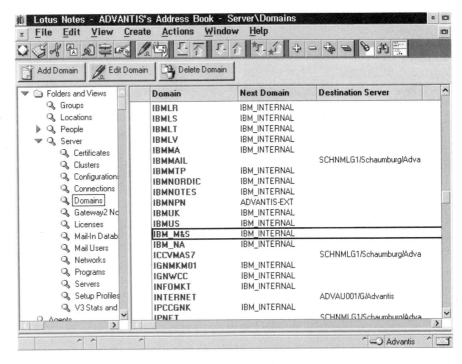

Figure 2.5 for an example of the screen for entering the non-adjacent domain).

Using Alternate Routes for Notes Network Backup

If your Notes domain has only one way to access another Notes domain, then a failure in any of the components in that route will cause access to be lost. Figure 4.5 shows an example. In this case, a failure in Notes hub server 1 in Domain A, a failure in Notes hub server 1 Domain B, or a failure in the wide area link connecting these two servers will cause access between Domain A and Domain B to be lost. Hub server 1 in Domain A, hub server 1 in Domain B, and the wide area link are all called *single points of failure* in the connection between Notes Domain A and Notes Domain B.

This section will discuss the different ways of setting up alternate routes in case any component fails in the primary route. The alternate routes can be established through another connection between two different servers in the two domains. The connection might use the same

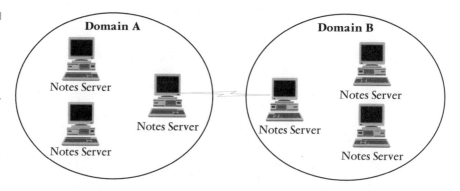

Figure 4.5
Example of a single route between Notes domains with single points of failure.

physical wide area connection, but it's a different logical connection as far as the Notes system is concerned. Also, any of the connections depicted could be leased lines or dial lines.

To backup a leased line, dial backup is often the most economical way to go. The dial backup is used only when there is a failure in the leased line, which might be very rare. Thus you might go months (or years!) without needing to use the dial backup. However, it's there and will automatically be used when needed.

Even with alternate routes established between your Notes domains, it is still easy to have single points of failure. Figure 4.6 shows an example. In this case, a single server in Domain A has routes to two different servers in Domain B. Thus a failure in one of the Domain B servers will still allow communications between the two domains. However, if the server in Domain A fails, no mail (or other communications) between the two domains will take place. Thus alternate routes are available, but the single point of failure in Domain A might be a problem. On the other hand, perhaps there is only one server in Domain A, and this type of backup route is all that is required.

The ideal way to establish alternate routes is to have no single points of failure that will cause loss of access between your two domains. Figure 4.7 shows an example of such a design. In this case, failure of a single server or failure of a single connection line will not disrupt communications between the two domains.

Once you've established the design for your alternate routes between domains, the implementation is then established with connection records in your N&A book.

Figure 4.8 shows examples of connection records in the IBM_EXTERNAL domain to reach the IBM_INTERNAL domain. This is a one-to-two type connection. The NOTES-GW server in the IBM_EXTERNAL domain is our SMTP gateway from the Internet to Notes (see "How Mail is Delivered" in chapter 8). NOTES-GW has connections to HCPNIBM

Figure 4.6
Example of an
alternate route
between Notes
domains with a
single point of failure.

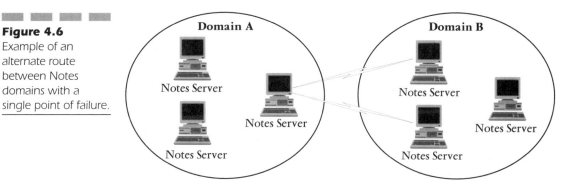

Figure 4.7
Example design
showing
connections
between Notes
domains with no
single point of
failure.

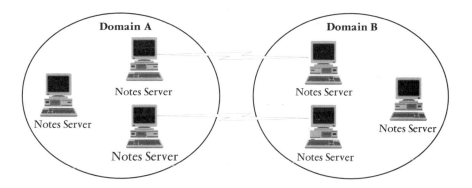

Figure 4.8
Example connection
records showing
alternate routes.

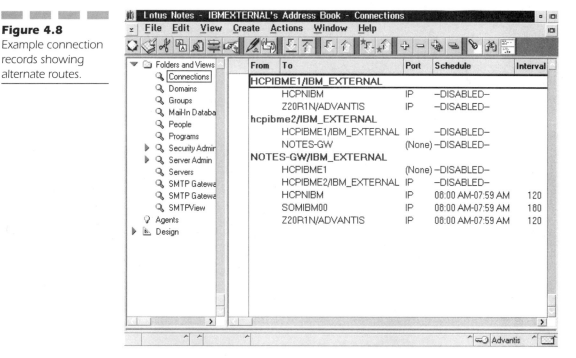

and SOMIBM00, which are two hub servers in the IBM_INTERNAL domain (at White Plains and Somers, NY, respectively). When NOTES-GW has mail for an IBM domain connected to IBM_INTERNAL, it will first try sending the mail to the HCPNIBM hub. If HCPNIBM fails to respond, after about five minutes, NOTES-GW will then reroute the mail to SOMIBM00. Any of the hub servers in IBM_INTERNAL (there are 10 of them) will be able to route to any of the 100+ Domains connected to IBM_INTERNAL. The alternate routing works quite nicely (we've seen it in action many times). Alternate routes between Notes domains are important both for mail routing and for replication.

Least Cost Mail Routing

Lotus Notes uses a "least-cost mail" routing algorithm to determine which alternate route to use. This section describes the logic used.

The mail router for WAN communication calculates explicit cost and attempts least cost mail routing. The Lotus Notes Release 3 route selection algorithm was enhanced to maintain a history of connection failures. It uses the failure history to weigh its routing choices. In this way, the router should be able to use alternate routes to re-route messages around network failures.

The routing table selects a least-cost path and tries to transfer the message to the next hop server. If the transfer fails (e.g., no answer or network timeout), the router adds an additional cost bias of 1 to the connection (or server) entry in the routing process.

When it retries the transfer, the cost via that connection will be higher than it was before. If there is an alternate route to the destination that was equal to the original route's cost, it will cost less than going through the original route, and the router will choose the alternate route.

Likewise, when a server receives an incoming phone call, the router sets a bias of −1 to the connection. This causes the new incoming connection to be used if it was previously equal to other connections. This form of bias is "one-shot" and will be deleted once the router has gone through its routing process once.

Because a routing failure is a locally detected condition, and because routing table updates via the Address Book is only replicated at a low rate, other servers will not be able to make the same routing choice that the local server has made. This lack of propagation of information can, in many cases, cause the message to not be routed properly or, in the worst case, cause insidious routing loops. Using only a bias of 1 means that, even if one server chooses a different route, it will never cause a misroute or loop because returning the message via the connection will cost at least a value of 1. In other words, because the algorithm is simply choosing a different path among equal cost paths, a loop is not possible (at least not one that didn't exist before the alternate route selection).

You should note the following:

■ The failure history (the cost biases) are kept in the routing tables. The routing tables are memory-resident and re-initialized when the server is restarted and the Address Book is modified. Therefore, the failure history is lost when either of these events occur. After one of these events, the router returns to the original routes until it "learns" to again use the alternate routes.

■ For the router to return to using the original route as soon as possible after it comes on line, the router will delete an individual connection's failure history whenever it receives a Link Notification Event (i.e., an incoming phone call). This will handle the case of the dead server coming back up and calling the original server. To handle the case of the dead server/connection coming back up but never calling the original server, all the failure histories will be deleted once an hour.

■ Only alternate paths of equal cost will be used. This means that there will be alternate paths that might exist but will not be chosen (see below for why this limitation is being imposed). Although this won't handle all network failures, it should solve many of the common ones, including:
  ■ Multiple dial-up servers between LANs and domains
  ■ Multiple servers within a LAN (i.e., bridges)

■ Preventing routing "loops" is a major problem with any kind of dynamic routing algorithm. One way to detect loops is to set a maximum hop "count" in each message and decrement it as it transits a server. A message is then returned or discarded when the hop count reaches zero.

Although a hop count prevents infinite loops, it doesn't avoid the following sort of loop. If a server chooses an alternate route, the next hop in the alternate route might decide that the least cost to the final destination is the server the message just came from. This could occur because the first server chose the alternate route, but the next server still thinks that the original route was best. (The servers did not make the same routing decision.)

Due to the complexity of solving this problem in the general case (e.g., via more frequent routing updates, sending the failure information along with the message, or sending a copy of the route already used with the message), the Notes R3 router will simply maintain the hop count (in a new item $HOPS). This will prevent routing loops when the Address Book does not properly converge via replication (e.g., the ACLs on one or more copies do not allow updates). Routing loops will be avoided in the normal case because, as mentioned previously, only alternate paths of equal cost will be chosen.

IBM's "New" Mail Routing Architecture

The IBM Mail domain strategy calls for a single mail domain by country. The choice of defining a mail domain by country was based on the following:

- It facilitates support for national language issues (for example, keyboard drivers and code pages).
- It is consistent with IBM's Internet domain structure and naming convention.
- High communicators tend to be within a country, and mail is routed most efficiently within a domain.
- Movement of individuals tends to be within country, thereby significantly reducing the administrative activities associated with user domain changes.
- It is a universally understood model.

The intent is to have one mail domain per country. In the event a country has a large number of IBM personnel and N&A book performance becomes problematic or other constraints are reached, a fallback alternative design of more than one domain per country is in place. By far, the largest single domain is for the United States. There is only one domain for the U.S. (IBMUS), and N&A book performance is not a problem. Release 5 of Notes and Domino significantly improves the performance of large N&A books, so it is very unlikely that there will be additional mail domains in the United States. IBMUS will include all 100,000 plus IBM employees in the United States. Having one mail domain per country also coincided with IBM's strategy for mail addressing with external companies, so that users have Internet and Notes addresses that are as similar as possible.

The end result is that each Notes user is a member of the mail domain representing his or her country, of which there are 131 to support all the countries in which IBM does business. Similar to the database domain model, mail servers are deployed within mail domains using a hub-and-spoke architecture. Within each mail domain is at least one hub that is responsible for routing all message traffic into and out of the domain.

Once received at the hub, the message is sent to the appropriate user mail server. There is also a single mail routing hub at each of our nine major Control Centers, which controls messages flowing into and out of each geographic region. The Control Center mail hub also serves as the domain mail hub for the country in which the Control Center is physically located. Each subordinate domain hub routes all outbound mail to its Control Center hub, which in turn directs it either to a remote Control Center hub or to another mail hub within the same geographic region. Control Centers are fully meshed to ensure that mail among the major population centers traverses no more than three hops and that no more than five hops separates any two users.

This topology concept is depicted graphically in Figure 4.9. Note that it would be very similar to a figure depicting intra-site replication. Just replace the caption "IBM Mail routing Topology" in Figure 4.9 with "IBM Database Replication Topology," and you have the replication topology figure!

Within each mail domain, all of the mail server spokes share a single Notes named network so that messages addressed to other users in the

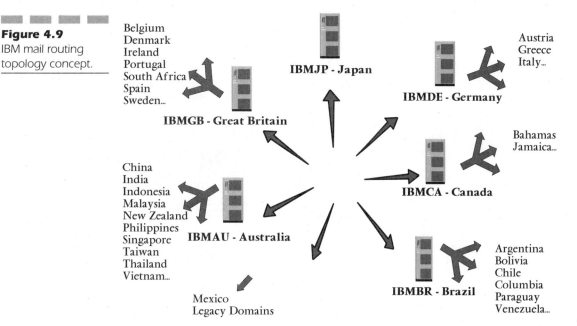

Figure 4.9
IBM mail routing
topology concept.

same domain are routed immediately and do not impact the hub. We have found that this structure is extremely effective in the IBM environment where the majority of mail is sent to people in the same country, and therefore the same mail domain.

How Big Should Mail Files Be?

A significant advantage of using Notes mail is that you can easily send file attachments of any kind. However, that's one of the reasons that it's very easy for users to have very large mail files. My mail file on the server is often over 100MB. When I go through and get rid of my mail with large attachments, I can reduce that file size by half. Even with today's large hard drives, allowing everyone 200MB for Notes mail files will create a problem. In the IBM roll-out of Lotus Notes, users are allowed up to 30MB for mail files. If your mail file grows larger than that, you will be subject to extra charges for the extra storage.

A feature of Notes Release 4 allows you to physically restrict the size of a database. You can apply this feature to mail databases, and it works well. However, you should be aware that the messages that are displayed to the user are not very descriptive. The messages indicate that the database is reaching its maximum size and that you shortly might not be able to update the database. However, you are not advised to delete some documents to fix this problem.

Database quotas work by allowing an administrator to set a size restriction or quota on a database so that the user cannot save anything to that database if it gets larger than the quota. In the case of mail, users can still send and receive mail but cannot save anything to their mail database. You can also set a warning just below this quota that advises them that the database is getting close to its maximum size.

From a "customer care" standpoint, it's probably better to have your Notes administrator advise users to reduce their mail rather than physically restrict the size of their mail files. Because mail is often a very critical part of a business process, not being able to save new mail might be more of a real business cost than the problem of having to periodically remind a few "pack rats" to reduce the size of their mail files.

The Notes R4 single-copy object store can be used to save the space required for mail. The shared-mail facility allows messages that are sent to several people to be stored in one central object store, with just the header of the message sent to the user's individual mail file. To the user, this is transparent. On the server, it can save up to about 35% of the storage used compared to storing that mail in individual mail files. This is because the number of duplicates of the same mail message is reduced. Shared mail is especially effective when files attachments are sent to multiple people.

Of course, you can only directly access the single-copy object store if you are connected to the network. So, if a user is working at home and not dialing in to connect, he would not be able to access this object store and so would not be able to access any documents that were sent to several people. However, the R4 replication process solves this problem. When a user replicates his mail to a local drive, whole documents are replicated from the single-copy object store, not just the headers of the documents from the user's mail database. Therefore, a user can work remotely and see all the single object store documents in his mail database. Of course, because he's replicated all the single object store documents to his local hard drive, he'd better have space available on his local hard drive!

Another way to keep mail files down in size is to send less Notes mail, especially routine mail to a large distribution list. Often, a discussion database can take the place of much of this mail. Instead of sending mail with a large attachment to everyone in your department, the item could be posted to your department's discussion database. This has much the same effect as a single-copy object store and doesn't use any storage in the mail files at all.

However, as good as this sounds, it often isn't done because the discussion database isn't as good an "attention grabber" as sending mail. Also, the readers of the discussion database might not be exactly the readers you want to receive the attachment. One alternative would be to post the item with the large attachment in a discussion database and then send out mail with a "doc link" pointing to that item. The point is that there are lots of ways to help reduce the amount of DASD required for mail files. Keep in mind, though, that Lotus Notes is meant to be a productivity tool, and it's best not to put any more restrictions on this productivity tool than necessary. So be liberal in your policy for allowing user storage on your Notes mail servers!

Named Networks

If you are going to implement one domain for applications and mail throughout your company, there is also the decision of how to divide your named networks. The rule is that servers can be grouped into a named network if they are running the same protocol, and they are constantly connected (i.e., not dial connected).

If all your servers worldwide are running TCP/IP, then they fulfill the conditions and can be put into one named network. This has definite advantages, because all mail will route at once and no connection documents are required. Also mail will route directly from the source server to the destination server, so mail routing will be efficient and mail routing problems will be easy to troubleshoot if mail is not delivered. Therefore, for R4 installations (and above) with the same protocol (usually TCP/IP) over a WAN, one named network is usually best. However, very large domains might be an exception. IBM, in its Notes R4 roll-out, has one domain for all users in the U.S., but it has a different named network for each of its large Geoplex Sites. See the section "IBM's 'New' Mail Routing Architecture" in this chapter for details on this mail routing architecture.

With R3 of Lotus Notes, there were some additional reasons for having different named networks. With R3 and one named network, when users did a File|Open Database, they would see a long list of all the servers in the domain. For that reason, under R3, Notes designers would often have one named network per major company location. However, starting with R4 of Notes, when users do a File|Database|Open, they no longer see a long list of all the servers in the named network. In fact, they only see servers they have previously accessed. This Notes method of only showing you the servers you have previously accessed is somewhat like your "bookmarks" or "hotlist" under your Web browser, except that you don't have to request that a Notes server get added to your "Notes bookmarks." That automatically happens as soon as you access a new Notes server for the first time.

The nice thing is that the servers that get saved in your "Notes hotlist" can be in any named network and in any Notes domain. This is a big improvement over the R3 policy of displaying all servers in your named network. For R4 and above, users don't see all of the servers in the named network, so this previous aspect of named networks does not apply when using R4 or R5 of Notes.

Mail Gateways

For Lotus Notes e-mail to be effective, you must be able to route the e-mail via gateways to other mail systems efficiently and with high reliability.

Gateways are devices that handle data translation between networks running different protocols. Network bridges or routers handle traffic between networks running the same protocol. If you need to send mail from Lotus Notes to a mainframe mail system, such as PROFS, a gateway is required. Gateways provide the message translation required. In fact, translation is the key word for a gateway.

Bridges are relatively simple devices that allow traffic to flow from one LAN to another with some filtering, but no translation. Routers are a step up in the network device hierarchy and allow more sophisticated filtering than a bridge. Routers also know enough about the network infrastructure in order to route messages the most efficient way over a wide area network. Gateways have the most "smarts" and can not only bridge between networks and route traffic, but can also translate data so that it can flow from one network type and be understood by a different type of network.

In any event, gateway function is very important in any Lotus Notes implementation. Gateways are used between Notes and the mainframe and between Lotus Notes and other mail services (e.g., cc:Mail, MS Mail, etc.). The following paragraphs provide information on some of the options to support gateway function between OV/MVS, OV/VM, and OV/400 and Lotus Notes. They are listed in increasing order of functionality, throughput, and cost.

OPTION 1: IBM MAIL LAN GATEWAY/2. This is a point-to-point gateway that runs on an OS/2 server (i.e., there is no host code to install) and allows mail to flow between all OV platforms and Notes, cc:Mail, and MS-Mail. IMLG/2 is ideal for low-volume situations (up to a few hundred messages per hour). In 1997, this product will be part of the SoftSwitch line of gateways.

OPTION 2: LOTUS MESSAGING SWITCH (LMS). This is a UNIX based e-mail backbone that originally ran only on DG Aviion hardware. It is now also available on systems running HP-UX and AIX. LMS supports far more platforms than just OV and Notes (e.g., SMTP/MIME, X.400, DEC, Wang, et al). (LMS was formerly called Soft*Switch EMX.)

OPTION 3: SOFT'SWITCH CENTRAL (OR LOTUS CENTRAL).
Central is like LMS but runs on an S/390 MVS or VM mainframe. This
is a heavy-duty, industrial-strength, e-mail network backbone infrastructure that is intended for the Fortune 500—and is priced accordingly.

These products also have options to do related things like synchroniz-
ing directories, connecting to calendaring systems, etc. They also offer
various APIs. Additional details on these gateways and others are given
in the following sections.

IBM Mail LAN Gateway (IMLG/2)

Internally, IBM has long used the IBM Mail LAN Gateway/2 (IMLG/2)
product to provide mail service between Notes and its mainframe mail
system. However, that function is in the process of migrating all its internal Lotus Notes gateways to LMS and SoftSwitch Central (see "IBM's
Internal use of SoftSwitch Central and LMS" later in this chapter). Also,
in 1997, the IMLG/2 product was taken over by SoftSwitch and will
serve the role of a low-end SoftSwitch gateway offering.

An IMLG/2 is simply a PC that runs the Notes server code and the
additional IMLG/2 code that interacts with the Notes server code. The
PC has a connection to a LAN in order to communicate with other
Notes servers and a connection to a mainframe, which is usually also via
the LAN. The IMLG/2 provides support for:

- Novell NetWare
- NJE (Network Job Entry)
- Lotus Notes to SNADS (SNA Distribution Services)
- cc:Mail to OV/VM (OfficeVision/VM) and PROFS (PRofessional
  OFfice System)
- Lotus Notes to OV/VM and PROFS

IMLG/2 Version 1 Release 3

IMLG/2 Version 1 Release 3 became available in June of 1995. This
release provides the following mail gateway features:

- Addresses mail to Microsoft Mail post offices, cc:Mail post offices,
  and Lotus Notes domains from a host office electronic mail system.
- Bridges LAN e-mail systems with other LAN and host e-mail systems.

- Reduces administrative address maintenance that exists between LAN office systems and host office systems.

- Is relatively easy to install and manage; all the code is resident on the LAN.

This gateway allows several popular LAN office electronic mail systems to communicate with other dissimilar LAN and host-based office mail systems. It resides on a LAN and is co-resident with the LAN office mail application. No additional software is needed on the host office mail system to enable communications.

In addition, Mail LAN Gateway/2:

- Performs the addressing mail format conversions required to send electronic mail to a dissimilar office system.

- Can be used to address mail to Microsoft Mail post offices, cc:Mail post offices, and Lotus Notes domains from a host office electronic mail system.

- Allows PROFS and OfficeVision/VM users to address multiple downstream Lotus Notes domains, cc:Mail post offices, and/or Microsoft Mail post offices as a single node.

- Offers connectivity support to Lotus Notes, cc:Mail, Microsoft Mail, PROFS, OfficeVision/VM, OfficeVision/MVS via DISOSS, IBMMail Exchange, OfficeVision/400, DISOSS, OfficePath/SNADS, NETDATA files produced by the SENDFILE and NOTE commands of the Conversational Monitor System (CMS) component of VM/SP, VM/SP HPO, VM/XA, VM/ESA, and EMC2/TAO with the Fisher InternationalSNADS Gateway.

Description of New Features in Release 3

IBM Mail LAN Gateway/2 Version 1 Release 3 has added connectivity support for Microsoft Mail. Existing users of the IBM Mail LAN Gateway/2 Version 1 Release 2 product supporting cc:Mail, Lotus Notes, OfficeVision/400, OfficeVision(R)/VM, PROFS, IBM Mail Exchange, OfficePath/SNADS, DISOSS, and OfficeVision/MVS via DISOSS can now upgrade to provide communication to Microsoft Mail.

IBM Mail LAN Gateway/2 Version 1 Release 3 reduces the administrative burden of maintaining addressing incompatibilities that exist between LAN office systems and host office systems. The added support for Microsoft Mail makes use of the IBM Mail LAN Gateway/2 to create the Microsoft alias names needed to communicate with the other incompatible electronic office systems. After the utility is run, new users are dynamically defined aliases when sending electronic mail through the gateway. This utility, combined with auto-aliasing, helps to reduce service administration time associated with managing the gateway.

Specified Operating Environment

HARDWARE REQUIREMENTS. IBM Mail LAN Gateway/2 Version 1 Release 3 requires 10MB of storage for the base code. Additional temporary storage will be needed for logging, addressing, and message buffering. The additional storage needed will be based on message frequency and message size. It is recommended that you have at least 30MB of additional storage available. The IBM Mail LAN Gateway/2 requires a communications adapter; for example, a token ring or multiprotocol adapter.

The IBM Mail LAN Gateway/2 operates on either token-ring or Ethernet networks.

SOFTWARE REQUIREMENTS. This licensed program requires the following operating system configuration:

■ Minimum OS/2(R) Version 2.11 with APAR PJ11462 or OS/2 Warp Version 3 and Communications Manager/2 Version 1.1 with service pack WR06150, or Version 1.11

■ IBM LAN Server 2.0 or higher, IBM LAN Enabler 2.0 or higher, or Novell NetWare Requester for OS/2

When supporting Microsoft Mail:

■ Microsoft Mail Release 3.0, 3.1, or 3.2

■ Microsoft Mail Message Services for IBM SNADS

When supporting Lotus Notes:

■ Lotus Notes Release 2.0 or higher

When supporting cc:Mail:

■ Any version of cc:Mail that supports cc:Mail Import/Export Version 3.2.1, 3.3.0, 3.3.1, or 3.3.2

This licensed program supports communications with:

■ OfficeVision/VM Version 1.1 or Version 1.2

■ OfficeVision/400 Version 2

■ PROFS Version 2

- OfficePath/SNADS
- DISOSS Version 3.4
- IBM Mail Exchange
- OfficeVision/MVS via DISOSS
- NETDATA files produced by the SENDFILE and NOTE commands of the CMS component of VM/SP, VM/SP HPO, VM/XA, and VM/ESA
- EMC2/TAO (Electronic Mail Communication/Totally Automated Office) with the Fisher International SNADS Gateway

COMPATIBILITY. The IBM Mail LAN Gateway/2 Version 1 Release 3 is upwardly compatible with IBM Mail LAN Gateway/2 Version 1 Release 2 Modification 1.

The IBM Mail LAN Gateway/2 is compatible with other IBM products: IBM OfficeVision/VM, IBM OfficeVision/400, PROFS and OfficeVision/MVS via DISOSS. The interface to OfficeVision/400 and DISOSS is based on SNADS document interchange architecture (DIA)/document content architecture (DCA).

Lotus Messaging Switch (LMS)

This gateway product from the Lotus Development Corporation performs very comprehensive gateway functions for Lotus Notes. It is more scalable for a large installation than the IMLG/2. However, this is also significantly more expensive. The LMS was formerly called Soft*Switch EMX.

In large enterprises, Lotus components will typically coexist with legacy systems and messaging components from other suppliers. To help integrate these legacy and third-party messaging systems into the Lotus Communications Architecture, Lotus will be enhancing the Lotus Messaging Switch to provide the following services:

- High-fidelity message switching among X.400, SMTP, IBM SNADS, IBM PROFS, DEC ALL-IN-1, and VMSmail, as well as among numerous other messaging environments.
- A high function "boundary MTA" acting as the node to connect internal networks to external networks for X.400 and SMTP. LMS has access controls, rules, and other features necessary for systems providing this boundary function.

- Directory synchronization among all major environments.
- X.500 support, including native support for the X.500 DAP (Directory Access Protocol), the LDAP (Lightweight Directory Access Protocol), and the Directory Systems Protocol.

LMS is currently available as an integrated hardware/software solution. In the future, LMS will be available on some of the same platforms as CommServer. LMS will not require a dedicated hardware system but will be able to be installed on the same physical hardware and operating system as CommServer and managed from the same management platform. Over time, LMS and CommServer will be integrated.

Message Capabilities in Notes Releases 4 & 5

Lotus Notes Releases 4 and 5 have a wide range of new enterprise messaging capabilities for users, administrators, and application developers. Starting with Release 4.0, integrated messaging and groupware provides you with the ability to connect simple electronic messages to documents stored in a shared discussion database or object store. This allows you to reference a document via e-mail, without having to attach or embed the document into the message. As a result, this reduces storage and communication system resources and ensures a recipient has the most recent copy of a document.

These capabilities are made possible by the distributed object store inherent in Lotus Notes. An object store is a flexible container that enables an object—whether a document, e-mail message, graphic, video clip, or scanned image—to be easily and securely stored, managed, and retrieved through links or direct user manipulation. Notes' object-based message store makes interpersonal mail more efficient and easier to manage.

SoftSwitch Central

SoftSwitch Central is an IBM VM or MVS application that provides multiprotocol message switching between host-based e-mail systems and directories and other mail environments. Integration and management of mission-critical line of business applications is made through open messaging APIs and toolkits.

IBM's Internal use of SoftSwitch Central and LMS

This section describes the IBM/Advantis use of SoftSwitch Central and LMS for mail gateway services between Notes and OV/VM, the Internet, and X.400 mail systems. Figure 5.1 shows a general diagram for the gateway design. The following sections provide descriptions of the components shown.

NOTES/AU. The Notes Access Unit code is a started task running on a Notes server. The Notes/AU application exchanges mail between Notes and LMS. The Notes/AUs run on IBM PCs on an OS/2 platform.

Central/CLC

This is the SoftSwitch Central component that runs on an MVS machine. It is used to exchange mail between LMS and OV/VM. The

Figure 5.1

IBM's design for Notes gateways using SoftSwitch Central and LMS.

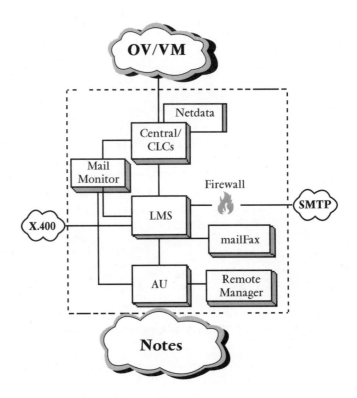

CLCs act as a SoftSwitch Network Application Programming Interface (SNAPI) client to Central and LMS.

LMS. The Lotus Messaging Switch works as a switching service in the IBM e-mail environment between Notes and OV/VM, X.400, and SMTP (for Internet mail). LMS is an MTA running on a UNIX processor that connects the MVS component (Central/CLCs) and Notes/AUs.

LMS NETDATA. The LMS Netdata is custom code that handles network data generated by host systems, such as the OV/VM Sendfile or CMS Note commands. It also processes PUNCH files generated from OV/VM systems. LMS Netdata consists of two started tasks running concurrently (named SSWNET1 and SSWNRV1).

SSWxND2 is another started task running as part of the Netdata component of SoftSwitch Central. Its purpose is to route mail items from Lotus Notes that have been sent to an invalid host node, back to the originating user.id. The sender receives a "Delivery Failure Report" outlining the address error and the original mail item sent. When Central (SSWxCEN) receives a mail item addressed to a node not found in its node table, the item is routed to the SSWxND2 task. SSWxND2 reverses the addressing (i.e., originating DGN and DEN are used as the target) and sends it back to Central. Central then processes it as a normal host to Notes item and sends it to the Notes user.

LMS SMTP. Mail flow between Notes and the Internet, by way of LMS, is controlled through a firewall. This firewall is strictly configured to act as an SMTP gateway to allow mail to flow in both directions. Each firewall directs inbound mail to a designated LMS to be forwarded to Notes. Mail destined for the IBM MPN Intranet is delivered to a mail hub for subsequent delivery to the user's OS/2 or AIX system. The LMS service supports MIME from Notes to the Internet and MIME/UUEncode from the Internet to Notes. Other protocol or services—such as Telnet, FTP, or TFTP—are not allowed to pass through.

MAIL MONITOR. This is a SoftSwitch Lotus product that sends e-mail messages to various network components. It then measures, records, and reports the time interval between the message being sent to a component and a response being received. From this timing information, it can perform the following operations:

- Detect and report failures of the mail network components
- Determine the quality of service provided to users (i.e., how long it takes to deliver mail)

Mail Monitor reports this information in a real-time graphical display and in the form of an event log that can be processed to generate reports about network performance, reliability, outages, etc.

PAGING. Initially, the Lotus Notes paging service used by IBM internally was based on the separate Lotus Pager Gateway product (see the section "Pager Gateway" in chapter 14). However, the pager service is being migrated to be part of LMS. Thus LMS will provide all of IBM's internal Lotus Notes gateway requirements.

REMOTE MANAGER. The Lotus AU/Notes can be managed either locally or remotely. Local management is done via the Lotus Notes console command interface. Remote management is done via the Lotus AU/Notes for OS/2 Remote Manager. The Remote Manager runs under the control of Hewlett-Packard OpenView in a Windows environment and communicates via Simple Network Management Protocol (SNMP) with the Lotus AU/Notes. The Remote Manager is composed of a Graphical User Interface (GUI) and an SNMP Manager. The SNMP Manager communicates with an SNMP Agent within the Lotus AU/Notes. This is shown in Figure 5.2.

ADMINISTRATION-BY-MAIL (ABM). Administration-by-Mail is a mail-enabled application that allows manipulation of the LMS Names

Figure 5.2
Access unit remote manager.

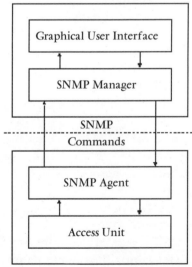

Remote Manager

Graphical User Interface

SNMP Manager

SNMP
Commands

SNMP Agent

Access Unit

Lotus Au/Notes for OS/2

Directory entries through e-mail. This allows authorized administrators to add, delete, modify, and list multiple Names Directory entries remotely.

QUERY-BY-MAIL (QBM). Query-by-Mail is a mail-enabled application that allows a user (usually an administrator) to issue queries against the LMS Names Directory by embedding the queries in the body of messages. QBM is accessible to any OV/VM user and to Lotus Notes users registered to LMS.

To use QBM, a request document is sent that contains one or more Names Directory queries to the QBM mailbox. Each query must indicate the attributes that are required to guide the Notes Directory search. QBM parses each query and searches the Notes Directory for matching entries. Information about each matching entry is written to a response document. If a query results in no matching entries or contains a syntax error, QBM will report the error in the response document. When QBM has finished processing all queries in the request document, the response document is mailed back to the user.

MAILFAX. MailFax allows e-mail users to have their mail messages delivered as faxes. MailFax controls the mail flow between Notes and a fax gateway interfacing directly to LMS via the SNAPI protocol driver. Mail flows in two directions depending on whether it is an outbound or inbound fax. The outbound fax flows from Lotus Notes, to the Notes AU, to LMS and then to MailFax. Inbound fax flows from MailFax, to LMS, to Notes AU, to Lotus Notes.

ROUTING CONFIGURATIONS. Figures 5.3 through 5.6 show mail routing scenarios with the use of the MVS Central and LMS gateways. Figure 5.3 shows the flow for Notes to OV/VM, while Figure 5.4 shows OV/VM to Notes. Figure 5.5 shows Notes to the Internet, and Figure 5.6 shows the Internet to Notes. The possible alternate paths are indicated. Note the "fire" icon used in these figures. This icon represents the firewall between the IBM network and the Internet.

SMTP Gateway

The Lotus Simple Mail Transport Protocol (SMTP) gateway provides mail translation services between Lotus Notes and the Internet. The SMTP gateway can also be used to provide much of the function that is provided by other mail gateways for Notes as long as the other mail systems have Internet addresses. Because Internet mail (SMTP) has become

Figure 5.3
Gateway routing
configuration for
Notes to OV/VM.

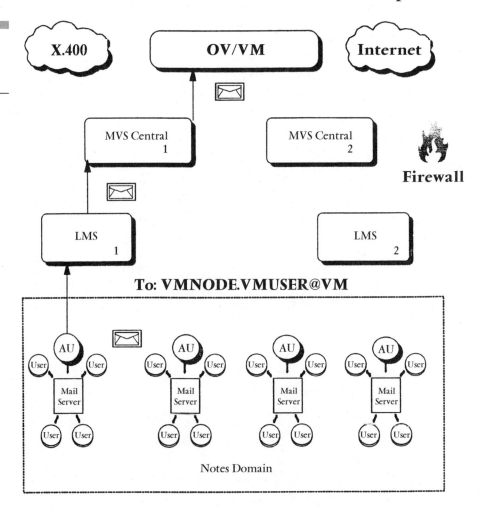

so popular, Notes with an SMTP gateway will allow you to send mail almost anywhere. One consideration, however, is formatting. A mail gateway, such as the IMLG/2, which is designed for processing mail between Notes and PROFS, will format mail arriving at PROFS so that it is basically customized for that system. On the other hand, IMLG/2 doesn't work for IBM Japan because of the 2 byte (Kanji) character set. IBM Japan is using SMTP and XAgent for this process. This again points to the benefits of having an Internet type (SMTP) mail gateway available.

This section describes the SMTP product. An example of the use of the SMTP gateway is discussed in the section "Sendmail" in chapter 8. Beginning with Release 4.1 of Lotus Notes, you get the Lotus SMTP MTA free (see the section "SMTP/MIME MTA: Release 4 and 5 changes" in this chapter for details). The Lotus SMTP MTA is installed on either an OS/2 or Windows NT Notes server (R4.1 or higher). The gateway logic

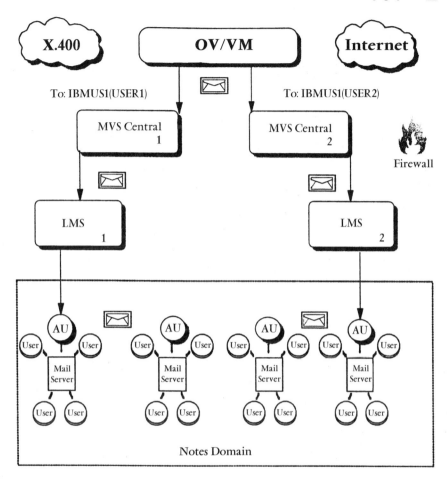

Figure 5.4
Gateway routing
configuration for
OV/VM to Notes.

described in the following paragraphs for the original R3 SMTP gateway is still valid for the latest versions of the Lotus SMTP MTA product.

The original Lotus Notes Mail Gateway for SMTP (OS/2) product was a Notes Release 3-based server add-in task that allowed Notes mail users to communicate with Internet mail users. The SMTP Gateway uses the Vendor Independent Messaging (VIM) interface to access Notes mail, and IBM TCP/IP for OS/2 SENDMAIL to access Internet SMTP mail.

The SMTP Gateway integrates diverse messaging environments into a platform for Notes mail and mail-enabled applications while offering easy setup and administration. The SMTP and other mail gateways offered by Notes contribute to the overall connectivity of Notes by ensuring that Notes mail users have access to different mail systems and can connect with vendors, customers, and business partners.

Figure 5.5
Gateway routing
configuration for
Notes to the Internet.

Figure 5.5 Gateway routing configuration for Notes to the Internet.

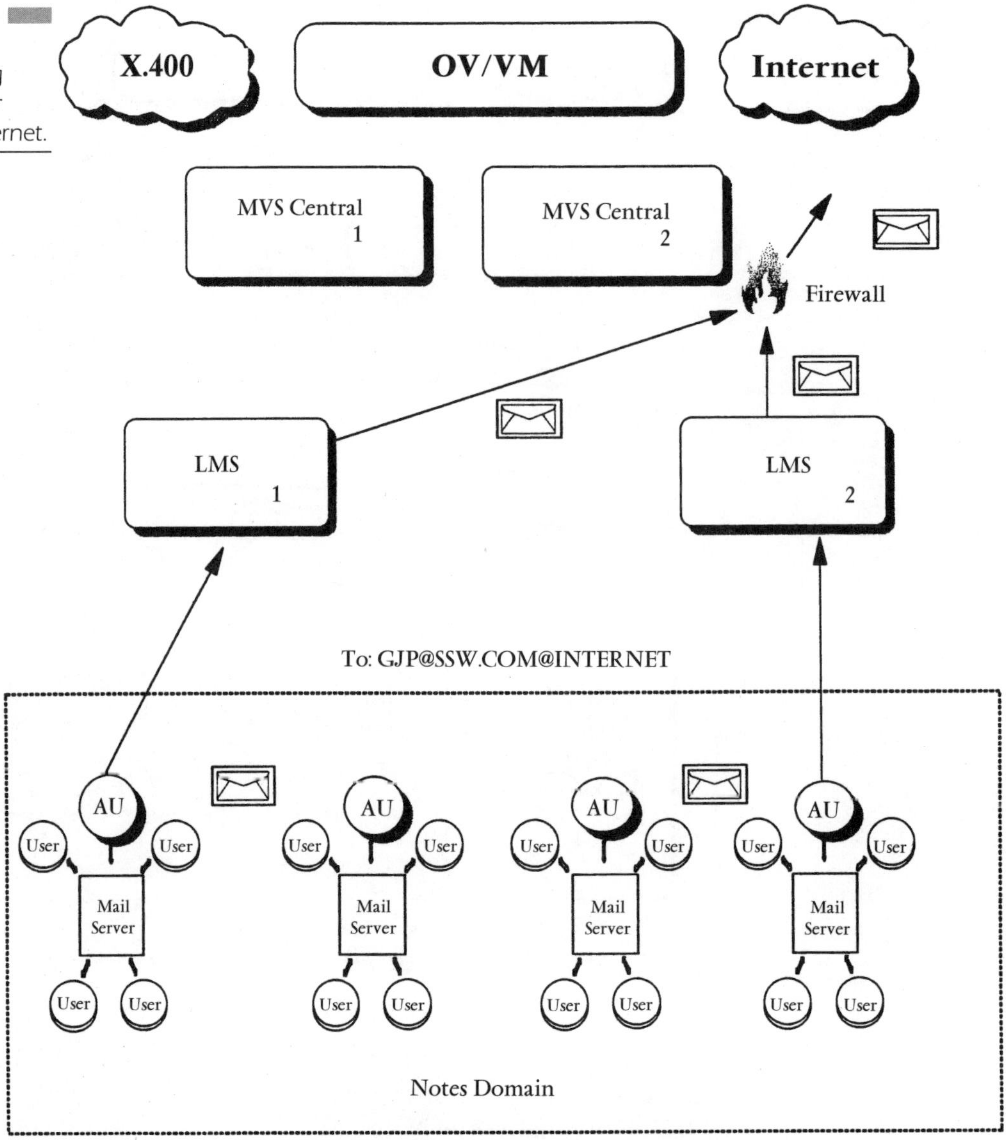

Required Hardware and Software

You need the following minimum hardware and software to install and use the original SMTP Gateway for OS/2 (see the section "SMTP/MIME MTA: Release 4 and 5 Changes" in this chapter for the SMTP MTA product):

■ One of the following:

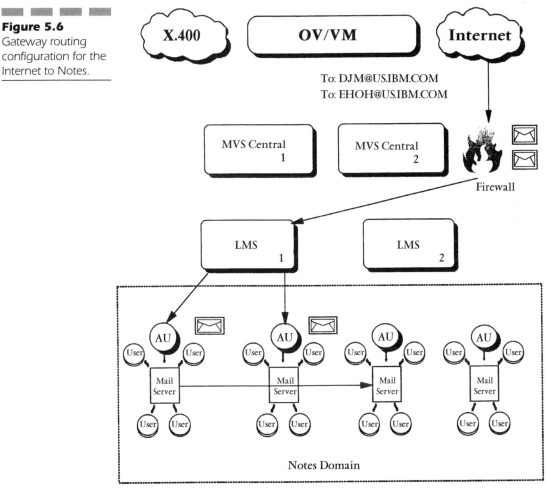

Figure 5.6
Gateway routing
configuration for the
Internet to Notes.

˜Notes R3 Server Software (or later) installed on a computer with OS/2 R1.3 and IBM TCP/IP Release 1.2.1 for OS/2 Base Kit with CSD# UN45351

˜Notes R3 Server Software installed on a computer with OS/2 R2.0 (or later) and IBM TCP/IP Release 2.0 for OS/2 Base Kit

■ The Lotus Notes Mail Gateway for SMTP (OS/2) disk

How the SMTP Gateway Works

This section describes how mail travels between Notes mail and the Internet via the SMTP Gateway and how messages are converted to allow this exchange.

MESSAGE FORMATS.   Notes and SMTP messages have different formats. When a message is sent from one system to the other, it must be converted to a format that can be read in the target mail system. This section describes the formats of Notes and SMTP messages and the differences between them.

A Notes mail message consists of a header, an optional body, and optional attachments, as shown in Figure 5.7. The Notes header consists of field name and value pairs—such as the To, From, Date, and Subject fields—that are analogous to the SMTP header. The Notes body can contain embedded textual information, such as color, italics, icon information, and document links. The Notes body can support rich text format (RTF). Notes attachments might be any native file system file.

There are several types of Notes messages. The Memo form is a user-generated message; examples of system-generated messages are DeliveryReport, NonDeliveryReport, and ReturnReceipt.

An SMTP message consists of a header and an optional body. The SMTP header contains fields of information about the message and its destination. The SMTP body is optional and contains short lines of 7-bit US-ASCII characters. The SMTP message body can include rich text through an ASCII representation.

MESSAGE CONVERSION.   The simplest case of message conversion occurs when a Notes message with a text body and no attachments is sent to SMTP. The headers and body convert directly to SMTP format.

When a Notes message is converted to SMTP format, the header fields are converted to equivalent headers. The body and any attachments are represented as the single SMTP message body.

Figure 5.7
Notes message
format.

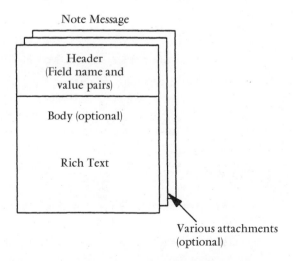

Note Message

Header
(Field name and
value pairs)

Body (optional)

Rich Text

Various attachments
(optional)

NOTE-TO-SMTP HEADER CONVERSION.   The SMTP Gateway converts all Notes headers to equivalent SMTP headers as in the previous simple case. Because not all Notes message headers convert to equivalent SMTP headers, the SMTP Gateway might generate the following special headers in the SMTP message:

- X-Date-Delivered—Only from Notes notice of delivery (generated by Notes mail router)
- X-Importance—Delivery priority associated with a Notes message
- X-Lotus-RTFAttachment—Flag an included rich text version of the body (via the SMTP Gateway configuration)

NOTES-TO-SMTP BODY CONVERSIONS.   The Notes message format is different than the SMTP message format. For example, a Notes message can have multiple attachments in addition to its body, while an SMTP message cannot have attachments. In order to make Notes messages readable in SMTP mail, the SMTP Gateway represents them in an SMTP-compatible format.

The SMTP Gateway uses the Internet Request for Comment (RFC) 1341 that describes the Multipurpose Internet Mail Extensions (MIME) recommendation to represent a body and attachments as a single text body. If necessary, the SMTP Gateway encodes any nontext Notes body and/or attachment using an encoding method. The default encoding method used by the SMTP Gateway is Basc64, as recommended in RFC 1341; however, you can alternatively specify that the uuencode method be used.

Many Notes databases can be mail-enabled by adding a "Send To" field to a database form. If the address placed in this field routes the message to the SMTP Gateway, the SMTP Gateway processes the message. The constructed SMTP message body is generated from a field called "Body" in the original Notes message. If the form that generated the message does not contain this field, then the ongoing SMTP message consists of only SMTP header fields and no body.

SMTP-TO-NOTES MESSAGE CONVERSION.   When an SMTP message is converted to Notes format, the header fields are converted to equivalent headers and the body becomes the Notes body and possibly attachments to the Notes body. This is shown in Figure 5.8.

SMTP-TO-NOTES HEADER CONVERSION.   There are more SMTP headers than Notes headers. SMTP headers that correspond directly to Notes headers (as in simple case header conversion) are converted to the

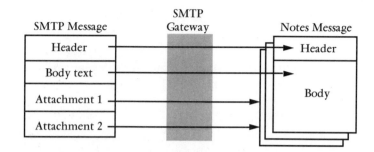

Figure 5.8
SMTP-to-Notes
message conversion.

same header field. SMTP headers that do not have a direct equivalent in Notes are discarded. You can configure the SMTP Gateway to retain the SMTP header information. If this configuration option is chosen, the SMTP headers become a separate text item of the Notes message. The item is called "SMTPHeaders" and is visible either by modifying your Notes message display form to include this item or by selecting *Design|Document Info*.

SMTP-TO-NOTES BODY CONVERSION.   SMTP messages entering the SMTP Gateway are checked for a MIME-Version header. If no MIME header is present, the entire body of the message is taken as the body of the resulting Notes message.

If the message has a MIME header, then that message is in the MIME format. The SMTP Gateway breaks MIME format messages into parts. The first part of the message becomes the body of the Notes message, and all other parts become attachments to the Notes message.

SENDING ENCRYPTED MESSAGES.   Encrypted messages can be sent from Notes. If the sender chooses the option to encrypt a Notes message, the public encryption key for each recipient is retrieved from the Notes Name and Address book and applied to the message for each recipient in turn. This produces a correctly encrypted message for each recipient.

If Notes cannot find a public encryption key for the recipient, it cannot send the message in encrypted format. You then have the option of sending the message in unencrypted format. Because recipients accessed through the SMTP Gateway are not in the Name and Address book, they can receive only unencrypted messages.

Message Routing

The SMTP Gateway is defined to Notes as a foreign domain. Mail sent through the SMTP Gateway is deposited in the gateway mail file of the foreign domain on the Notes server. The SMTP Gateway retrieves mail from that file, converts it to SMTP format, and submits it using Sendmail. This is depicted in Figure 5.9.

All SMTP mail addressed to the Notes OS/2 server is redirected to the SMTP Gateway. This is done by reconfiguring Sendmail during the SMTP Gateway Installation (done automatically by the installation program) to use the delivery agent program (GWMAILER.EXE) to transfer mail to the SMTP directory. The SMTP Gateway retrieves the mail from this directory, converts it to Notes format, and submits it to the Notes mail router, as shown in Figure 5.10.

NOTE: *Because all SMTP mail for the SMTP Gateway is addressed to the Notes OS/2 Server, you might want to configure the server with a name suggesting its gateway function. An example might be "Notesgw."*

To do this, use the Install and Configuration Automation Tool (ICAT) on IBM TCP/IP for OS/2. Select Option 5 - Configure Services from the configuration menu and change the computer's name.

If you do change the name, make sure that this change is known throughout the rest of the network.

SMTP/MIME MTA: Release 4 and 5 Changes

The latest version of the Lotus SMTP gateway is called the Lotus Notes SMTP/MIME MTA. This is also a Notes add-in task that you install into the Notes executable directory (e.g., C\NOTES). It is installed on a

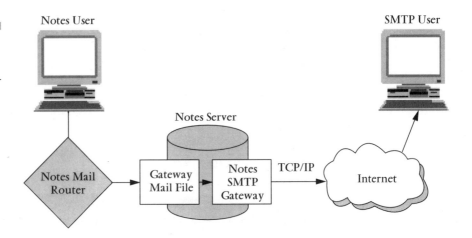

Figure 5.9
Notes-to-SMTP message routing.

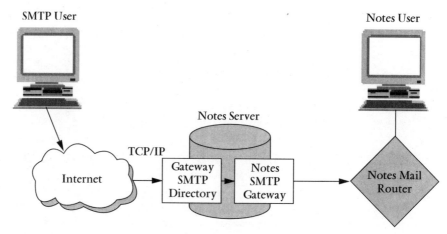

Figure 5.10
SMTP-to-Notes
message routing.

Release 4.1 (or higher) Notes server (OS/2 or Windows NT). It provides the following functionality:

■ Direct access to the Internet

■ Full implementation of a MIME-compliant SMTP

■ Extensive support of MIME content types for better end-to-end message fidelity. This means that all IANA media types are handled by the SMTP MTA. For example:
  ■ Text: Plain and enriched
  ■ Multipart: Mixed, digest, alternative, and related
  ■ Application: Octet stream, postscript, extension, and more
  ■ Image: jpeg, gif, ief, and tiff
  ■ Audio: basic
  ■ Video: mpeg
  ■ Message: rfc822, external-body, and partial

■ MIME-conformant as specified in RFC1521 to assure a high-level quality of message presentation

■ Tight integration with Notes; any MIME attachment can be viewed from the Notes client

■ Supports the MIME Header Extensions for Non-ASCII text

■ Extensive international character set support

■ Preserves Notes-to-Notes message fidelity over an SMTP backbone, using encapsulation

■ Can act as a peer on a multi-vendor SMTP environment

■ Transparent connectivity with Notes, cc:Mail, VIM, and MAPI clients

■ Wide range of address conversion options and flexible addressing capabilities

▪ Built-in diagnostic tracing facility that can be turned on for any of the SMTP MTA components to assist in isolating message error symptoms

▪ Simplified management and administration of the messaging backbone

Table of Mail Gateways and Message Transfer Agents for Lotus Notes

There are many mail gateways and Message Transfer Agents (MTAs) that can be used to transfer mail between Lotus Notes and other mail systems. We have discussed the IMLG/2, LMS, SoftSwitch Central, and SMTP Gateways. Table 5.1 gives a list of many of the mail gateways available for Lotus Notes. The other mail systems (in addition to Lotus Notes) and the operating systems supporting the mail gateway are listed. The table is printed courtesy of the Lotus Development Corporation.

What is the difference between a mail gateway and an MTA? Originally, the term Message Transfer Agent was used only within the X.400 community. It described a piece of software that was able to move mail messages from one system to another. The term *gateway* also refers to software that moves mail messages between systems, but the term is usually used only when the systems in question are of different types. Lotus has adopted the use of the term Message Transfer Agent to describe its Notes-SMTP, Notes-X.400 and Notes-cc:Mail software. This is to highlight certain advanced features of its products that provide benefit to the user:

▪ The Lotus MTAs can all be managed as part of the Notes system.

▪ They are multithreaded and designed to be scalable.

▪ They function as fully capable servers in both the environments they are serving. Thus the Notes-cc:Mail MTA appears to cc:Mail users as if it were a cc:Mail router and to Notes users as just another Notes server.

Functionally, however, both MTAs and gateways achieve much the same end result.

TABLE 5.1

Lotus Notes Mail Gateways

Vendor	Product	Other mail systems supported	Operating system platform
ACC VAX	MBlink	DEC All-In-One, VaxMail, SMTP, MHS,	OS/2 Notes Server and DEC
			VAX for Hub
CMS	VBridge for Notes	Banyan Vines Mail	OS/2 Notes Server and Vines Server for Hub
Control Data	Mail'Hub	X.400 and/or SMTP/MIME. Full X.500 directory and synchronization utilities. Mail'Hub supports a large list of messaging protocols like LMS.	Unix
David Goodenough Associates	DGAteway	SNADS (OV/MVS)	OS/2 Notes Server
IBM	Mail Gateway/2	cc:Mail, PROFS, OV/VM, OV/MVS, DIOSS, SNADS, OV/400, MS-Mail, IBM Mail Exchange	OS/2 Notes Server
InterNotes	InterWin	SMTP, NNTP	Windows Client
ISOCOR	Notes Router for X.400	X.400	OS/2 Notes Server and Unix or DOS ISOPLEX X.400 MTA
Lotus /Soft'Switch	SoftSwitch Central	X.400, SMTP, IBM SNADS, IBM PROFS, OV/VM, OV/MVS, MHS, cc:Mail, MS-Mail, IBM Mail Exchange	S/390 MVS or VM Mainframe
Lotus	Lotus Mail Exchange Facility	cc:Mail	OS/2 Notes Server
Lotus/Soft'Switch	Lotus Messaging Switch (LMS) (Formerly called Soft'Switch EMX)	X.400, SMTP, IBM SNADS, IBM PROFS, DEC ALL-IN-1, VMSmail, cc:Mail, MHS, MS-Mail, FAX messages, etc.	Unix on Data General Aviion Platform. Plans to include HP-UX and AIX.
Lotus	Notes MHS Gateway	MHS	OS/2 Notes Server
Lotus	Notes SMTP/MIME MTA Gateway	SMTP	OS/2 Notes Server and IBM OS/2 SendMail Windows NT Notes Server

Vendor	Product	Protocols Supported	Platform
Lotus	Notes VaxMail Gateway	VaxMail	OS/2 Notes Server and DEC VAX for Hub
Lotus	Notes Gateway for SkyTel	SkyTel Alpha Pagers and other receivers	OS/2 Notes Server
Phoenix	Mail Express	Microsoft Mail	OS/2 Notes Server
Retix	Notes OpenServer Gateway	X.400, SMTP, MHS, MS-Mail, cc:Mail	OS/2 Notes Server and Unix, OS/2, or DOS OpenServer X.400 MTA
Wingra	Missive-Notes (Access Unit for Notes)	DEC All-In-One, VMSMail, PROFS, OV/400, MHS, SMTP, cc:Mail	OS/2 Notes Server and DEC VAX for Hub
Worldtalk	Worldtalk 400	SMTP/MIME, X.400, cc:Mail, MS-Mail, WP-Office, MHS	HP/UX or SCO Unix

Dial Access for the Mobile User

"ISDN: Dead Again" was the title of a September, 1996 Forrester report written soon after 56-Kbps analog modems received a lot of press. Forrester noted that ISDN was still "hard to buy, hard to find, and hard to make work." On the other hand 56-Kbps analog modems should be "high-speed and simple, provide easy upgrades for dial-up pools, and be lower cost than ISDN." Their recommendation was to upgrade to 56-Kbps analog modems rather than ISDN and keep ISDN in a niche for things like video-conferencing.

Forrester Report, September 20, 1996.

An increasingly important aspect of Lotus Notes for corporations is dial access from anywhere in the world. Ideally, this access would be a local telephone call, and the protocol would be TCP/IP. Users require dial access to Lotus Notes Servers that might be on the Internet or on the company's internal network.

One of the most appealing aspects of Lotus Notes has always been its support for the mobile user. Notes lets the mobile user work very effectively without network access. However, when it is time to send e-mail, replicate the latest changes from the company Notes databases, and perhaps surf the Internet via the Notes client, dial access is usually required. Often the mobile user needs dial function that will let him access the Notes servers back at his main office. This chapter discusses the different dial options available for the mobile Lotus Notes user.

Figure 6.1 shows a general network design for providing Notes access.

There are many ways to have dial access to your Notes servers. The simplest way is to use XPC dial. XPC is the dial protocol that comes with every copy of the Lotus Notes software (client or server). No additional software is required for the customer to use this dial access to the Notes server. Of course, each Notes client workstation needs a modem, and the Notes server needs to have a number of modems, enough to handle the demand for concurrent dial-in sessions. In addition to XPC dial, there are several other ways to obtain dial access to your Notes server. The following sections describe some of these access methods and also discuss other aspects of dial access to Lotus Notes.

Dial Access to an Internal LAN

A good way to get to a Lotus Notes Server via dial access is to use a product that supports dial access to a company's LAN. Two IBM products that support this access are LAN Distance and LAN Hop. With this type of dial access, you usually can do all the things you can with your

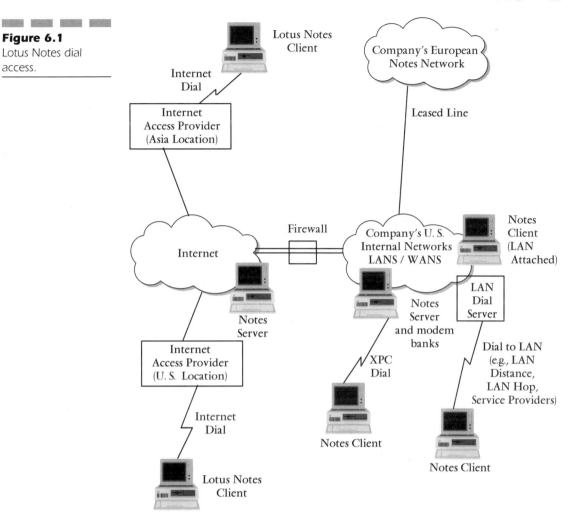

Figure 6.1
Lotus Notes dial access.

direct LAN access from your office. That is, in addition to getting to your Notes server, you have access to your file servers, your mainframe, etc. The only difference is the access speed.

Dial Access via the Internet

Several Internet Service Providers (ISPs) provide worldwide access to the Internet via dialing to local numbers. Then, if your company has a Notes server on the Internet, this Notes server could be accessed from anywhere in the world via local dial access to the Internet. Mail databases for employees could be replicated from the company's intranet to the

Notes server on the Internet. Then these users could obtain their Notes mail either by accessing their Notes server on the company's internal LAN or via dialing the Internet. Release 4.5 of Notes is "socksified" and allows replication through a socks server (Internet firewall). That's one way to get your Notes mail and databases from your company's internal LAN to your Notes server on the Internet.

Dial Access with ISDN

ISDN (Integrated Services Digital Network) is an international standard architecture for integrating different types of data streams on a single network, such as data, video, and voice. It is now being offered by telephone carriers in the U.S. and many other countries.

The standard service includes two 64-Kbps B Channels that can be used either for data or voice and one 16-Kbps D channel used for signaling. B Channel bonding allows the combination of the two B Channels to create one 128-Kbps data channel.

ISDN can be used as a native protocol by Lotus Notes with appropriate drivers; however, if ISDN adapters are placed in Dial Gateways, such as LAN Distance, then the TCP/IP protocol could be used by Notes clients with switched ISDN capability. Switched ISDN using the LAN Distance Gateway for Notes access with the TCP/IP protocol has been tested at IBM Global Network with very good results. It should be noted that different telephone companies (for example, in Europe) have not implemented ISDN in exactly the same way. Therefore an ISDN adapter that works in the UK might not work in Germany. If possible, it's a good idea to use the same adapters on both sides of the dial link.

Dial Access from Service Providers

The different Lotus Notes network service providers offer many dial options. One comprehensive service is from the IBM Global Network. This service provides worldwide local dial access to your private network, to your virtual private network provided by IBM Global Network, or to the Internet. Security is provided by login with password, user profile for restricting network access, routing restrictions, and restrictions on Domain Name Services. All users must have an account, user id, and nontrivial password that must be changed periodically. User profiles are set up that define which networks can be accessed by each individual user. Therefore, only users from your company can get to your network. Also, you cannot get to other company networks. Users cannot use Domain Name Service to search for "interesting" networks to access.

Using a dial provider eliminates the need for you to manage your own banks of modems 24 hours a day, 7 days a week. Also, with the service provider, you don't have to worry about busy signals and dial resource capacity because the service provider takes care of that. The service providers utilize "hunt groups" where a call to a local telephone number rolls over to another modem until the next available modem is reached. In the unlikely event that all the modems for your local number are busy, 1-800 service is available. Also, in very remote areas, 1-800 service is always an option.

Another advantage of using a dial service provider is that you will benefit from technology updates very quickly. For example, service providers have swiftly moved from 28.8-Kbps to 56-Kbps modems. They will also rapidly offer ISDN service as it becomes available from the Regional Bell Operating Companies in different parts of the country. Similarly, TCP/IP dial protocols have been enhanced (for example, SLIP to PPP).

Using a dial service provider can also be cheaper than providing your own dial service because a service provider has economy of scale for hardware, software, and help desk support.

The "New" Technologies

There are several new technologies that are becoming available that could significantly affect our access to Lotus Notes in the near future. These technologies are :

- *Cable-TV modems*—A typical cable modem would offer 10-Mbps data rates over existing cable-TV wiring. Such devices, which include an Ethernet connection to that you attach your PC, connect to the coaxial wiring that delivers your cable-TV signal.

- *ADSL*—Asymmetric Digital Subscriber Line is a new technology that will soon be offered by the telephone companies. ADSL will allow megabit access to Lotus Notes from your home or office over twisted-pair telephone lines and will be a strong contender to cable modems. With ADSL, the downstream channel speed ranges from 1.5- to 6.1-Mbps, while the upstream speed goes up to 640 Kbps. This works well for Web browsing and many Notes applications where most of the traffic is from the server down to the client. Full-scale ADSL deployment is expected in 1998.

■ *Wireless access*—PCS (Personal Communications Services) is the digital technology that is starting to replace cellular phones in the area of wireless communications. This wireless technology is much better for data transmission than the current cellular phone technology. PCS is quickly becoming a reality in many parts of the country.

Dial Recommendation

From a cost standpoint, Lotus Notes users/designers should leverage Internet dial capability as much as possible. All mail stored and transmitted on the Internet should be encrypted using the encryption function of Lotus Notes. The next section gives the author's experiences using this type of dial access to Notes while traveling in Europe.

Experiences of a Mobile User on European Trips

This note summarizes the author's April, 1995, and August, 1996, experiences accessing Lotus Notes by local dial to the IBM Global Network local access nodes in the UK, France, Sweden, and Italy. Overall, after I switched over to workstation-based mail, I found that it's an excellent way to access Notes when out of the country. Most of this section was written as e-mail while I was on the train from Copenhagen to Stockholm. The e-mail was then completed and sent when I dialed the local IBM Global Network number from a hotel in Stockholm.

One nice thing about workstation-based mail is that the speed of the dial connection is of no real concern. Mail is read and written from the mail file on the local hard disk, so there is no delay (it's faster than using a mail server on the local LAN). Then when a dial connection is made, new mail is replicated to the laptop's local mail file and sent to the home server in the background. I've been using the Tools|Replicate command after I've highlighted my local mail file and preselected to have my mail file replicated in the background, my outgoing mail transferred, and my "home server" in White Plains, NY. Then, while that's going on in the background, I can look at any databases I don't have replicated to my local hard disk.

The different telephone jacks were a problem in the UK, before I bought an adapter, and in Sweden where I also bought an adapter. In

Paris, the hotels I stayed at used the familiar RJ11 jack. In Sweden, the telephone jack required for the wall looked like a big four-prong power plug. Some of the telephones themselves have RJ11 jacks, but it seems most of the hotels have Swedish-built phones where the wiring from the telephone is permanently connected. The biggest problem in Sweden is that many of the hotel rooms have digital lines. Of course, IBM locations have analog lines available, so that's no problem.

This note was sent from a "guest office" at the Royal Viking Hotel in Stockholm, where they had assured me they had an analog line. The code was "99" to get an outside line. The problem was that, when I set up the INET dialer to dial 99,internet#, I could hear the operator saying "Allo" over the PC speaker. The line was analog all right, but it required an operator to set up the outside line. The INET dialer does not have a manual dial option, so I coded the number as 99,,,,,,internet# and asked the operator just to give the next call an outside line without waiting for a voice request, and that worked! Actually, it wouldn't be much fun if there were no challenges! Differing power plugs and telephone jacks provide challenge enough.

Our IBM Notes colleague in Sweden told me he's solved the problems with hotels not having analog lines, etc. by traveling in Europe with a cellular phone that has access anywhere in Europe. He hooks his laptop to the cellular phone and then always has a consistent dial interface (i.e., he doesn't have to be concerned about hotel codes to get an outside line, if he has to go through an operator to get an outside analog line, etc.). He also has no concerns as to whether a hotel only has digital lines. It is possible to have a corporate cellular phone (or phones) with European access that could be used during European trips for laptop dial access.

In August, 1996, I traveled to Southern France and to Northern Italy and accessed Notes via local Internet dial access. This following note was sent from Lucca, Italy, with dial connection to IBM's Internet access in Florence. This was the first time I tried dial connection in Italy, and I had some adventures.

My dial connection kept giving me the error "no dial tone." When I first tried this from a hotel in Florence, it seemed to make sense, because I'm quite sure the hotel had a digital system (at least the phone system at the hotel reception area said "digi vox"). However, when I tried the port at the reception desk used for their fax machine, I got the same problem. I was afraid I had burned out my modem trying to connect via a digital phone system. The same problem occurred at this residence in Lucca.

Yesterday, I went to a computer store in Florence. The technician there told me that it's necessary to use an "X3" in the modem command string in Italy (my Italian was about as good as his English, so we had a good time trying to communicate). That worked! I'm using a ThinkPad 360 for my laptop. It has an APEX 14.4 PCMCIA modem. The modem commands always used for this machine were (which are the default for the APEX modem in the dialer):

```
AT&F
ATE0&K3
```

This was changed to:

```
AT&FX3
ATE0&K3
```

The "X3" command tells the modem to dial without a dial tone. Anyway, I just tried dialing again without the "X3," and I got the "no dial tone" error. There don't seem to be any "CompUSA" type computer stores in Italy. The two I visited were very small, with only two people working in each. At least the "language" seems to be universal. They used the words "modem" and "computer."

One more thing, and that's Notes security. When I access Notes on the Internet, I see the message "network traffic is being encrypted at the server's request." I suspect that slows the response time somewhat, but it gives a certain sense of security, because the server is on the Internet. From our earlier tests, encrypting network traffic didn't seem to add much to the Notes response time. I am using workstation-based mail, so I can read and write notes without a network connection. Then, when I replicate in the background, the network encryption overhead shouldn't be noticeable at all.

An International Dial Mobility Kit Used by IBM Executives

This section describes a dial mobility kit made up for IBM executives who travel abroad. The sections that follow are the instructions that come with the kit.

Information All Global Dial Users Need to Know

The following sections cover the things you need to know about international data dialing.

TELEPHONE "PLUGS" ARE DIFFERENT IN MOST COUNTRIES. The U.S. RJ11 plug, available with most Laptop modems, is not used in other countries. To connect, you must have the correct telephone adapter for the country you are in. Use the following procedure to connect:

1. Remove the telephone plug from the wall.

2. Use the modem Saver Line Tester to ensure the connection is to an acceptable analog line. If the test fails, then it is a digital connection and you cannot use your modem. In this case or if the telephone was hard wired into the wall, you will need to use an acoustic coupler to connect.

3. Plug in the correct country telephone adapter.

4. Connect your RJ11 MODEM cable to the adapter.

5. Connect the telephone plug into the back of the adapter. While you can't use the telephone at the same time you are using your laptop, this will enable you to make telephone calls when your laptop is not in use. There are a few adapters that don't allow you to connect the telephone in addition to the laptop.

POWER PLUGS, VOLTAGE, AND HERTZ VARY BY COUNTRY. While laptop computers operate on batteries, their power supplies must be able to handle the countries voltage and Hertz as well as mechanically plug into the countries power outlet to recharge their batteries. These power supplies must be the "universal" version and rated for 50/60 Hertz and voltages between 100 V and 240 V. Ensure that you have the required power plug adapters for the countries that you will be visiting, as well as a "universal" power supply. The rating is written on the power supply and should state that it supports inputs of 100/240 V and 50/60 Hertz.

DIALING FROM A HOTEL ROOM IN A FOREIGN COUNTRY CAN BE CHALLENGING. Frequently you need to enter special digits, or you might need the hotel operator initiate the call. If you are dialing to a server in another country, you must use the appropriate country code. U.S. "800" numbers cannot be dialed from other countries.

Prior to attempting to have your laptop connect to the dial server, dial the number directly from the telephone to ensure that you hear the familiar squeal of a fax machine or modem. If the digits you dialed don't connect you with the server, check with the hotel to ensure that you are dialing correctly. If you can't dial the number using the telephone, you know the problem is not with your laptop or dialer program.

Procuring Telephone and Power Adapters

IBM has a national contract with AR Industries Road Warrior Outpost to provide IBM employees with telephone adapters and power adapters at substantial discounts. Anyone can browse the adapters available from their Web site (http://www.Warrior.com/IBM.HTML).

Encryption Considerations when Traveling Abroad

Large North American companies, or any company with people who travel abroad frequently, should standardize on the worldwide edition of Lotus Notes, which uses the Lotus Notes international encryption standard. This also makes sense for companies that are global.

Typically, a large global company would have its North American Notes users, employing the North American edition and its employees located outside North America using the International edition. This creates a couple of problems. First, with this arrangement, it is not possible to encrypt mail and send it from a North American user to an International user because the encryption levels are different. Secondly, if a North American user wants to travel abroad and carry a laptop with Lotus Notes installed, to be legal, the version of Lotus Notes must use the international level of encryption. To do this, he would have to install the International version of Notes on the laptop, *and* he would not be able to use his existing user.id, because that is an encrypted file, encrypted with the North American level of RSA encryption.

NOTE: *Due to an agreement announced in January of 1996, between IBM/Lotus and the federal government, Lotus will be able to standardize on the 64-bit RSA encryption level (current North American version) worldwide (24 of the bits can be decrypted easily by the U.S. government). This will solve the problems mentioned earlier for companies with users of Notes with different encryption levels. In the meantime, the following section describes how you can "legally" use the North American version while traveling abroad.*

Foreign Travel with North American Version of Lotus Notes

The following is a statement from the Lotus legal department regarding the January, 1996 "Personal Use Exemption" to the ban on export of data encryption technology by the U.S. government. It deals with the Lotus Notes Client code installed on portable computers. The author used this regulation in traveling during August, 1996, with the North American version of Notes to France and Italy.

The U.S. government recently issued a regulation permitting the temporary export of laptop computers that contain encryption functionality, provided certain conditions are met. The Lotus product that benefits from this new regulation is the North American version of Lotus Notes.

The new regulation, referred to as the "Personal Use Exemption," permits the export of Lotus Notes North American version without an export license (1) on a temporary basis, (2) by U.S. citizens and resident aliens, (3) for personal use and not for sale, demonstration, or further distribution, and (4) provided a record of travel is kept by the exporter. Export is permitted to any destination except banned or embargoed countries (currently Cuba, Iran, Iraq, Libya, North Korea, Sudan, and Syria).

The new regulation permits employees who are U.S. citizens or resident aliens to travel with North American Notes on their laptops without first obtaining an export license, provided they comply with the requirements stated in this e-mail.

Travelers must keep a record of each trip, which is a simple document (a form of this document is included in Notes Release 4). The document must include: dates and countries visited; dates of leaving the U.S. and re-entry; cryptographic products that are hand carried; and a statement that all conditions have been met and no products were stolen, transferred, or compromised.

Your user.id on a Mobile Workstation

If you need Notes on different machines, then you can make copies of your original user.id and use that on other machines such as mobile workstations. I have Notes on my PC at work, my PC at home, and a ThinkPad I travel with. All three machines have a OS/2 Notes client installed on the hard disk. I also have the Windows client of Notes installed on my workstations at work and home (at work, I sometimes have both the OS/2 client and Windows client, under WIN-OS/2, up at the same time; sometimes accessing the same database). Each copy of the Notes client (three OS/2 and two Windows) has its own NOTES.INI, which was built automatically when I installed the code. However, I use the same user.id for all five clients (actually just copies of my user.id from my office workstation that I have in the NOTES directory on each machine). Therefore my user.id on each machine has the same certificates as the original I have at work. This has worked very well.

The point is that you can use your original user.id with all its certificates on as many machines as you want. Of course, you could have your user.id on a diskette and use that wherever you use Notes. Then you could use any Notes workstation with your user.id. Anybody else, with your user.id and password, could also use any Notes workstation. Therefore your user.id and password should be closely guarded!.

One other point on multiple user.ids: With Notes R4.5 and above, it's easy to switch between two different user.ids that you might have for two different domains. I've had a user.id in the ADVANTIS domain for several years; however, during the spring of 1997, we started to migrate users from the ADVANTIS domain to become part of IBM's "IBMUS" Notes domain.

When I received my user.id for the IBMUS domain, I immediately tried it out. My IBMUS mail server is on the SP2 in Southbury, CT, and it was at that time an R4.1 server. I was able to use my mail in the IBMUS domain, but switching back to the ADVANTIS domain required switching both my user.id and my location. However, starting with the R4.5 client, there is an easier way. With my R4.5 client I was able to define two locations by going into my personal N&A book and clicking on the Add Location button. Actually, I just renamed my "office" location definition to "Advantis," and I created a new location "IBMUS."

My Notes 4.5 client let me define a user.id to switch to with the new location. So, for the "IBMUS" location definition, I was able to enter my new IBMUS user.id. Then, with one "click" in the location area at the bottom of the Notes desktop screen, I can switch between being in the ADVANTIS domain and in the IBMUS domain.

During our migration period, I didn't put in any forwarding addresses in either N&A book (ADVANTIS or IBMUS). I just looked at mail both in the ADVANTIS and IBMUS domains, because it was so easy. Of course, cross certificates, along with mail forwarding, would have given me much of the same function (and that was part of our migration process), but the switch location feature that came with R4.5 is so nice I've stuck with that. My Advantis user.id is a North American version, and my IBMUS user.id is an International version. Therefore, the location switching only worked if I used a North American version of the Notes R4.5 client. With the International version of the Notes client, I could only use an International user.id.

Using the International Edition of Lotus Notes

Lotus offers two different editions of Notes code, based on your location in the world. The major difference is the type of encryption level used. The North American edition uses an RSA encryption level that, by Federal law, is not to be transported outside the United States. The second type of Notes code is the International edition. This version of Notes uses the International level of public key encryption (a shorter RSA key).

North American keys can decrypt information encrypted with an International key, but International keys cannot decrypt information encrypted with a North American key. This means that, if a North American server and an International server want to communicate, they must use an International certificate to cross-certify. Lotus did offer a third edition of Notes called the "World-Wide Edition of Lotus Notes," but this was basically the International edition with American English. The International edition of Notes offers "American English" as an option, so it becomes the equivalent of the previous World-Wide edition.

An International ID file can contain and use a North American certificate (or cross certificate) and use the certificate under either a North American or International version of Notes to access servers with either a North American or an International ID file. Also, when you use an International version of Notes, you can only run the International version of Notes with an International ID. (To create an International ID, choose File|Admin|Register New User; in the Other User Settings box, under license type, choose International.) The user can use this International ID on domestic or International versions of Notes. However, a domestic ID (which is a license type of North American) that is embedded in the NOTES.ID file cannot be used with an International version of Notes.

You can still use a North American certifier to create International IDs and thereby have a North American certificate on an International

ID. However, someone using an International version of Notes in a different domain would need a common International certificate or cross certificate depending on the naming convention (flat versus hierarchical).

When users travel overseas, they must ensure that they have International IDs (unless they use the Personal Use Exemption discussed earlier). Then they don't need to worry about whether their certificates are North American or International. For these reasons, IBM has standardized on the International edition of Lotus Notes for all servers and clients in the roll-out of their new Notes architecture. Going to the International edition has made it more difficult to migrate IBM "legacy domains" to the new architecture; however, in the long run, using the International edition as the worldwide company standard will simplify administration and Notes management.

Table 6.1 gives an overview of the two Notes edition differences and considerations needed to determine appropriate use.

TABLE 6-1

Lotus Notes English Language Editions

	North American	International
Description	North American encryption with North American defaults	International encryption with International defaults
Where can it be sold and used?	United States and Canada only	Worldwide
Who is most likely to use it?	Customers in North America (United States and Canada)	Customers outside North America and North American customers standardizing on International encryption and/or while traveling outside North America
How is it identified on the "Help About Notes..." screen?	[release number] [release date] For example: Release 4.5 February 22, 1997	International [release number] [release date] For example: International Release 4.5 February 22, 1997
Encryption	North American	International

Description of the Three Versions of Notes Clients

There are three different types of client licenses in terms of functionality. These are the "Full Lotus Notes License," the "Lotus Notes Desktop License" and the "Lotus Notes Mail License."

Full Lotus Notes License

The full Notes license gives you all the features Notes has to offer. This includes design and administration capability. Many organizations will distribute this kind of license throughout, as it allows people to design databases on their local drives. It also means that administrators or developers can use their ID on any workstation and still be able to perform the more advanced functions that are restricted by other types of Notes licenses. Most mobile users of Notes would find this version of Notes very valuable because, when you are disconnected from the network, you'll be able to use your workstation as a Notes server and create and modify databases. This license is, of course, the most expensive.

Lotus Notes Desktop License

If you want to be more restrictive about the functionality available to your users, you can give them a desktop license. This takes out some of the design capability and so restricts the changes users can make to databases, even if they have Designer access in the ACL. The desktop license is a run-time Notes client. It has the ability to run any Notes application. This license is cheaper than the full client, so on a large scale, it could save you a lot of money. Many users do not need application design capabilities.

Lotus Notes Mail License

The mail license has full mail and Web navigation capabilities. In addition, this license allows you to use some standard Notes applications—such as document libraries and discussion databases—but not very many more. If you are going to implement Notes just for mail and Internet access purposes, then this would be the right client to use.

The Future of Dial Access for Lotus Notes

Network dial technologies continue to improve at a rapid rate. 33.6-Kbps modems are now the norm, and PC manufacturers are now standardizing on 56-Kbps modems. Much of the demand for ever-increasing dial bandwidth is fueled by users of the Internet and World Wide Web. Notes dial becomes a beneficiary of this rapidly improving dial technology. In fact, Lotus Notes with the Domino server is very much a part of the Internet/World Wide Web demand for increased dial bandwidth.

Modem speed on plain old telephone service (POTS) continues to improve. 56-Kbps modems are becoming popular. 56 Kbps is the "downstream" speed (from the server to the client). It's still 28.8-Kbps upstream (client to server). However, the downstream speed is by far the most significant direction for Web browsing and many Notes applications.

The advent of 56-Kbps modems prompted a Forrester report titled "ISDN Dead Again" because ISDN's basic speed of 64 Kbps is only slightly better than the 56 Kbps you could get with the new modems over your existing dial lines. Megabit access speed for the mobile user will first come through cable modems and ADSL. These are not dial access systems but will be available from your home and some hotels as the technology catches on. Web browsing and groupware applications such as Lotus Notes will continue to grow rapidly and so will the demand for easy high speed network access for the mobile user.

Security Aspects of Notes Over the Network

Notes has so many components that the Internet lacks. Authentication [checking the identity of users logging in], for example, is not built into the Web, but it's handled by Notes. At some point the Web will have all these things, but it will be challenging to integrate them.

Bill Wilson, senior vice president of Johnson & Higgins,
a New York City insurance brokerage
that has 6,000 employees on Notes.
(Taken from July 8, 1996 issue of Fortune magazine)

The Security issues for Lotus Notes are:

■ User authorization and authentication

■ Server access control

■ Database access control

■ Data privacy and integrity

■ Verification of originator

■ Physical access control

■ Internet firewalls

Lotus Notes itself includes a substantial amount of security features regarding the first five issues. These features must be supplemented with the security procedures discussed in this chapter. Security within the Lotus Notes product itself can be looked at as seven layers from the outside in.

The outside layer is simply the security involved in having any access to the Notes domain. The inner-most layer involves security at the field level within a Lotus Notes document. Figure 7.1 depicts these seven layers.

Consider the (somewhat contrived) example of Human Resources databases stored on a dedicated Notes server. Only Human Resources staff have server access. Some of the staff members have access to the Resume database, but not to the Job Offer database. Within the Job Offer database, there are further restrictions on access to executive Job Offer versus non-executive Job Offer. Because different forms are used for executive and non-executive Job Offer, access to all executive Job Offer documents (remember that a document is a filled-in form) can be controlled at the form level. Of the staff members who can access the executive Job Offer documents, only senior staff members can access the Job Offer documents for Vice Presidents and above. These senior staff members can fill in the base salary and health benefits portion of the document that are based on formulas. Only the executive staff members of the Human Resources department can fill in the restricted section on the incentive bonus plan. Finally, only the Director of Human

Figure 7.1
Seven layers of Lotus
Notes security.

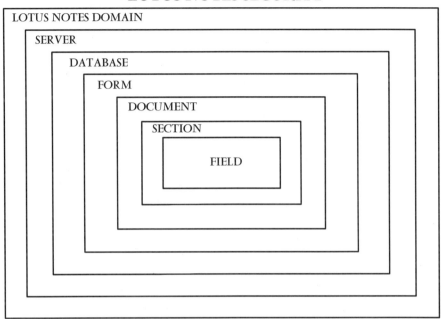

LOTUS NOTES SECURITY

Resources can fill in the field in the restricted section for the number of stock options.

User Authorization and Authentication

The Notes Administrator authorizes a user (in this document, meaning a person or a server) to access a Notes Domain by creating a user ID file and a Person document in the Name and Address Book database. The user ID file is kept privately with the user and the Person document is available publicly. The user ID file contains:

- User name
- License number
- Private key
- Public key
- Password
- Encryption key(s) (optional)
- Certificate(s)

The license number and public keys are unique, are assigned to the ID file when it is created, and cannot be changed. The user name, the license number, and the public key together identify a unique user. The certificate validates the association of the given user name, license number, and public key with a digital signature using the Certifier ID. The digital signature uses a public key encryption method to prove that the certificate was created by the Certifier ID (is official) and that the certificate has not been altered (not forged, data integrity maintained). (See the section "Cryptography" later in this chapter for a full discussion of encryption and digital signatures.) Whenever a new user ID or server ID is created by the Notes Administrators, a certificate is issued using the Certifier ID (called the Cert ID or certifier, for short).

A Notes certificate is somewhat analogous to a government-issued passport. The United States government has validated the association of my photograph, my name, birthdate, and other vital information. The passport is stamped and contains other security measures (holograms, special ink, etc.) to demonstrate that the passport is official, has not been altered, and has not been forged (data integrity maintained). A passport can be used for identification when moving about within the United States. (A Notes certificate is all that is required to move about in your own Notes Domain.)

Of course, a passport is more often used to travel to other countries. The United States passport is only recognized with those countries that have reciprocal diplomatic relations with the United States, so I can only travel to those countries. Likewise, passports from countries that do not have reciprocal diplomatic relations with the United States are not recognized by the United States, so citizens of those countries cannot enter the United States.

Exchanging diplomatic relations is somewhat analogous to Notes cross-certification with other Notes domains. Notes certificates must be exchanged (with digital signatures of the respective Certifier IDs) before any user or server from one domain can communicate with ("enter") the other domain. International travel requires a passport from your government and diplomatic relations with other countries. Inter-domain Notes communications requires a certificate from your domain and cross-certification with other domains.

A user must prove his identity each time he accesses a server. The authentication process begins with a check that the user's certificate (and thereby the user's public key) is trusted by the server. Once trust of the public key is established, the public keys of the user and the server are used for an encryption/decryption test to verify the user's private key. Once authenticated, the user can then access the server and its databases subject to controls described in the following sections.

Certificate Strategies

The role of the Certifier ID has profound security implications. Whoever has access to the Cert ID (or a copy of the Cert ID) can add a user or server to your Notes domain. There are still other security measures to limit what this impostor user or server can do, but creating the IDs can be a big step to breaking into your data. The point is that you need to guard your Cert ID and closely restrict access to it!

The Cert ID is a file with an associated password. Never store the Cert ID on a hard disk! Only store the file on a floppy disk, and keep a copy off-site. Lock the floppy disks in a safe place! Some people store the Cert ID floppy in a fireproof safe (and I do not consider this to be an extreme measure). Choose a nontrivial password that is at least 8 alphanumeric characters or more. Make sure you remember the password, but do not make it readily available to non-administrators. The Cert ID is critical to the operation of your Notes and Domino network. You need it to create new users and servers. You need it for cross-certification to communicate with other Notes and Domino Domains. I cannot over-stress the importance of the Cert ID and its careful, secure handling. Enough said.

Control of certifiers and your overall certificate strategy is a very important part of Lotus Notes security. There is no security feature in Notes that allows you to restrict access to a whole domain. Domain security is enforced entirely through certificates, which is actually done on a per server basis. Although it is common to have one domain throughout the organization sharing a common organization name, this is not always the case. IBM's Global Notes Architecture is a good example of this, where there are multiple domains (one for each country) that all share a common high-level organization identifier ("/IBM").

At the user level, with hierarchical certificates, if a user shares a common ancestral certificate with a server, then the user can authenticate with that server. This is the advantage of having one organization certificate across the enterprise. At the server level, cross-certificates allow communication between organizations. The cross-certificates are held in the Public Address Book. Cross-certification can occur between users, servers, organizations, organizational units, or any combination of these. For example, server-to-server, server-to-organization, or user-to-organizational unit.

If you want to distribute data to users in another organization, you should cross-certify at the server-to-server level. This means that only one of their servers can communicate with your server. You are not giving their users access to your servers, which would put extra load on that server and would also be much harder to monitor for security purposes.

Server-to-organization and organization-to-organization are much less secure because you are either giving access to a much wider group of people, giving a server access to more of your organization, or both. It is much harder to monitor who is accessing your system this way and so lessens the security of your Notes network.

When you are dealing with flat certificates, you must remember that these are stored in the ID file rather than the Public Address Book. Usually, domains that only have flat certifiers are those that started off with Release 2 of Lotus Notes, when flat certifiers were the only certifiers available. In the author's experience, newly established domains are hierarchical, with hierarchical certifiers.

At the user level, with flat certificates, authentication is only performed in one direction: the server authenticating the user. Users can access any server with which they share a common certificate, provided that the certificate is trusted by the server. At the server level, with flat certificates, authentication is performed in both directions, as both servers have to authenticate each other.

If your server and an external server share one common certificate, they can only authenticate if both servers trust that certificate. This means that one of the servers has to trust a certificate that does not belong to their organization. If this is you, the result is that any other server that holds that external certificate can access your server. This is a huge security risk! It means that any users or servers that the external organization has also given their certificate to can now access your server. This is therefore not a viable option if you want to restrict who can access your server. It is far more secure to have two certificates in common, one from each organization, and to only trust the certificate your organization owns. It is usually a good idea to only trust a flat certificate if it is controlled by your organization.

Server Access Control

Server access is controlled by fields in the Server form in the Name and Address Book or through variables in the NOTES.INI file. Figure 7.2 shows a server record in the ADVANTIS domain for a replication hub server named ADVRH001.

Server access can be explicitly granted or denied to replica creation, to file creation, to specified port(s), or to the entire server. For example, whenever an employee leaves the company, the employee's user ID is added to the DenyAccess group. This group is explicitly denied access to all of the corporate servers. The configuration variables can restrict

Figure 7.2
Server record in the ADVANTIS N&A Book.

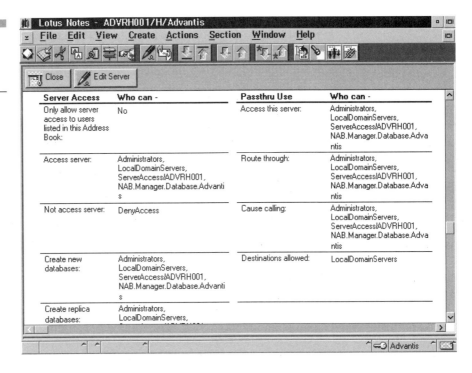

which groups of users can access the server via the dial-in COM port. This variable is often used when the Access Server variable is not used or when dial-in access is restricted to a subset of the users or servers who can access the server over the LAN/WAN.

Port security can be used for LAN or COM ports. They can also be used to deny access to specific ports. For any access level changes to take effect, the server must be restarted.

NOTE: *A user working at the server console can potentially bypass all access control measures including user authentication. Therefore, physical access control to the server must be closely restricted. Password-protected keyboard locks should be used on all Notes servers. However, the server.id should not have a password so that the server can be rebooted automatically, if necessary. For the same reason, Notes servers should not have password protected power on locks.*

The Administrators group will have access to all servers and is the only group that can create new databases or new replica databases. Consequently the Administrators group should be listed in the "Access server," "Create new databases," and "Create replica databases" fields (see Figure 7.2).

Local domain servers normally need access to other servers in the same domain. In Figure 7.2, the local domain servers are given as the default name LocalDomainServers. In the ADVANTIS domain, a special group NAB.Manager.Database.Advantis was created for a few people who need to administer the ADVANTIS Name and Address Book on all servers. Other domain servers get a default group name of OtherDomainServers. It is generally a good policy not to use this default name. In Figure 7.2, the other domain servers group is called Server/Access/ADVRH001. These are servers in other domains that you need to replicate with.

Port Server Access Lists

Access to the server from dial-in COM ports can be further restricted through Port Server Access Lists. There is no entry in the Name and Address Book to make these changes. They are enabled by defining specific variables in the NOTES.INI file on the server. The following is the IBMGN-EXT standard configuration for these variables:

```
ALLOW_ACCESS_COM1=<accesscomgroup>,Administrators
ALLOW_ACCESS_COM2=<accesscomgroup>,Administrators
```

ALLOW_ACCESS_COM1. This configuration variable restricts that groups of users can access the server via the dial-in COM port. This variable is often used when the Access Server variable is not used or when dial-in access is restricted to a subset of the users or servers who can access the server over the LAN/WAN.

ALLOW_ACCESS_COM2. See above.

Port security can be used for LAN or COM ports. They can also be used to Deny Access to specific ports with the DENY_ACCESS_<port-name>.

Database Access Control

Database access is controlled by Access Control Lists (ACLs). Every database has an ACL. An ACL is a list of people, groups of peoples, servers, and groups of servers who have access to a database and specifies the document privileges they have, such as Read, Compose, and Design. The database manager maintains the ACL. It is highly recommended that the Default ACL setting be set to No Access. Anybody with Manager access can change the ACL (for example, add or delete users with Manager access), so the set of people with Manager access must be closely restricted.

The creator of a database automatically has Manager access, so the set of people who can create a database must be controlled with the Server Access Control described earlier. See "Database Privacy and Integrity" later in this chapter for more details.

━━ ━━ ━━ ━━ ━━ ━━ ━━ ━━ ━━ ━━ ━━ ━━ ━━ ━━ ━━ ━━

NOTE: *A user working at the server console can potentially bypass all access control measures including user authentication. Therefore, physical access control to the server must be closely restricted.*

Table 7.1 shows the Access Control Levels for Lotus Notes.

TABLE 7.1

Access Control Levels for Lotus Notes.

Access level	Read?	Write?	Edit documents?	Edit design?	Administer?
1. No Access	No	No	No	No	No
2. Depositor	No (not even documents user wrote)	Yes	No	No	No
3. Reader	Yes	No	No	No	No
4. Author	Yes	Yes	Yes (but only documents user wrote)	No	No
5. Editor	Yes	Yes	Yes	No	No
6. Designer	Yes	Yes	Yes	Yes	No
7. Manager	Yes	Yes	Yes	Yes	Yes

To look at the database access control list for a Notes database, click on the icon representing the database. Then select a File|Database|Access control to display the access control. A typical access control list for a discussion database is shown in Figure 7.3. Notice that, for this database, the default access is "author" that is typical for a public discussion database in a Notes domain.

Database Privacy and Integrity

The Notes databases must be kept private and protected from tampering during data storage and transmission. This is accomplished by encrypting information stored on disk and during network communications.

Figure 7.3
Looking at the
database access
control list.

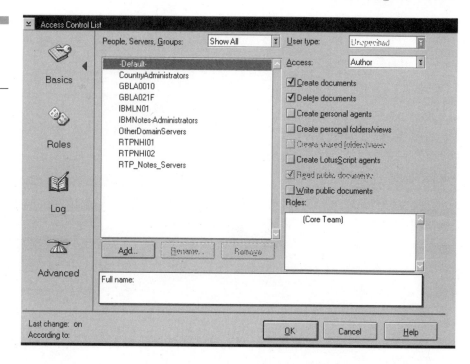

The Notes Administrator enables network data encryption with the Port Setup dialog box. Mail encryption is selectable by the user. Encryption maintains database privacy and integrity even if physical security is breached. The drawbacks of encryption are:

- Minor performance degradation
- Encrypted mail cannot be read if the user ID file is lost
- Encrypted items are skipped when creating a full text index

Verification of Originator

Originators are verified by electronic signatures based on the user's private key and a 128-bit fingerprint constructed from the data being signed. Entire documents or single fields can be signed. Signing mail assures the recipient that the person who sent the letter is who the user claims to be.

NOTE: *If an ID is stolen and is not password-protected, it can produce a forged signature. User ID files must be physically secured. A copy is as good as the original. If users store their user ID file on the hard drive of their workstation, password-protected keyboard locks and power-on locks must be used to ensure copies of the user ID file cannot be made when the workstation is unattended. Nontrivial passwords must be used.*

Physical Access Control

Physical access control of both servers and workstations is vital for the overall security of the use of Lotus Notes. Physical access to servers and workstations can potentially circumvent all of the above security measures.

Internet Firewalls

Internet firewalls restrict the traffic flow between the corporate network and the Internet. Traffic can be restricted based on routing filters so that only authorized network addresses are allowed through the firewall. Traffic can be restricted by sockets so that only authorized applications (such as SMTP mail) are allowed through. These are some of the tools available to protect the corporate network. The concept of packet filtering is further discussed in the following paragraphs.

Packet filtering provides security for Internet connections by permitting or denying packets into the corporate network based on IP address and TCP protocol port. Certain applications use reserved "well-known" ports. For example, Telnet (remote login) uses port 23, SMTP mail uses port 25, and Lotus Notes uses port 1352. Thus Telneting into the corporate network is denied by filtering out all packets destined for port 23.

Likewise, a very restrictive connection could permit only Lotus Notes packets to pass between the corporate network and the Internet. The only way a person could "enter" the corporate network would be as a Lotus Notes user. At this point, the Lotus Notes user would be subject to the Lotus Notes security described previously.

The combination of the packet filtering and Lotus Notes security protects the corporate information from unauthorized access. Most routers are capable of implementing packet filtering. In other instances, a firewall such as IBM's NetSP can be situated by the router to provide this function. This method of security requires no change to the standard Lotus Notes software.

User Authorization: Name and Address Book

Lotus Notes servers and the information contained on these servers must remain secure. The Notes system's security is built on the premise that all servers are physically secure, database ACLs are properly set, server access is properly set, and the IBMGN-EXT Name and Address Book ACL remains secure. The security standards for local Notes servers outlined here must be met before a Notes server is allowed in the IBMGN-EXT Notes network. It is the responsibility of the local Notes administrator to maintain the security of the servers and the databases that they administer. Also, it is the responsibility of the local Notes administrator to work with the Central Administration Group, when necessary, to ensure the security of the IBMGN-EXT Notes network.

Central Administration's Responsibility

Information stored within the IBMGN-EXT Name and Address Book is extremely critical and must remain secure. The IBMGN-EXT Name and Address Book ACL must be properly set for the database to successfully replicate throughout the IBMGN-EXT Notes network. Central Administration is responsible for the management of the IBMGN-EXT Name and Address Book. Local Notes administrators should not modify the IBMGN-EXT Name and Address Book ACL. Central Administration is also responsible for certification and recertification, the creation of Notes server IDs and user IDs, and the creation of associated certificates.

Local Notes Administrator's Responsibility

Termination of users, administrators, and vendors is also a security consideration. To ensure the security of the IBMGN-EXT Notes network, local Notes administrators must work with Central Administration to ensure that users, administrators, and vendors are properly terminated. Also, remote, dial-up access is another security consideration.

Remote, dial-up access encompasses communication with other companies outside the IBMGN-EXT organization who have integrated Notes into their environment, dial-up access for IBMGN-EXT Notes users, and engagement access. It is critical that the integrity and security of the IBMGN-EXT Notes environment is maintained during such types of communication. Local Notes administrators must adhere to the standards and procedures for these types of communications in order for the security of the IBMGN-EXT Notes environment to remain intact.

Database Security

Notes provides database-level security by allowing an Access Control List (ACL) to be defined for each database. An ACL defines the people, servers, and groups who can gain access to a database and specifies the kind of access and document roles they have. Each database must have at least one database manager who maintains the database ACL. This document describes the IBMGN-EXT standards for setting database ACLs based on the requirements for sharing and replicating data throughout the firm.

It is important to note that the information stored within most databases is highly confidential and should remain very secure. Therefore, the standards described here are intended to assist the database manager with setting the databases ACLs properly. This is important if the databases are to successfully replicate throughout the IBMGN-EXT Notes network topology. The recommendations here are not intended to provide open access to data by any group of users.

ACL Guidelines for Nonreplicated Databases

The ACL for databases remaining on a single server need only include the group and user names that will have access to the database and the RemoteReplicationServers group. For each group or individual name listed in the ACL, define their specific access level. For example, one group of users can be defined with Author access allowing them to create new documents or edit ones they had previously authored. Another group can be defined with Reader access allowing them to only view the information in the database. One or two individual names must be entered with Manager access allowing them to change the structure of the database and manage the ACL list.

It is highly recommended that the Default ACL setting be set with No Access. It is also important to note that the server access variables will restrict which users have access to a specific server before they could try opening a specific database.

NOTE: *Roles can be defined on forms and views and provide additional levels of securing information in databases. Roles defined in the application are administered by the database manager in the ACL for each group, server, or individual.*

ACL Guidelines for Replicated Databases

Once a database is replicated with other servers, the guidelines for managing the ACL list become more critical. The ACL for databases that will replicate with other servers must be set up to ensure that all servers, including hub or intermediate servers, have the proper access levels in order to replicate all necessary changes.

It is not adequate to define only the source and final destination servers with the proper ACL levels. All intermediate servers that the database is replicated through must be defined with an ACL level high enough to replicate all changes to the destination server. For example, if a change was made to a form or view in the source server, then all intermediate servers would require at least Designer access to replicate it to the next server.

To assist the database managers and server administrators with defining database ACL, a firmwide hub server group has been defined. It contains a list of all servers throughout the firm that will be used as a replication hub. Figure 7.4 gives recommended server ACL settings. See Table 7.2.

TABLE 7.2

Recommended Server ACL Settings.

Recommended server	ACL settings
Server or group name	ACL setting
LocalDomainServers	Manager
<source server name>	Manager
OtherDomainServers	Editor or reader
RemoteReplicationServers	Manager

NOTE: *The settings for users of the database can vary depending on what type of access they require. However, all servers must have their ACL levels high enough to replicate user changes from the origination point to the destination point.*

LocalDomainServers. This is the group name containing the list of servers that are used as replication hub servers and are managed by the Central Administration group.

<source server name>. This is the name of the server where the application originated. It is the server where most or all database design changes are made as well as changes to the ACL.

OtherDomainServers. This group contains the names of all servers that are at the bottom of the replication hierarchy. This group is included in database ACLs to provide for proper database replication across the worldwide network. This ensures that new, updated, and deleted documents flow back to the source server database.

RemoteReplicationServers. This group is used to ensure that databases will replicate properly to remote users' PCs. The Remote ReplicationServers group should be given Manager access in the Access Control Lists (ACLs) of all databases that are replicated to remote users.

The previous example applies only to the management of ACLs for replication purposes between servers. The use of roles in databases can affect database replication. Therefore, it is critical that the source, hub, and spoke servers have all necessary roles or replication might not work properly. For more information on roles, see the Lotus Notes Database Design manual. The management of the ACLs for individuals or groups is dependent on the requirements on those individuals or groups.

Cryptography

Cryptography, the art of writing text in secret code, is one way to secure data sent over a network. Cryptography is one of the most important aspects of security for electronic commerce, including transactions over the Internet. Two of the newest ways to secure transactions over the Internet involve use of SHTTP (Secure Hypertext Transfer Protocol) and SSL (Secure Sockets Layer). Both of these techniques encrypt data between a Web browser and a Web server.

Encryption Example

Let's examine a few examples to appreciate the strength of encryption. One of the simplest encryption algorithms is to shift each letter of the message by one letter in the alphabet. Romeo might use this algorithm to send Juliet a rendezvous message about when and where to meet tonight. Of course, Romeo and Juliet don't want anybody to know about their tryst, especially their parents! Romeo's plain-text (un-encrypted) message is "THE CHURCH AT MIDNIGHT." The encrypted message is "UIF DIVSDI BU NJEOJHIU." Offhand, the lovers' parents would not be able to read this encrypted message. Another version of this method is to shift each letter of the plain-text message by two letters of the alphabet. In this manner, the encrypted message is "VJG EJWTEJ CV OKFPKIJV," equally mysterious as the first version.

The algorithm used in the previous two examples is the same algorithm: Shift the letters of the plain-text message by so-many ("N") letters of the alphabet. There are 26 letters in the alphabet, so there are 26 different versions of this algorithm. Each of these versions can be associated with a unique private key.

Before Romeo can send Juliet encrypted messages, he must send her the private key that he is using so that she can decrypt his messages. This private key must be kept private so that nobody else can decrypt and read the message. Romeo sends his private key to Juliet via his trusted courier. In this case, Romeo sends key 9, so Juliet will shift all letters of the message by 9 letters of the alphabet.

Well, Romeo wants to secretly send Juliet flowers, so he sends key 24 to the flower shop owner via trusted courier. When the flower shop owner receives the encrypted order from Romeo, she will shift letters in the encrypted message by 24 letters of the alphabet. Similarly, Romeo sends key 17 to the candy shop owner via trusted courier.

Juliet uses key 9 with Romeo because he already sent that key to her with the implicit agreement that this is the key to be used for all communications between the two of them. Juliet needs a different key with the candy shop owner than Romeo or else she can read Romeo's orders (which she intercepted), won't be surprised to receive the candy, and their romance will be in jeopardy! Juliet sends the candy shop owner key 21 via her trusted maiden, and all is secure.

There are a few problems with this simple algorithm. It is too easy to break the code. Shifting letter the same amount is too simple. Furthermore, there are only 26 different keys, so it does not take long to try every possibility. Each pair of senders and receivers must have different keys. Thus, the candy shop owner has a one key (17) to communicate with Romeo and a different key (21) to communicate with Juliet. Romeo and Juliet, meanwhile, use yet another key (9).

Keys get used up quickly as the number of users increase. For n users, $(n \times (n\text{-}1)/2)$ keys are required. Thus, 1000 users will need almost 500,000 keys! Distributing keys prior to sending encrypted messages by courier is inconvenient at best. Distribution of such a large number of keys becomes nearly unmanageable.

Strengthening Encryption

Encryption is strengthened by making more keys and making each key longer. Romeo sends Juliet the key 5 22. She shifts the first letter of the encrypted message by 5 letters of the alphabet, the next letter by 22 letters of the alphabet, the next letter by 5, the next by 22, and so on. Now there are $26 \times 26 = 676$ keys.

A longer key that Romeo could send Juliet is 7 19 13 24 6. Juliet shifts the first letter by 7 letters of the alphabet, the next letter by 19, the next by 13, the next by 24, the next by 6, the next by 7, the next by 19, and so on, cycling through the key. Now there are $26 \times 26 \times 26 \times 26 \times 26 = 11,881,376$ keys! Even if Juliet's mother obtained the encrypted message and knew the encryption algorithm, it would be past midnight before she could try all possible keys to decrypt the message (assuming Mother's computer was down and she had to decode by hand). Romeo and Juliet would have finished their midnight tryst before Mother could figure out when and where they were meeting.

We have increased the numbers of keys and made a more involved algorithm. These basic examples can give you a feeling of how to strengthen encryption.

Private Key Cryptography

The encryption and decryption that Romeo and Juliet used is an example of private key cryptography. Private key cryptography uses only one key to encrypt and decrypt a message. This key is private (secret) and must be distributed in a secure manner to each sender/receiver pair of users before sending encrypted messages. Because a particular key is used for only one pair of users, different keys must be used for each different pair of users. This distribution of keys is not convenient.

Despite its drawbacks, private key cryptography is still very popular. An example of private key cryptography is the Data Encryption Standard (DES) that was published as a standard by the U.S. government in 1977. DES is still widely used, for example, by the U.S. government and by the American Banking Association. In fact, the confidence in DES is so high that Electronic Funds Transfers have relied on DES encryption. DES uses a series of substitutions and transpositions to scramble the message. There are 72,000,000,000,000,000 (72 quadrillion) keys. Assuming that a computer can try one key per microsecond, a brute-force attack would take about 2000 years to try every possible key. This is indeed quite strong encryption. The problem of key distribution still exists with DES, however, because DES is private key cryptography.

Public Key Cryptography

Public key cryptography solves the problem of pre-arranged distribution of keys. Public key cryptography methods assigns a pair of keys to each user: a private (secret) key and a public key (hence the name). Each pair of private and public keys are mathematically related so that one private key corresponds to one and only one public key.

The private key must be kept private—kept by the user only. No copies of the private key must ever be made and given out. The private key is the heart of message privacy (encryption/decryption), message integrity (digital signatures), and, in many cases, data access and user authentication. The public key, on the other hand, is made public—known by everybody. The public key should be available in a public directory for all to access. This eliminates the need for pre-arranged distribution of keys prior to sending encrypted messages. Messages encrypted with one key are decrypted with the other key. Here are the basic elements of public key cryptography:

- Each user has a public key and a private (secret) key.
- Each pair of public and private keys is mathematically related.
- Private key is secret—kept by user only.
- Public key is made public— known by everybody.
- No pre-arranged distribution of keys needed.
- Encrypt with one key; decrypt with the other key.

The next three examples will help clarify how public key cryptography works.

Message Privacy: Encryption

Romeo has decided to use public key encryption instead of private key encryption. Today he encrypts his rendezvous letter using Juliet's public key (available from the local phone directory) and sends the encrypted message to her. Juliet can decrypt the message because she has the corresponding (her own) private key. In fact, because only Juliet has her own private key, she is the only person who can decrypt the message. Anybody can encrypt messages using Juliet's public key, but only she can decrypt these messages using her private key. (To re-emphasize, if anybody has a copy of your private key, they can read your encrypted mail. *Don't allow your private key to be copied, with or without your permission!*)

Message Integrity: Digital Signature

Digital signatures assure the receiver that the sender is the actual person who sent the message. This is analogous to a handwritten signature or a person's wax seal used for the same purpose with hardcopy documents. In fact, a digital signature is an improvement over the handwritten signature because a digital signature can also verify that the message has not been modified.

For example, King Ferdinand writes an authorization for Columbus to provision three ships and to hire mates. The King signs the authorization by encrypting with his (the King's) private key and sends the encrypted message out. Anybody receiving the authorization can decrypt the message using the King's public key (which everybody has access to) can see the King's digital signature to verify that the King did approve the authorization and that Columbus did not modify the authorization. Columbus cannot go to the shipyard and ask for four ships when he is only authorized for three ships. (To re-emphasize, if anybody has a copy of your private key, they can impersonate you and forge your signature. *Don't allow your private key to be copied, with or without your permission!*)

User Authentication

User authentication is similar to digital signatures. Suppose Marco Polo wants to check on his caravan database. How does the server know that Marco Polo is who he says he is? Marco Polo logs in to the server with his user ID and password. The server sends him a random number.

Marco Polo encrypts the random number with his private key and sends the encrypted message back to the server.

The server uses Marco Polo's public key to decrypt and compares with the original random number sent. If the two numbers match, then Marco Polo's identity is verified. Only Marco Polo has his own private key, so only Marco Polo could send back the correct response. (To re-emphasize, if anybody has a copy of your private key, they can impersonate you and forge your signature. *Don't allow your private key to be copied, with or without your permission!*)

The Overall Picture

Table 7.3 gives an overall picture of mail and signature security.

TABLE 7.3

Encryption for Mail, Signature, and User Authentication

Security Type	Encryption	Decryption	Sender	Receiver
Privacy (mail encryption)	Use receiver's public key	Use receiver's private key	Any	Only 1
Integrity (digital signature)	Use sender's private key	Use sender's public key	Only 1	Any
User authentication	Use sender's private key	Use sender's public key	Only 1	Any

The RSA Algorithm

The RSA algorithm is an example of a public key cryptography algorithm. The algorithm was published in 1978 and has been named using the initials of the last names of its developers: Rivest, Shamir, and Adleman. The algorithm is based on the fact that it is easy to multiply prime numbers to create a large number and it is very, very difficult to find which prime numbers were multiplied together to initially create the large number. Among mathematicians, this is the well-known, difficult problem called *prime factorization* of large numbers.

This asymmetry is ideal for cryptography! It's easy to create keys and is very hard for unauthorized users to determine the key. Furthermore, the RSA algorithm is very strong. Using a 1024-bit key, the number of keys is 1 followed by about 300 zeros. This number is so large that there is not even a name for it! Assuming a computer can try one key per microsecond, a brute-force attack would take about 2 billion years to try every possible key! This is indeed very strong encryption—much stronger than DES. RSA is public key cryptography, so the distribution of keys is easy. Lotus Notes uses RSA cryptography as the basis for its security functions.

Internet Mail Security

RSA has been also popularized by providing the basis for security in PGP. PGP, which stands for Pretty Good Privacy, is a program that provides message encryption/decryption, digital signature, and message integrity on the Internet.

These functions are provided with Lotus Notes; however, when communicating with users on the Internet, it cannot be assumed that the recipient is a Notes user. The Notes public and private keys thus cannot be generally used with Internet mail because the public keys and the security algorithms are not publicly available on the Internet. PGP public keys and security algorithms are publicly available on the Internet, and PGP is becoming more widely used.

The mail text is encrypted with PGP, then sent with any SMTP mail system, such as the UNIX mail systems ELM or PINE. The recipients receives the encrypted mail with their favorite SMTP mail system, such as Ultimail. The recipients then decrypts the mail text with their PGP private key.

Contributing to the widespread use of PGP is the fact that the public keys and security algorithm is free and freely available for noncommercial use from various anonymous FTP (file transfer protocol) sites on the Internet, such as net-dist.mit.edu. A commercial version is also available from the company ViaCrypt (Phoenix, AZ).

Security Risks with Ethernet and Token Ring Networks

PROBLEM 1.   Unused network ports can permit network snooping with "Sniffers."

SOLUTION 1.   Disable all unused network ports

Nearly all local area networks (LANs) are Ethernet or Token Ring networks. Ethernet and Token Ring networks are both broadcast networks, which means that all systems attached to the network get a copy of every transmission. (This fact and its security implications are not emphasized or well-publicized.) An individual system will keep the data packet if it is addressed to that system.

For example, King Arthur, Queen Guinevere, and Lancelot all have their computers connected to an Ethernet network (the same applies to Token Ring networks). When Lancelot sends a message to Guinevere, Arthur's and Guinevere's Ethernet adapter cards (also called a *network interface card*) on their computers each receive the message; however, because the message is addressed to Guinevere, only Guinevere's adapter card will recognize the address and forward the message on the mail application program.

Meanwhile, the wicked Modred plots to expose the dishonorable love affair between Guinevere and Lancelot. He knows the castle at Camelot has been prewired, so there are spare, unused Ethernet ports available. The sly Modred connects an instrument called a Sniffer to an unused Ethernet port, copies all data packets sent on the Ethernet, and reads all of the mail Guinevere and Lancelot (and everybody else) sends. Nobody even knows that the evil Modred has copied the packets because this is the normal operation of an Ethernet and everybody has received their mail without a problem.

A Sniffer is intended for use by network administrators to monitor, detect, and correct network problems. In the wrong hands, a Sniffer can be a nefarious weapon. A Sniffer can generate packets to test the response of network equipment; however, mail can also be "forged" and appear to come from "impersonated" senders. To foil Modred's Sniffer attempts, Merlin, the network administrator, disables all unused network ports. (Sometimes it seems as if sorcery causes network problems and requires wizardry or magic to fix the problems!)

PROBLEM 2. "Sniffer" swapped for a computer on an enabled port.

SOLUTION 2. Physical security. Control physical access to network ports in use.

Modred finds that unused network ports have been all disabled, so while Galahad is on a knightly quest, Modred sneaks into Galahad's chamber to use Galahad's enabled port. He disconnects Galahad's computer and connects the Sniffer for the night. Merlin learns that somebody is tampering with people's computers, so he enforces the policy that all chambers must be locked when unattended.

PROBLEM 3. Network adapter cards acting like a network "Sniffer."

SOLUTION 3. Sign and encrypt all transmissions.

Modred applies to connect his own computer to the Ethernet network, so Merlin's assistant enables the port in Modred's chamber. Modred conspires with the Ethernet adapter card expert, Morgan le Fay, to configure the card to act as a Sniffer. The card normally receives all packets, and only those packets addressed to it are forwarded on to the application program. All other packets are filtered out.

Adapter cards can be configured to not filter out by address, so *all* packets are sent on to an application program. The computer with this modified adapter card then acts just like a Sniffer. (This fact and its security implications are not emphasized or well-publicized.) This situation requires a bit more expertise and is not documented (as opposed to a Sniffer that you can easily buy and get a written manual on how to operate it). This case is not so common but is more subtle and requires more resources to counter. The best way to ensure secure transmissions is to electronically sign and encrypt your transmissions with the tools provided with Notes or with PGP. Then, even if Modred does get a copy of Lancelot's mail, Modred will have a very difficult time to unscramble or change the message or to impersonate Lancelot.

Physical Security Requirements for Routers and Other Network Equipment

PROBLEM 1. Routers and other network equipment are susceptible to "sniffing."

SOLUTION 1. Physical security. Control physical access to routers and other network equipment.

Sniffers can be readily attached to routers and other network equipment. The best way to prevent unauthorized sniffing is to tightly control physical access. Large, established network service providers certainly understand the need for tight physical security and the risk of not having it. Other institutions encourage openness and might find some security measures too restrictive or confining. At a minimum, it is strongly recommended that routers and other network equipment be kept in a locked room with restricted physical access. Small amounts of equipment are often mounted on a standard 19-inch-wide rack and locked in a wiring closet. Large amounts of equipment are often placed in a temperature-, humidity- and power-controlled room with access controlled by a round-the-clock security guard.

Security Smart Cards

There are a variety of credit card-sized security smart cards that enclose a computer chip and typically have a one-line display. One approach is to encode a complex algorithm on the chip to calculate and display a different number once a minute.

When I use my modem to dial the access computer and enter my user ID and password, the access computer looks up which card is registered to me so that it can use the same algorithm and starting point as the chip on my card. The access computer then asks me to read the number currently displayed on the security smart card to synchronize the clocks on my card and on the access computer.

Now that the computer knows which card I have and what "time" I have on the card, it can predict what numbers will appear next. The access computer prompts me for the new number now displayed on my card and matches that number with its own calculation. If there is a match, I am authenticated and allowed to use the network and servers I am authorized for.

Security smart cards use the principle of having "something I have" and "something I know." The "something I have" is the security smart card, and the "something I know" is my user ID and password. Neither alone is sufficient to gain access. If thieves stole my security smart card, they could not gain access because they would not know my user ID and password. If spies obtained my user ID and password, they would not be able to gain access without the numbers displayed on the smart card. Even if sniffers were used to obtain my user ID, password and numbers displayed on the smart card (all of which I transmit over the network), the numbers are only good for that one session. What is needed to gain access then is the smart card algorithm, the "current time" on the card, the user ID, and the password. The smart-card algorithm and the "current time" might possibly be obtained by opening the smart card and reverse engineering. Most cards are tamper-resistant and disable or destroy the card when opened. (I do not know of any cards that will self-destruct a la *Mission Impossible*, but nevertheless, I would not recommend opening a security smart card! At best, you will lose your money for the price of the card.)

The benefit of "something I have" and "something I know" is sometimes viewed as a liability. The security smart cards have to be registered and distributed. The cards must be carried by users whose wallets are already bursting at the seams. The users must be vigilant to notice if the card is missing and report the loss immediately to the central administrator (similar to losing a credit card). The administrator must then disable the missing card and issue a new card.

All told, there is more administration overhead and this is the price for added security. Security smart cards are physical and cannot be copied. These cards are a good option for providing access to networks and computers more securely than a method only using a user ID and password. The cost and administration overhead must be balanced with the convenience and security required.

The numbers generated by the security smart card are used for authentication to gain access to networks and servers. In a rough sense, this can be compared to the use of the Notes private key for authentication. One important distinction is that the smart card is material and something people can hold. It is like a physical key used to unlock a regular door lock. People are familiar with physical keys and understand that it must not be stolen or copied.

On the other hand, the Notes private key is contained in the user ID file stored on a diskette or on the hard drive of a workstation. Private keys and computer files are much more abstract. People are not as familiar with private keys and computer files and usually do not appreciate how easy it is to copy the private key or how serious it is for somebody else to have a copy of the private key. When you make a copy of your house key, you know because you have given your key to somebody. Your Notes private key, however, can be copied from your diskette or hard disk in the time it take you to get a drink of water or go to the bathroom and you will not even know your private key was copied!

If you store your Notes user ID file on your hard disk, you should:

- Use a password-protected power-on lock and a password-protected keyboard lock.

- Lock your keyboard whenever you leave your computer, no matter how long you will be gone.

If you store your Notes user ID file on diskette, you should:

- Never leave the diskette in your disk drive when you are not at your computer.

- Never leave the diskette lying around in the open.

(These items are not emphasized or well-publicized, so they are worth repeating throughout this book.)

The Internet's SSL and SHTTP

Secure Sockets Layer (SSL) is the security protocol usually used by Web pages on Internet servers. The URL for these types of pages uses the https://format. You usually encounter this type of Web page when you need to provide confidential information in a fill-out form on the Web page, such as credit card information.

Domino 4.5 and Notes 4.5 support the data encryption, server authentication, and message integrity features that are part of SSL Version 2.0.

Using a Notes client, when users come across a Web page that has been secured using SSL, a TCP/IP connection is established, and an authentication process begins. First, the Internet server sends a certificate to the Notes client. The Notes client then uses the information provided in a Certificate Authority (CA) certificate to validate the certificate of the remote Internet server. If the Notes client trusts the certificate, a secure data channel is opened and all data sent between the client and the server is secured. The data can only be read by the authenticated server that has the private key issued by the CA that matches the public key sent in the server's certificate.

Secure Hypertext Transfer Protocol (SHTTP), the Internet protocol used for creating secure HTTP connections using public key technology, is also used in the previous scenarios.

C2 Security

The U.S. government has established criteria for security for anybody doing business with the government. These criteria are described in the December 1985 document "Trusted Computer System Evaluation Criteria," which is popularly known as the "Orange Book" because of the color of the book's cover.

There are four main groups of security: A, B, C, and D. A is the highest level of security. Within these groups, there are numbered levels such as C1, C2, B1, B2, and B3, where C2 offers higher security than C1. D is the level assigned to those systems submitted for evaluation but that fail to meet even the C1 rating. There are no systems with this rating, although DOS for the IBM-compatible computers and the Apple Macintosh would be expected to get this rating.

To obtain a rating, a system developer must test all security-feature requirements, document the results, and submit the document to the National Computer Security Center (NCSC), which is part of the National Security Agency (NSA), which is part of the Department of Defense (DOD). The NCSC will then do its own testing. The NCSC maintains the Evaluated Products List of which products have which ratings. The ratings apparently apply to operating systems and not to applications. The NCSC has this information about IBM products:

- IBM MVS/ESA rated B1 in 1990
- IBM OS/400 rated C2 in 1995
- IBM OS/2 not listed
- IBM AIX not listed

AIX version 4.x has strong security features and, if evaluated, would probably have a B1 rating.

NCSC does not have Lotus Notes on the list. Lotus representatives indicated that Notes itself is not rated although Microsoft Windows NT is on the list. A recent (October 1995) news item indicated that Windows NT is rated C2.

The Orange Book does not address network security, so the NCSC published the "Trusted Network Interpretation of the Trusted Computer System Evaluation Criteria" (the "Red Book") in 1987. It is not clear that a corresponding C2-level security rating exists for networks.

It is recommended for government contractor customers that all Notes data communication be encrypted using the encryption that comes with Lotus Notes. This encryption is based on the RSA algorithm, which has been extremely difficult to crack.

Domino Security Practice for Intranet Servers

Domino comes with many security options, and decisions must be made about how to implement selected features. This section is one possible scenario of security practices for intranet servers to illustrate one set of choices. This might not suit your intranet, extranet, or Internet Domino servers, but the discussion will point out where choices must be made.

Table 7.4 summarizes the example Domino Security Practice. There are five access modes, and your first choice is to decide which modes you will support. We will only allow anonymous Web browser or standard

Notes client access. Details are described later. Next, decide if a dedicated server is required for the public information (no confidential information) and an isolated server for the private information (where security is required). We will mix public and private information on the same Domino server and control access with the database Access Control List (ACL).

TABLE 7.4

Domino Security
Practice

Web browser/ Notes client mode	Public information— no confidential information	Private information— security required
Web browser—Anonymous	Can access	No access —controlled by database ACL
Notes client—Standard	Can access	Can access —controlled by database ACL
Notes client—Anonymous	Not supported	Not supported
Web browser—User id/ password login	Not supported	Not supported
Web browser—SSL secure login	Not supported	Not supported

Web Browser—Anonymous

Information hosted on the Domino server will be either public or private (sensitive or requiring security). Anonymous Web users can access the Domino server but will only be able to access the public databases. Anonymous Web users will not have access to any database that is private, is sensitive, or requires security. A Notes client must be used to access any information that is private, is sensitive, or requires security. Database security will be handled at the database level using Access Control Lists (ACLs). Only Notes Administrators can modify ACLs.

ADMINISTRATOR DETAILS. Server access by anonymous Web users is set in the Server document of the Domino server in the Public Name and Address Book. In the Security section of the Server document, Yes will be selected for Allow anonymous HTTP connections:. Information access is then controlled at the database level using the ACLs.

Web users will not be required to provide a user ID and password to access the Domino server and will be known by the name Anonymous. (This is a special Notes group, just as Default is a special Notes group.)

With no explicit action by the Notes Administrators, all anonymous Web users would be given the Default access, which might be higher than desired. Therefore, it is important that the Anonymous group be explicitly added to the ACL for all databases and be given the default access level of "No Access." This is a requirement for anybody creating a new database (Notes developers, Notes Administrators, etc.).

If the database is to be made public, the access level for Anonymous can be made higher. In this manner, Anonymous access to a database must be given deliberately. ACLs are controlled by the Notes Administrators only. Database owners can modify Group membership in the Public Name and Address Book, but only Notes Administrators can change the database ACL for the Groups.

Another security feature used in conjunction with the explicit ACL lists is the Maximum Internet browser access option in the Advanced section of the ACL. Access by Web users is limited by this setting. For example, if Anonymous is listed explicitly in the ACL with "Reader" access and Maximum Internet browser access is set to No Access, the Anonymous Web users will have no access to the database.

The Maximum Internet browser access setting does not raise access levels, but only limits Web access levels. For example, if Anonymous is listed explicitly in the ACL with "Reader" access and Maximum Internet browser access is set to Author, the Anonymous Web users will have "Reader" access to the database. The "Reader" access level does not get raised to "Author" access level.

The default maximum Internet browser access is "Editor" access, which means that Web users can create documents and edit all documents, including those created by others. This is too high for our infrastructure. The maximum Internet browser access should be set no higher than "Author," which means that Web users can create documents and edit documents, but only ones they created. Editor, Designer, and Manager functions should be done with a Notes client in a secure manner. It is the database developer who sets the Maximum Internet browser access to Author.

Actually, there is an inconsistency with giving all Web users the Anonymous group membership and Author access. Because all users appear to have the same name, any user can edit a document "Anonymous" has created. In effect, any user can edit any document and essentially has Editor access. Depositor access does not allow users to read documents. Reader access does not allow users to edit. A possible workaround is to allow users to submit documents and read documents, but not edit any documents by hiding the fields in edit mode or using roles and field controls in creative ways.

The Maximum Internet browser access setting is more relevant to the situation where Web users other than anonymous Web users access the Domino server, which is not the case here. In our situation, the Maximum Internet browser access setting and the ACL for the Anonymous group should be the same for a given database. In situations where additional Web user have user IDs and passwords, different people will have different access levels and usually higher than the access level for Anonymous users. The Maximum Internet browser access setting is more useful in these cases.

Notes Client—Standard

The standard Notes client will have access to the Domino server in the normal manner. Server access lists, cross-certification, database ACLs, and all normal Notes security features apply.

Notes Client—Anonymous

Anonymous access by Notes clients (a Notes 4.*x* feature) will not be supported. The reason for requiring the use of a Notes client is for security. Anonymous access by a Notes client defeats the whole purpose of using a Notes client, in this case. Anonymous access by a Notes client is useful for a Notes server on the Internet. In this case, the Notes server behaves similar to an anonymous FTP server where you allow anybody to download your content.

ADMINISTRATOR DETAILS. Server access by anonymous Notes users is set in the Server document of the Domino server in the Public Name and Address Book. In the Security section of the Server document, No will be selected for Allow anonymous Notes connections:.

Web Browser—User ID/Password Login

Domino supports the use of a Web browser with a user ID and password to login to a server and/or a database. Although this provides some measure of security that is greater than having no user ID and password (anonymous access), this is not very strong security and is not easy to administer. Mainframe systems (VM, MVS), UNIX systems, and Windows NT systems are usually set up to deny login after a given number of unsuccessful attempts. The Domino Web browser login allows unlimited attempts without any disincentives. If secure access is required, using a Notes client will be required. Using a Web browser with a user ID and password will not be supported for the time being. This feature is desired by end users, but more work must be done to ensure security.

ADMINISTRATIVE DIFFICULTIES. Every person in your Notes Domain would need a Person document in the public Name and Address Book with a user ID and HTTP password.

It is difficult to register every person in other Notes Domains that you communicate with who needs access to the Domino server. Each user needs a Person document in the public Name and Address Book with a user ID and HTTP password. This is an administrative burden.

The HTTP password (which is independent from the Notes password) needs administration. The HTTP password must be initialized and distributed to the user. The password will need to be changed at times. This is a burden if the Notes administrators are the only ones who can change the HTTP password in the Name and Address Book. This is a security concern if the end user is allowed to change the Name and Address Book. It is possible to design the Person document with Author roles to allow each user to update only the HTTP password field in the Public Name and Address Book. This is a decision you have to make.

There is a registration application that is available for download from the Domino Web site. This allows the Web user to register their name and password themselves and to change the password themselves. This solves the administration problem, but creates a security problem. I can register as the Chief Executive Officer (the CEO, the head of the company) with a password I know. Then I can have whatever access the CEO has. Not good. The user identity cannot be verified independently.

The user ID format is not user-friendly. Using the fully distinguished name, such as Peter Lew/White Plains/Advantis/US, is desired from a security viewpoint. This is used in the Notes ACLs and guarantees uniqueness. If there are two people in an organization with the same common name, such as Peter Lew, they would have the same access rights if only the common name was used in the ACL. This would be analogous to using a flat Notes domain structure. The drawback of using the fully distinguished name is that more typing is required when accessing the Web site. Users also need to be educated as to what their fully distinguished name is and where to look it up if they forget what it is. It should be noted that entering the user ID and password is required only once per session (connection with the Domino Web server with a Web browser).

The HTTP password is *masked,* not encrypted. This might be okay if only the Notes administrators can edit the Person documents.

Web Browser—SSL Secure Login

Secure Web browser access can be provided by using the Secure Sockets Layer (SSL) protocol. Like Notes, SSL is based on the RSA public key/private key encryption method. Like Notes, SSL also uses certificates to vouch for the identity of clients and servers. The problem is that SSL is like Notes—using SSL requires an administration effort that parallels Notes administration. SSL keys and certificates must be administered like Notes keys and certificates (in the user ID and Name and Address Book directory). Cross-certification must also be administered. We are already using Notes for secure access, so providing secure Web access (to our intranet Domino servers) is not necessary. SSL is useful for extranet communication between businesses on a private network and, of course, on the Internet. SSL is not so useful for the intranet used internal to a company.

Disable Web Users from Seeing all Databases on the Server

In the "Security settings" section of the Server Document in the Name and Address Book, there is an option called Allow HTTP clients to browse databases. The default setting is Yes, which allows Web users to use the ?OpenServer URL command to see a list of all databases on the Domino server. This URL would look something like http://www.Mythical Site.com/?OpenServer. This is equivalent to a File|Database|Open command in Notes.

Set this option to No so that the Web users will not see a list of databases. Hiding the name of database files is no substitute for true security, and all of the Notes security features available should be applied as necessary. Nevertheless, showing names of database files can cause people to become curious and to try harder to access information they are not authorized to access. It is better to consciously control access to the list of database shown instead of giving access to everything to everybody. Even though Web users cannot see the list of databases, they can still open individual databases for which they have access.

The Internet
and Lotus Notes

Then came the Internet. Microsoft CEO Bill Gates called it the "Internet tidal wave." After the initial hype by the popular press, it soon became clear that the Internet's biggest immediate impact would be on the corporate workplace. Internet technology makes it easy to publish and share information with colleagues, whether in the same office or in branches overseas. Intranets, as these internal networks are called, seem to promise all the capabilities of Lotus Notes, and they are easier to set up and cheaper to maintain.

However, in fact, the Internet might have been what Notes needed all along. Since it launched Notes in 1989, Lotus's biggest challenge has been getting people to understand its purpose. The Internet, the World Wide Web, and intranets give people a window into the world of collaboration across a network. Eric Schmidt of Sun says: "Notes will offer more of the pieces for intranets than any other solution."

Fortune magazine, July 8, 1996.

One very cost-effective way to use Lotus Notes globally is to use the Internet as your wide area network. This chapter discusses this topic.

What is the Internet?

The Internet is a collection of networks. Internet service providers (ISPs)—such as IBM, NETCOM, Delphi, SURAnet, CERFnet, or PSINet—have their own networks and connect their customers to these networks. The Internet service providers' networks are in turn connected to each other via the Internet backbone network.

For example, IBM's network (called OpenNet) might connect an insurance company's network and a bank's network, while SURAnet might connect a university's network. When an insurance company employee sends mail to the bank's employee, the mail will be sent over the IBM OpenNet. When the bank employee sends e-mail to the university employee, the mail will go from the bank's network, across the OpenNet to the Internet backbone, to the SURAnet network, and then to the university network. In both cases, nonetheless, the e-mail has traversed the Internet.

The IBM OpenNet and the SURAnet network are privately owned but are part of the Internet—indeed being connected to the IBM OpenNet is being connected to the Internet. The IBM OpenNet is IBM's piece of the Internet. Thus, to make the initial statement of this section a bit more precise: The Internet is the collection of networks attached to the Internet backbone.

How Data is Sent Reliably Over the Internet

The Internet is a packet-switched network, so data is sent as packets. To understand the context of packet-switched networks, it is helpful to revisit the original goal of the Internet.

In the 1960s, the Department of Defense desired a communications system that would continue to function even if part of the system was destroyed. Connection-oriented networks are not adequate because, once an intermediate switch is destroyed, the connection is broken and no further communication takes place. This is analogous to having a phone conversation end when a telephone central office switch goes down. To address this problem, the Advanced Research Projects Agency (ARPA), the research arm of the Department of Defense, began funding research for packet-switched networks.

For packet-switched networks, the user's data (such as mail) is broken into small packets, each packet containing the source address, the destination address, a sequence number, and a portion of the user's data. Each packet is sent from one router to another router until the destination is reached. The routers use the destination address to decide which router to send to next, based on its routing table. The routing table changes dynamically to bypass network congestion or outages.

Packets from the same message might take different paths across the network and might arrive out of order. The receiver uses the sequence numbers to put the packets back in order. Packets might get lost, so the receiver sends an acknowledgment for each packet received. If the sender does not receive acknowledgment for a packet, then the packet will be resent.

Once all the packets have arrived, the original message is reconstructed and the mail is delivered—ta da! In this manner, reliable communications is achieved over an unreliable network. This was the paradox that ARPA was initially confronted with and resolved by developing the TCP/IP protocol suite.

Today, users of the Internet and TCP/IP benefit from the military's desire for reliable communication over an unreliable network.

A Brief History of the Internet

ARPA not only pioneered the development of TCP/IP but also fostered its widespread use. ARPA provided the well-known ARPANET backbone of the Internet beginning in 1969, which connected universities across the country allowing research in packet-switching technology as well as communication via e-mail and file transfer. ARPA then funded the incorporation of the TCP/IP protocols into the University of California at Berkeley's version of UNIX: the Berkeley Software Distribution UNIX, or BSD UNIX for short. BSD UNIX was adopted by more than 90% of the university computer science departments.

The National Science Foundation (NSF) became more involved with the Internet in 1985. The NSF established supercomputer centers at six universities across the country. These supercomputer centers were so expensive that the NSF wanted to provide widespread access to remote users. Of course, NSF turned to the Internet and TCP/IP.

In 1986 NSF funded a new backbone network, the NSFNET backbone, which initially connected to and then replaced the venerable ARPANET. In 1990, IBM, MCI Communications Corporation (MCI), and Merit (a consortium of universities in the state of Michigan) formed a nonprofit organization, called Advanced Network and Services, Inc. (ANS), to build and operate the NSFNET backbone network of the Internet. IBM used its expertise to build the high-speed routers and network interface cards and to write the routing and management code required. MCI provided the phone lines and Merit provided the Network Operations Center (NOC) to manage, monitor, and operate the network, routers, and other associated hardware and software.

The original NSFNET backbone used 56-Kbps connections. (This is frightfully slow by today's standards in the United States where an individual can get 33.6 Kbps over a plain old telephone line, and 64 to 128 Kbps over a ISDN line. However, consider yourself lucky. The fastest connection for an entire corporation in developing countries is 64 Kbps! This is important to consider when you design your global Notes network!)

In April of 1995, NSF funding was discontinued and ANS was dissolved. (As an interesting historical note, ANS formed a for-profit part of the organization called ANS CO+RE Systems, Inc. in June of 1991, which was later bought out by America On-Line (AOL) in 1995.) The NSFNET backbone has been replaced with Network Access Points (NAPs). The United States NAPs are located in San Francisco, Chicago, New York, and Washington, D.C. National Service Providers (NSPs) connect to the NAPs, second-stage service providers connect to the NSPs, third-stage services providers connect to the second-stage providers, and so on. There is still some government funding in the United States for the NAPs, but each service provider is now responsible for their own funding and billing for their part of the network.

In 1995, the NSFNET backbone operated at 45 Mbps—much faster, though sometimes it still does not seem fast enough! (Fact of the

Universe: You can never have too much computer memory, too large of a disk drive, or too fast of a network connection!) In 1997, the backbone of the Internet had speeds up to 155 Mbps.

The current booming Internet can trace its origins back to the ARPANET. During and just after the Vietnam war, ARPA was criticized for funding research that was very destructive, irrelevant, or too far-out. While this might have been true in many cases, we should give credit to ARPA for giving birth to TCP/IP and the Internet, which will have a direct, creative, and very profound impact on all of our lives for years to come.

Internet Addresses: Naming Convention

Users new to the Internet might be somewhat confused with Internet addresses. This section should help clear up some of this confusion about the naming convention used on the Internet. My Internet address is pwlew@vnet.ibm.com. "com" is the top-level domain, "ibm.com" is the domain, "vnet.ibm.com" is my subdomain, and "pwlew" is my user name. Originally, there were seven top-level domains that Internet address ended in: com, edu, gov, mil, net, org, and int. Here are some examples of what these domains refer to and how Internet addresses incorporate these "suffixes":

- com—Commercial organizations ("com" is pronounced "kahm")
  - ibm.com—International Business Machines Corporation (IBM)
  - lotus.com—Lotus Development Corporation
  - att.com—American Telephone & Telegraph Bell Laboratories
- edu—Educational organizations ("edu" is pronounced "ed-ju")
  - stanford.edu—Stanford University
  - caltech.edu—California Institute of Technology (Caltech)
  - berkeley.edu—University of California, Berkeley
- gov—Government organizations (United States) ("gov" is pronounced "guv")
  - whitehouse.gov—White House Public Access
  - doe.gov—Department of Energy (DOE)
  - nsf.gov—National Science Foundation (NSF)
  - ca.gov—State of California
- mil—Military organizations (United States) ("mil" is pronounced "mill")

- army.mil—U.S. Army
- navy.mil—U.S. Navy
- net—Networking organizations, including Internet service providers ("net is pronounced "net")
  - mci.net—MCI Telecommunications
  - cerf.net—California Education and Research Federation (CERFnet)
  - oar.net—Ohio Supercomputer Center (OARnet)
  - psi.net—Performance Systems International, Incorporated (PSI)
- org—Noncommercial organizations ("org" is pronounced "org")
  - npr.org—National Public Radio (NPR)
  - red-cross.org—American Red Cross
  - sierraclub.org—Sierra Club
- int—International organizations ("int" is pronounced "int," rhymes with "mint")
  - nato.int—North Atlantic Treaty Organization (NATO)

New domains are registered on behalf of organizations by the central authority, the Internet Network Information Center (the InterNIC). The seven top-level domains listed earlier are broad enough to include all types of organizations. Indeed, you are required to choose one of the seven top-level domains when you apply for a new domain.

One item to point out is that it is very unlikely that I could have an address pwlew@my-school.edu.com. The InterNIC would not register the domain "edu.com" because it does not make sense. The organization is categorized as either an educational organization or a commercial organization but not both. Because the organization is a school the correct address would most likely be pwlew@my-school.edu.

(I had once spent an extended discussion with a user who could not get his mail delivered to his school, which had an address like professor@his-school.edu.com. I tried to explain that his address was probably incorrect, but he insisted that he had successfully sent mail there before and that I had a problem with the Notes SMTP Mail Gateway. Well, now that all of you have read this, I will never have that type of problem again!)

As the Internet expanded internationally, the set of seven top-level domains expanded to include 2-letter country codes as top-level domains. For example, the code "fr" is for France, "de" (short for Deutschland) for Germany, "it" for Italy, "uk" for the United Kingdom, and so on. (These codes are based on an international standard: ISO 3166.)

Some countries follow the convention of com.fr and edu.fr, while others follow the convention of co.uk for commercial organizations and ac.uk for educational organizations (academic community). Many commercial organizations are international but choose to use a single domain worldwide. For example, zurich.watson.ibm.com is a subdomain for the IBM Research Center in Zurich, Switzerland. A geographic domain referring to Switzerland was not used because people first consider the organization as a part of IBM rather than a part of Switzerland.

There is some ambiguity in addresses for those using online services such as Prodigy, America On-Line, or the IBM Internet Access Kit. From the (fictitious) addresses magician@prodigy.com, wonder@aol.com, and miracle@ibm.net, you might think that Maggy Magician works for Prodigy, Wilma Wonder works for America On-Line, and Michael Miracle works for IBM. However, these (fictitious) people are subscribers to the online services that provide e-mail facilities and Internet addresses for the subscribers. You have to be a little careful, but equipped with these tools, explanations, and examples, new users to the Internet can now decipher Internet addresses from around the world.

How Mail is Delivered

You process your mail using a Mail User Agent (MUA), such as Notes, cc:Mail, MS Mail, or elm or pine in the Unix world. With a MUA, you can send, receive, reply to, store, and delete mail items. The MUA provides the user interface on the client workstation that users are accustomed to dealing with, and everything else happens behind the scene.

How mail gets delivered to another person on the same server, or on a server across the world, is taken care of by the Mail Transfer Agent (MTA). The MTA looks up the address of the destination server in the directory, then the MTAs on the source and destination servers begin a dialogue to transfer the mail using the appropriate protocol. The Notes MTA is the "Router" program, and the directory is the Name and Address Book for Notes networks. The Notes MTA does not have its own network protocol; instead, Notes uses its own database functions to accomplish the same goals. For TCP/IP networks, the most common MTA is a program called "sendmail," the directory is the Domain Name System (DNS), and the protocol is called Simple Mail Transfer Protocol (SMTP).

Running a Notes SMTP Gateway requires a working knowledge of

- sendmail
- Domain Name System (DNS)
- Simple Mail Transfer Protocol (SMTP)

sendmail

sendmail's main role is to deliver the mail. This appears to be a one simple and straight-forward task, but actually involves many activities including (1) deliver to local host, (2) deliver to nonlocal host (over the network), (3) queue the mail, and (4) manage aliases.

sendmail determines from the address if the mail is for a local host or nonlocal host. For the local host, sendmail delivers the mail by appending the mail to the recipient's mail file. For a nonlocal host, sendmail queries the DNS to obtain the TCP/IP address of the nonlocal host and establishes an SMTP session by which the mail is transferred. The sending host often holds a copy of the mail in a mail queue until the mail is actually transferred. This provides for more reliable service in case the destination host is down, the destination host is unreachable because the network is down, or the destination host is not responding because it is busy. Delivery will be re-attempted and, if delivery fails, the mail will be returned to the sender. (The size of this queue should be closely monitored. If it is growing too large, it is a sign of trouble. Mail should not stay in queue for a long time.)

Aliases allow for shorter mail addresses, mail distribution lists, and mail forwarding for those who have changed e-mail addresses. Internet addresses are intentionally kept as short as possible so that they are easy to remember and do not take too long to type. (This is especially important if you are like me; my memory gets shorter as I get older! I am fortunate to be a touch typist, but I sympathize with two-finger typists. Still, I appreciate short e-mail addresses.) For example, you would not be happy if your Internet address was John_Patrick.somlan01.IBM_Internal.IBM_External@hcpibme1.ims.advantis.com. You would probably prefer something like patrick@gemini.ibm.com or even jrp@gemini.ibm.com. This is what a sendmail alias can provide.

You might want to go even further and use jrp@ibm.com. This would be fine if IBM were a company of 100 employees; however, with over 200,000 employees, IBM is too large to use just the domain ibm.com, and it makes more sense to use subdomains to organize the various parts of IBM. The gemini.ibm.com subdomain is used for Internet mail to IBM to be delivered as Notes mail or to stay as SMTP mail (for UNIX users, for example). The vnet.ibm.com subdomain is used for Internet mail to IBM to be delivered to mainframe mail systems such as PROFS or OV/VM. The subdomain watson.ibm.com is used by IBMers at the Thomas J. Watson Research Center, and zurich.ibm.com is used by IBMers at the Research Center in Zurich, Switzerland. From these examples, you can appreciate the value of subdomains to organize the functions of a large organization.

The ability of sendmail to provide an alias is a wonderful feature for end users. The price is more work for the sendmail administrator. Each alias is just a line in the /etc/aliases file, which does not seem to be a big deal. This might be fine for 100 aliases, but I do not want to manually type in 10,000 aliases! In addition, all aliases must be checked for uniqueness. For example, entries in an /etc/aliases file might look like:

```
George_Washington:    georgie
John_Adams:           johnnie
Thomas_Jefferson:     tommy
James_Madison:        jimmy
James_Monroe:         jimmy2
John_Quincy_Adams:    johnnieq
```

Here, even with just six entries, we have run into some minor conflicts. We have two people named James. They both cannot use the alias "jimmy," so the second was assigned "jimmy2." Perhaps we should use last names as an alternative. Then James Monroe could use the alias "monroe." Applying this logic, we have two people named Adams and cannot use the same alias. In fact, they are both named John, so we cannot use first names either. We can use "johnnie" and "johnnieq" to distinguish the two John Adams.

The Notes Name and Address Book can be used to register a large number of Notes users in a batch method. The users' Full Name and Short Name can be extracted from the Name and Address Book, exported to a file, reformatted, and appended to the /etc/aliases file. At IBM, when the migration was made from mainframe mail to Notes mail, the Short Name was taken to be the OV/VM (PROFS) user ID. This is a good default and is usually unique, but there is a limitation of 8 characters, which leads to unusual looking user IDs. For example, because of the multiple David Frank's in the company, they could not be all "dfrank." One user ID became "dfran005."

sendmail aliases are not limited to 8 characters, so David's alias could be "dfrank5." In fact, to avoid confusion with all the other David Franks, one of them might choose "tiger" or "surfer" as their alias. The alias does not have to be directly related to the person's name; although, for business communications, it might be more appropriate.

One possible automated registration form would ask for a person's top five choices for an alias. After the form is submitted, a program will determine whether any of the choices are already in use. If all five choices are already taken, the program will prompt the user for more choices. The program can also check against other rules, such as no special characters, no profane or other disallowed words, or any length limitations. Users should be able to modify their own, and only their own, alias entry (change, add, or delete).

A policy must be established for how many aliases a single person can have. For example, Fortunato "Tony" Cusato might use "cusato" as his alias. However, for historical reasons, most of his friends and colleagues know his e-mail address to always have been "fortunc." Mail sent to fortunc@gemini.ibm.com will bounce (be returned to the sender and marked undeliverable), so the system administrator might give Tony a second alias to allow this mail to be delivered.

Domain Name System (DNS)

Domain Name System (DNS) is a practical necessity to using the Notes SMTP Gateway to send mail to Internet recipients. A DNS server translates the sender's and receiver's Internet name (TCP/IP name) to the corresponding Internet address (TCP/IP address). This TCP/IP address is then used by the network communications equipment (routers, bridges, gateways, switches, etc.) to deliver the mail to the recipient's workstation or mail server. For example (using a fictitious name and address), the TCP/IP name odysseus@ibm.com translates to 9.1.2.3. Routers understand numbers, but people can remember names better than numbers; thus, DNS makes addressing easier for people.

DNS is configured to distribute updates across the Internet automatically. Therefore, if you have to change your TCP/IP address because you moved to a different city, you can update your DNS record with the new TCP/IP address while keeping the same TCP/IP name. This avoids having to notify all of your current (and future) correspondents. If you publish your Internet address on your business card, DNS helps you avoid printing new business cards each time you change your TCP/IP address. I move offices quite often (IBM means "I've Been Moved") and usually change TCP/IP addresses with each move. I love this feature of DNS!

Typically, an organization with Internet connectivity will operate its own DNS server. In fact, two DNS servers are required for redundancy to provide high reliability and availability. Each DNS server requires a computer (usually a UNIX server) running TCP/IP and the DNS application. Of course, this also means that a UNIX administrator is required. The data used by DNS frequently changes (over the period of days or weeks in large organizations), so in practice, DNS requires a full-time administrator.

If the organization uses TCP/IP internally, then two DNS servers would provide DNS service for the internal network in addition to the two DNS servers for the external (Internet) network. (The same DNS servers are not used for both internal and external networks to avoid advertising internal TCP/IP addresses to the external users who might potentially attack a computer. Not advertising the address does not prevent attacks, but only makes attacks one step more difficult.) The DNS administrator would then administer the four DNS servers. Some organizations do not use UNIX or TCP/IP internally and find it more cost effective to outsource the Internet connection and related services.

Using the Lotus SMTP Gateway Product

As the popularity of Notes and the Internet grow, so does the need for the SMTP Gateway. The Lotus SMTP Gateway product was described in chapter 5. This section will discuss how the SMTP gateway is being used within IBM.

In a large organization, you would replace the mail relay and the Notes SMTP Gateway with a messaging switch, such as the Lotus SoftSwitch LMS switch. This switch would be a gateway between Notes mail, SMTP mail, cc:Mail, mainframe mail, fax, pagers, and more. The alternative to one large switch is to have several smaller gateways. This might be preferable if the volume of mail is low, if only a few functions are needed, or if the costs do not justify the performance level. The LMS switch is what is currently used within IBM.

These are the problems and solutions discussed in this section:

PROBLEM 1.   Need to send Notes mail to the Internet.

SOLUTION 1.   Connect a Notes SMTP Gateway directly to the Internet.

PROBLEM 2.   Security exposures—the Notes SMTP Gateway runs sendmail, which is not secure.

SOLUTION 2.   Connect a mail firewall (such as the IBM NetSP) directly to the Internet and to the corporate network and put the Notes SMTP Gateway behind the firewall.

PROBLEM 3.   Using other mail programs in addition to Notes, such as cc:Mail, MS Mail, etc.

SOLUTION 3.   Add a server like the Gemini Switch Unit between the firewall and the Notes SMTP Gateway. This is a UNIX system running sendmail and is often referred to as a mail relay.

The IBM Gemini Mail System routes mail from the Internet to any of several mail systems used internally by IBM. Figure 8.1 shows this system.

Mail destined for a Notes user arrives from the Internet to the NetSP firewall, is transferred to the Gemini Switch Unit, is then transferred to the Notes SMTP Gateway, and is then delivered using normal Notes mail routing (see Figure 8.1). Mail from Notes to the Internet takes the reverse path.

The Notes SMTP Gateway is necessary to convert Notes mail to SMTP mail, which is what is used on the Internet. The Notes SMTP Gateway is a Notes server that allows mail transfer to other Notes servers. The Notes SMTP Gateway also runs the sendmail program, which allows transfer of SMTP mail to other servers running sendmail. The Notes SMTP Gateway could be attached directly to the Internet without a firewall, but this is a security risk and is *not recommended* because the Notes SMTP Gateway runs sendmail, which is notorious for security exposures.

sendmail is very flexible and configurable, which makes it very complex. sendmail has evolved and grown over the years, which makes it very large. A complex and large program is much more likely to have bugs and security exposures than a simple small program, and such is the premise behind mail firewalls like the NetSP. The Internet side of the NetSP firewall will only accept SMTP dialogue with Internet computers. If Internet mail is given to the NetSP firewall for delivery, the NetSP will deposit the mail in a certain directory on its disk drive. The corporate network side of the NetSP will pick up the mail from this directory and transfer the mail on to the Gemini Switch Unit. In this manner, there is no direct dialogue between the Internet computers and the corporate network computers. Also, the program running on the NetSP to handle this work is simple and small, so the program can be tested and inspected to assure there are no security defects.

Figure 8.1

Example SMTP GW design showing connection between the Internet and a corporate network.

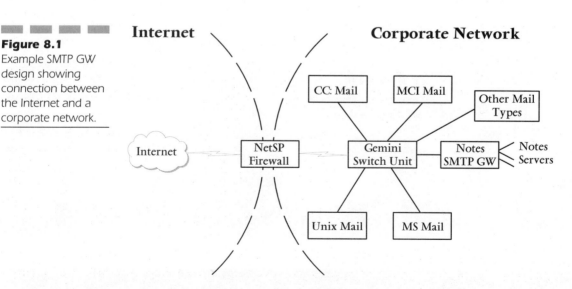

It is now clear why the NetSP firewall is needed. Why is the Gemini Switch Unit needed? The Notes SMTP Gateway can be connected directly to the NetSP, and this will work fine for organizations purely using Notes. For organizations with a mix of mail users, mail needs to be switched to the proper gateway, such as cc:Mail, MS Mail, MCI Mail, mainframe mail (e.g., PROFS), or straight SMTP mail to UNIX users. This is the role of the Gemini Switch Unit. Using the sendmail configuration file, the Gemini server maps a user ID with the proper address for the appropriate mail system.

Let's follow Internet mail sent to pwlew@gemini.ibm.com to see how this works. Whenever an Internet computer sends mail for the subdomain gemini.ibm.com, the computer will query a Domain Name System (DNS) server and learn that all mail for this domain (that is, all mail for somebody@gemini.ibm.com) should be routed to the NetSP. The NetSP will only accept mail for this domain and will transfer all such mail to the Gemini Switch Unit. The Gemini server will look up pwlew in a table and rewrite my address as Peter_Lew.Advantis@notesgate .gemini.ibm.com. The Gemini server will query a DNS server and learn that all mail for the server notesgate.gemini.ibm.com should be routed to the Notes SMTP Gateway. sendmail will then deliver my mail to the Notes SMTP Gateway where the mail is converted to Notes mail and my address becomes Peter Lew@Advantis, where Advantis is my Notes domain. The Notes mail goes happily on its way in the Notes network. If I were a cc:Mail user, the Gemini server would translate my pwlew@gemini.ibm.com alias to something like Peter_Lew@ccgate.gemini.ibm.com and transfer my mail to the cc:Mail SMTP Gateway and so on. (No, we don't have any server called Watergate!)

As you can see, operating a Notes SMTP Gateway requires a working knowledge of sendmail, DNS, firewall operation, UNIX and UNIX system administration (most firewalls run on UNIX, and the UNIX sendmail is the most stable for the Gemini server), and of course, Lotus Notes. This combination can become quite involved, and many organizations consider outsourcing this part of the operations. As an example of the material to be mastered, the classic book on sendmail by Costales is more easily measured in inches rather than pages. (Well, I exaggerate. . . . The book is about 1½″ thick or about 800 pages. To paraphrase a Chinese saying, "The journey of 1000 pages begins with the first page!")

The World Wide Web (WWW)

The World Wide Web is a collection of servers network-connected to the Internet that are used to present multimedia content. The ownership and responsibility for the Web servers are distributed around the world with no central authority.

The World Wide Web servers and the associated Web browser clients are the latest answer to the question: How do I find information on the vast Internet?

There is so much content on the Internet, how can it be organized usefully and easily? Prior to the Web, "gopher" servers and clients were the most popular method to access information. Gopher was a text-based system that presented the user with a series of menus. The Main Topics menu might offer News, Weather, Sports, and Travel. The menu item, say Weather, is selected by entering the number of the item at the command line. A menu for Weather is then presented offering United States, Europe, Asia, and so on. The individual states or countries can be selected from the next menu, and finally the individual cities or regions can be selected, and the weather forecasts will then be presented. The weather forecasts for different parts of the world are contained on different servers around the world but are accessed from the same gopher menu. The user does not need to know the address or location of the servers.

The World Wide Web servers and the Web browser are backward compatible with gopher servers and clients and add a graphical user interface. The first addition to notice is hypertext. This text is marked in a different color (for example, blue) and/or is underlined with a solid line. Clicking on this text is similar to selecting an item from a gopher menu—you are connected to another Web "page" that can be on a server literally on the other side of the world. When you return to the original "home" page, the blue hypertext will change to purple and/or is now underlined with a dashed line. This helps you keep track of what you have already seen.

The most important addition of the World Wide Web is graphics. The previous gopher example of the Weather information could be presented on a Web page as a graphical image of the world. Clicking on the graphics can rotate, translate, or zoom to select the desired region of the world. The forecast can be shown in text, rainfall amounts can be shown in a plot, the current satellite image can be shown, a video can be shown of the latest hurricane movements, and an audio can be heard of an eyewitness describing the hurricane with the wind howling in the background.

The introduction of the graphical user interface to the Internet has been a profound change. This is analogous to changing from the text-based, command-line-oriented DOS operating system for the IBM-compatible personal computer (PC) to the graphical user interface of Windows or OS/2. The use of the Internet is enriched and is easier with the graphical user interface as is the use of the PC. The introduction of the World Wide Web and its graphical user interface is one significant reason for the recent explosion of the Internet. (Of course, life is not so simple. More powerful and cheaper computers that can handle the graphics and faster and cheaper network connections to transmit the data were also responsible for the explosion.)

Type of Content

Here are examples of the type of content you will find on the World Wide Web:

- The Vatican Library
  - Art, manuscripts, and maps from centuries ago
  - The URL is: http://www.software.ibm.com/is/dig-lib/fvatican.htm
- Weather
  - Color graphics showing temperatures, fronts, rainfall, snowfall, etc.
  - The latest satellite images
  - Hurricane formation and paths of past hurricanes
  - The URLs are:
    http://wxp.atms.purdue.edu/maps/satellite/sat_ir_east.gif and
    http://web.usatoday.com/weather/wfront.htm
- News
  - The latest headlines from Reuters
  - The URL is: http://www.yahoo.com/headlines/current/news
- Stock Quotes
  - Delayed 15 minutes
  - Plots for the past 6 to 12 months
  - The URL is: http://update.wsj.com (from the Wall Street Journal)
- Real-estate listings
  - Text, photographs, and videos
  - Check out houses before you drive around town
  - Very handy for people relocating across the country (or across countries!)
  - The URL is: http://www.netprop.com

How Businesses Can Take Advantage of the World Wide Web

- Advertise.
  - Text, graphics, images, audio, and video.

- Let users read the script and see still images from television ads.
- In fact, let users download the same commercials shown on television. (I missed some of the witty IBM commercials so I caught up by accessing the IBM home page at www.ibm.com!)

- Distribute product and service catalogs.
  - Let customers see and read descriptions of products and services.
  - Product line and price list always up-to-date.
  - Provide phone numbers, fax numbers, and e-mail addresses to get more information.
  - Make it easy for customers to order products and services.

- Collect customer feedback on products and services.
  - Provide phone numbers, fax numbers, and e-mail addresses to get feedback.
  - E-mail feedback automatically provides addresses to contact customer.
  - Utilize feedback for product and service improvements.
  - Utilize feedback for marketing research and development.
  - Utilize feedback for sales leads.

- Distribute press releases.
  - Always up-to-date.
  - Saves the large cost of printing and distributing hardcopies worldwide.
  - Ensures that everybody gets the press release at the same time.
  - Makes it easier for the press to cut and paste quotes into their articles.

- Distribute bug fixes.
  - Saves the cost of producing and shipping diskettes and/or CD-ROMs.
  - Better than bulletin boards (BBSs).
  - Don't have to maintain your own modem banks. (Users dial their Internet service provider's modem banks, which is usually a local call.)
  - Can support many more simultaneous users more cheaply.
  - Easier to provide worldwide service.

- Provide store hours and directions.
  - Always up-to-date.
  - Saves employees' time on the phone giving routine information over and over again.
  - Keeps phone lines open for more "important" phone calls.

- Post job opportunities.
  - What better way to find people who are computer and network savvy!

■ The mere fact that the applicant found the want ad on the Internet prescreens applicants.

Some Web Sites for Lotus Notes

There are many Web Sites on the Internet related to Lotus Notes and network security. This section lists several of these. The number of World Wide Web sites offering interesting information on Lotus Notes will continue to expand. Some hints on "surfing the 'Net" to find these sites are discussed at the end of this section.

■ Lotus Home Page—http://www.lotus.com

■ Lotus Domino Home Page—http://domino.lotus.com

■ Lotus Notes FAQ Web Site—http://metro.turnpike.net/kyee

■ Lotus InterNotes Home Page—http://www.internotes.lotus.com. This Web page contains forums with Iris developers, downloads of Domino and Notes code, technical papers, step-by-step guides, interviews with Iris developers, etc. You can connect to this site with your Notes client.

■ Iris Associates (Developer of Lotus Notes)—http: //www.iris.com. Iris was founded in 1984 and became a wholly owned subsidiary of Lotus Development Corporation in June, 1994.

■ IBM Home Page—http://www.ibm.com

■ IBM Global Network Home Page—http://www.ibm.com/ globalnetwork/

■ Interliant Home Page—http://www.interliant.com

■ CompuServe Home Page—http://www.compuserve.com

■ CompuServe Remote LAN access services for Lotus Notes— http://www.compuserve.com/look@/network/notes.html

■ Network Security Information
 ■ RSA Home Page—http://www.rsa.com. RSA is the data encryption standard used by the Lotus Notes product for all encryption requirements. "Internet Mail Security" in chapter 7 gives some details on the RSA encryption algorithm.
 ■ RSA users—http://www.rsa.com/AboutRSABrief.html
 ■ http://www.yahoo.com/computers/Security_and_Encryption
 ■ http://www.yahoo.com/computers/PGP_Pretty_Good_Privacy
 ■ http://www.yahoo.com/computers/Firewalls/
 ■ http://www.yahoo.com/computers/Kerberos/

Here are some hints on ways to "surf the 'Net" to find additional material on Lotus Notes and network security. There are thousands of Web sites on the Internet, and new Internet sites are cropping up at a staggering rate. Web search engines are often the best way to find what you're looking for. Yahoo is a comfortable place to use as your home base to start the search. Then, if you can't find what you want on the Yahoo search page, choose another search engine. I usually click on the "Net Search" button on my Netscape Navigator browser and that lets me choose from more than a dozen search services.

Get to the Yahoo home page with the URL http://www.yahoo.com. From this home page, enter the words "Lotus Notes" in the request area, and then click on the Search button. This will find entries with both words together (i.e., "Lotus Notes," not just "Lotus" or "Notes").

After trying the Yahoo search, try the WebCrawler. You can get to that by going to the WebCrawler home page directly with the URL http://webcrawler.com. The WebCrawler is not as all-inclusive or as easy to use as some of the other options—especially Lycos—but it's a great tool to get the job done quickly.

If WebCrawler doesn't get you what you want, call in the slow-but-thorough heavy-hitter: Carnegie Mellon's Lycos (http://lycos.cs.cmu.edu). Lycos searches are the most thorough of all—right down to the word level through almost 3 million Web documents. That includes Web, FTP, and Gopher sites. I've found that Digital's Alta Vista is also one of the fastest and most comprehensive search engines. Get to Alta Vist with the URL http://www.altavista.digital.com.

Another good search engine, developed by IBM Research, is available with the URL http://www.ibm.com/Search. Try finding entries for "Lotus Notes" with this search engine, and see how the results compare to those from Yahoo, WebCrawler, Alta Vista, and Lycos. You'll find that the results are quite different.

Lotus InterNotes Web Publisher: Tying Notes and the Web

The Lotus InterNotes Web Publisher is a tool that ties Lotus Notes with the World Wide Web. The Lotus InterNotes Web Publisher is much more than just an authoring tool to convert documents from the Notes format to the HTML format used by the World Wide Web (the Web). The Lotus InterNotes Web Publisher functions include three important areas:

■ Converting from Notes format to HTML and from HTML to Notes.

■ Extending powerful Lotus Notes functions to the World Wide Web.

■ Managing changes to Web site hyperlinks, utilizing the security and replication strengths of Lotus Notes.

Conversion between Notes and World Wide Web (HTML) Formats

The Lotus InterNotes Web Publisher can convert an entire Notes database to the World Wide Web (HTML) format. Any Web server can then publish the HTML files as Web pages. Any Web browser can be used to access the Web pages. Graphics and fonts are maintained as close as possible within the limitations of HTML. Notes Views are presented on the Web as hypertext lists.

Beginning with Lotus InterNotes Web Publisher version 2.0, not only can Notes documents be converted to Web pages (HTML format), but Web pages can also be converted to Notes documents. This is normally accomplished by first publishing a Notes form to the Web (converting Notes to HTML format), then converting the completed Web form from HTML to Notes format. In this manner, information can be input from a Web browser and entered into a Notes database. This feature offers a convenient method to collect customer requests for information, to collect customer feedback on products and services, and to interact directly with customers in general.

Extending Notes Functions to the World Wide Web

Filling out a "Notes form" with a Web browser is like composing a Notes document with a Notes client. This function is now extended to the World Wide Web. The data entered in the Web form is entered into the respective field in the Notes database. The full set of Notes database capabilities are then available, such as views, sorts, searches, and workflow.

For example, if customers enter their name, address, company, and product feedback, the corresponding database could be viewed by name, ZIP code, or company to help analyze where the most problems are coming from. In another case, say a customer wants to submit a request to buy a book (like this one!) using a Web browser. After submitting the completed Web form to the Notes database, Notes functions will scan the form to determine that it is an order form for this book and route the form to the appropriate approver, along with inventory, billing, and shipping information collected from other Notes database. Extending Notes workflow functions to the Web is quite a significant accomplishment. Additional features extended to the Web include radio buttons, check boxes, scrolling lists, and free-format text fields. Again, the information submitted from the Web form will be entered in the respective fields of the Notes form.

The Notes formula validation function is another powerful extension to the Web-formatted Notes forms. For example, if I entered "   peter lew   " in a name field, the @Trim and @ProperCase functions could be applied to convert my entry to "Peter Lew" by removing extra spaces and by properly capitalizing. The set of Notes formula validation functions is rather extensive and includes text parsing, applying logical operators, checking if text entry is in a specific list or database, checking if data is in range, performing arithmetic functions, and operating on date-time entries.

The Notes full-text search function is another powerful extension to the Web-formatted Notes forms. The Notes full-text search engine is recognized as a sophisticated tool to search an entire Notes database. The Lotus InterNotes Web Publisher allows a Notes full-text search query to be initiated from a Web page. The search is performed on the Notes server just as if the request was initiated from a Notes client and is performed on demand (that is, searches do not wait for the next scheduled publication period). This is another example of how valuable Notes functions can be extended to the World Wide Web.

Managing a Web Site Using the Lotus InterNotes Web Publisher

The Lotus InterNotes Web Publisher is a powerful tool for managing a Web site. Using the inherent feature of views in Lotus Notes, hypertext links can be easily modified for large Web sites (e.g., sites with hundreds and thousands of Web pages). Using the strong replication feature of Lotus Notes, the creation of Web pages for a Web site can be distributed geographically and organizationally. Using the strong security features of Lotus Notes, a company can control who can modify specific Web pages.

The beauty of the World Wide Web is the hypertext links that helps you navigate to other local Web pages or other Web sites with the click of a button. The set of links becomes increasingly involved as the number of Web pages and Web sites grows. For each new Web page added, at least one existing Web page must be changed to link to the new page. For each Web page deleted, all the Web pages (say a dozen pages) linked to the deleted page will need to be changed. Tracking and updating links quickly becomes unmanageable for all but the smallest of Web sites. Notes databases and views are ideal for managing the dynamic nature of Web sites. Views are lists of documents (which become published as Web pages) and are automatically updated each time a document is added or deleted. The Lotus InterNotes Web Publisher publishes the views as hypertext links, and voilà! You never have to worry about maintaining the links again! The links are now managed by the inherent workings of Lotus Notes databases.

The Lotus Development Corporation's Web site (http://www.lotus.com) was created and is managed by using the Lotus InterNotes Web Publisher and contains over 25,000 documents. The various departments within Lotus are dispersed geographically and have their own responsibilities. Yet the Internet Products department, the Accounting department, and the Human Resources department can each contribute to their own local Notes database, and all changes will be consolidated via replication to the Lotus InterNotes Web Publisher. The changes are then published to the Web site. In fact, the publishing of the Web pages is very much like replication between a Notes server and the Web server. The first time, publication is set so that all Notes documents are converted to Web pages. Subsequent publications convert only updates. Publication is scheduled by day of the week and for a given time frequency. Publication can be forced to occur immediately, similar to forced replication.

The security features of Notes, such as Access Control Lists (ACLs), control who can modify the Notes databases. In this manner, a department can limit who can modify the Notes databases that will be published as Web pages. It is undesirable to have everybody updating Web pages, as this situation would be too chaotic and sensitive information might be unintentionally leaked by unauthorized individuals. The combination of replication and security allow decentralization of authority and responsibility while maintaining a central clearing house for the information.

Who is Using The Lotus InterNotes Web Publisher?

It is natural to expect, given the close integration of Lotus Notes and the World Wide Web offered by the Lotus InterNotes Web Publisher, that many users of the Web Publisher were previously Lotus Notes users—in fact, this is true. It is interesting that many users of the Web Publisher are not Lotus Notes users. These users are attracted by the broad combination of functions and features described earlier. Lotus Notes has its strong points, and the World Wide Web has its strong points. Combining the two further expands the broad capabilities of Lotus Notes and gives a much more potent Web site.

Now that Lotus Domino combines a Notes server with a Web server, you might expect that the Lotus InterNotes Web Publisher is no longer needed. Some people prefer using a different Web server from the one bundled with Lotus Domino. In this situation, you would use the InterNotes Web Publisher in conjunction with your favorite Web server.

Additional information on the Lotus InterNotes Web Publisher is given in chapter 14. See the section "Lotus Notes: The Enterprise-scale Intranet Server" for sample screen captures of a Notes database displayed using the InterNotes Web Publisher.

Firewalls: Packet Filtering, Socks Server, and Proxy Agents

Firewalls typically have three possible modes of operation: filtering packets, acting as a socks server, and using proxy agents. Using proxy agents is the most secure mode, but it comes with the price of being the hardest to implement and placing the heaviest load on the firewall. Packet filtering is the easiest to implement and places the lightest load on the firewall; however, it is the least secure. A socks server is moderately difficult to implement, places a moderate load on the firewall, and is moderately secure. This comparison is summarized in Table 8.1.

The decision over which mode to use depends on the requirements of the situation. Is the highest security required? How much time, skill, and effort is available for implementation? Will the firewall server (that I can afford) be powerful enough to handle the load (this is a cost/performance issue)? If the highest security is required, I will need to buy, deploy, manage, and support many firewalls. How will I do this, and how much will it cost? Based on the answers to questions like these,

TABLE 8.1

Comparison of
Firewall Types

	Packet filtering	Socks server	Proxy agent
Degree of security	Lowest	Moderate	Highest
Ease of implementation	Easiest	Medium	Hardest
Load on firewall	Lowest	Moderate	Highest
Scalability to many users	Good	Moderate	Not very good
Logon to firewall required?	No	No	Yes
TCP/IP layer used	IP	TCP	Application

Table 8.1 can help with the decision. The following sections give technical details on how the different modes work.

Packet Filtering

Data to be sent from one TCP/IP network to another is first divided into packets, and it is these packets that are actually transmitted. Each packet contains the source (sending) IP address, the source's port, the destination (receiving) IP address, and the destination port, where, in general, a port identifies an application. For example, over a TCP/IP network, port 1352 is used for Lotus Notes, port 23 is used for telnet, and port 25 is used for SMTP mail. (Certain ports, such as these, are called *well-known ports* and are registered with the central authority to avoid conflicting usage.)

A table is maintained on the firewall that contains the allowed four-part combinations of source IP address, source port, destination IP address, and destination port. Each packet is checked against this table. If the four-part combination is not allowed, the packet gets dropped (filtered). If the four-part combination of IP addresses and ports is allowed, the packet gets transmitted. This is the basic operation of a packet-filtering firewall. Many routers are capable of packet filtering, so purchasing a full-fledged firewall might not be necessary in all cases. End users are generally not directly aware of the packet-filtering firewall (unless they can't reach their destination) and do not have to logon to the firewall.

Socks Servers

Socks servers also use a table of allowed four-part combinations of source IP address, source port, destination IP address, and destination port like the table used for packet filtering. There are two sides of a socks server: an internal side and an external side. The internal side communicates only with the internal network, whereas the external side only communicates with the external network. An internal client packet sent to an external server is intercepted by the internal side of the socks server.

If the four-part combination of source IP address, source port, destination IP address, and destination port is not allowed by the table entry, the packets are discarded. If the four-part combination of source IP address, source port, destination IP address, and destination port is allowed by the table entry, the external side of the socks server attempts to communicate (establish a session) with the external server using the destination IP address and destination port. If the external server responds, the socks server gets out of the way and lets all subsequent packets flow uninhibited (having a wild time!) from the internal client to the external server.

From this point on, the socks server just appears to be another router and does not have to do much work. The work is done on the client and server, and the socks server is not involved. For example, using a Web browser on an internal client to access a Web server on the Internet through a socks server appears as though the client was directly connected to the Internet. The client's workstation handles the graphics display and caches the Web pages in the client's workstation memory (in contrast to proxy agents discussed later). End users are generally not directly aware of the socks-server firewall (unless they can't reach their destination) and do not have to logon to the firewall.

Applications (such as Web browsers, telnet, and FTP) make calls to TCP-level programs. Most applications are not designed to work with socks servers (are not "socksified"). Because socks servers operate at the TCP level, it is good practice to not "socksify" the application itself, but to socksify the TCP/IP stack. Vendors of TCP/IP stacks are beginning to socksify their products, and this approach is becoming more widespread. A good source of information and assistance regarding socksifying applications or TCP/IP stacks is Trusted Information Systems, Inc. (TIS). Their Web site can be accessed at http://www.tis.com.

Proxy Agents

Firewalls using proxy agents operate at the application layer. There are actually two sides of the proxy agent: an internal side and an external side. The internal user first telnets to the firewall and logs on to the firewall with a user ID and a password. The internal user then starts the application (say a Web browser) using the internal side of the proxy agent and attempts to begin communication with the external server. The internal side of the proxy agent can only communicate with the external side of the proxy agent and no further on the external network. The external side of the proxy agent receives all commands and data from the internal side of the proxy agent, then acts on behalf of the internal user (acts as a "proxy") and re-issues the commands and data to the external server. Likewise, the external server can only communicate with the external side of the proxy agent and no further on the internal network. The internal side of the proxy agent receives all commands and data from the external side of the proxy agent, then acts on behalf of the external server (acts as a "proxy") and re-issues the commands and data to the internal user. This sequence continues until the internal user ends the application on the internal side of the proxy agent.

Using proxy agents is very secure because the workstations and servers on the internal networks never communicate directly with workstations and servers on the external networks. The proxy agents are always at work on the firewall. The flip side (and downside) is that the firewall gets loaded down as more people use the firewall. Furthermore, the application is running on the firewall, which further bogs down the firewall. The Web browser, for example, actually runs on the firewall, processing the graphics and caching Web pages in the firewalls memory. (The internal user's workstation is used to display the graphics, but the firewall performs the actual computing work.) Well, this just does not scale well for a large number of users, but it is secure. The only way to scale up is to add more computing power, more memory, more disk space, and so on. There is a trade-off between security and cost. There is a similar trade-off between security and user convenience. Users must always be authenticated via user ID and password; this adds security but is not convenient.

Using proxy agents is the most secure of the three methods described here; however, proxy agents also require the most effort to implement. Proxy agents operate at the application level, which means that the actual application source code must be rewritten, which is a significant effort. End users usually do not have access to application source code, so this means that end users cannot write the proxy agents themselves.

Owners of the application source code are usually driven by market demand. If there are enough users of proxy-agent firewalls, then the application owners will develop a proxy agent version of their application. Many organizations choose not to use proxy-agents firewalls given the trade-off between high security and ease (difficulty) of implementation, loading of the firewall (performance impact), scalability issues (cost), and logon requirements (inconvenience).

It is important to understand the security requirements of your situation and to understand the trade-offs of the three main modes of firewall operation. There are situations where packet filtering is the right solution, where a socks server is the right solution, and where using proxy agents is the right solution. The information presented here is intended to help you decide on the right solution for your situation.

Replicating and Web Navigating through a Socks Firewall

Replicating from an internal network to the Internet via a Socks firewall can work very well. Performance is as good as over the internal network. This is much faster than replicating securely via XPC modem dial (1.5 Mbps vs. 33.6 Kbps). This allows us to more quickly replicate databases from internal Notes servers to external InterNotes Web Publisher/Web servers (Domino) on the Internet. This is a significant solution for internal use and for external customers.

Using the Web Navigator on the internal network to access the Internet WWW via a Socks firewall works well. Performance is good—as fast as using a Web Browser via a Socks firewall. This is much faster than today's solution of using the Web Navigator with a proxy Web agent. This is a significant solution for internal use and for external customers.

From our office, we can use Notes to access a Notes server on the Internet via a Socks firewall. Then, we do not need to dial the Internet. Currently, our Notes administrator dials the Internet to administer the Notes server on the Internet. He can now use the Socks firewall instead. In the same way, he can access anonymous Notes server on the Internet, such as the NotesNIC operated by Lotus. This way, we will not all need analog lines, modems, and dial programs on our office workstation.

This is a viable solution for connecting IBM Notes servers on the IBM internal MPN network to Notes servers of existing Advantis customers on IBM's commercial network, LAN Internetworking 1.1.

The Notes server on the internal network uses the socksified TCP/IP stack, which is part of OS/2 Release 4 (Merlin).

Notes Release 4.5 is also socksified. However, previous versions of Notes can call a socksified TCP/IP stack, and Notes behaves as if it is socksified. Any application that calls the socksified TCP/IP stack will behave as if the application itself is socksified. Socksifying the TCP/IP stack is much more efficient than socksifying each and every application. The work only has to be done once for the protocol stack instead of for every application. Furthermore, IBM can socksify its TCP/IP stack because IBM owns the code. IBM cannot socksify everybody else's code without access to the code (and it is inefficient, does not make sense, and IBM would not want to anyway). The TCP/IP stack that is part of OS/2 Release 4 (Merlin) was socksified by IBM.

Your security staff might be concerned about using Notes through a Socks firewall and downloading data from the Internet. Using Notes through a socks firewall is analogous to using an FTP client or Web browser through a Socks firewall. In each case, the session can only be initiated from the internal network. Also, in each case, data can be downloaded form the Internet. In fact, Notes is more secure than FTP because the Notes servers use a secure challenge/authentication method to validate identities. You know with whom you are communicating. (This is true unless you access a Notes server that allows anonymous Notes access. This case is then similar to anonymous FTP and is subject to some risk.)

More security is obtained by using Notes network encryption of all data communications between the two Notes server. This method encrypts data at the application layer and is quite strong security. Web browsers and servers have similar, though not as extensive, security features as Notes, but only if SSL or S-HTTP protocols are used. Most sites are not using these secure protocols, so typical Web browsing is not secure. Your security staff must decide what risks are acceptable and determine the appropriate security policy.

The replication process can be used to allow Notes mail to flow through a Socks firewall. The most straight-forward way to do this is to replicate the mail.box between Notes servers on the outside and inside of the firewall. Of course, to get replication to work, both mail.box files would need the same replica ID. That's no problem.

When the mail.box on the outside server is empty, just delete it and replace it with an empty replica of the mail.box from the inside server. The mail.box is just another Notes database; it can be replicated, etc. Deletions from the mail.box get replicated in the normal way. So when you allow a two-way replication of the mail.box, then as mail is sent out from the mail.box, those deletions would replicate and both copies of the mail.box would be empty once all mail is sent out from both the inside and outside Notes servers. For this to work, you need to disable normal Notes mail transfer between the internal and external servers and rely on the replication of the mail.box to "transfer mail."

This method of mail transfer has been tested by ISSC Australia for mail transfer between IBM's internal networks and the Internet. While this method does work, it means that the internal server needs to be a server dedicated to the same purpose as the external server—in order to reduce the risk of mail not destined for external transfer being replicated outside the firewall (that mail won't go anywhere, but there is a certain risk by just replicating it outside the firewall).

The Lotus Web Navigator Internet Browser

The Personal Web Navigator is part of every Notes client. It allows a Notes user to harness the power of Domino to share Web content directly with colleagues. You can mail active Web pages (complete with graphics and working URLs) to other Notes client users, integrate Usenet newsgroups with Domino discussion databases, monitor online information sources for the latest information, etc. If you prefer, the Lotus Web Navigator can be replaced with either Netscape Navigator or Microsoft Internet Explorer. They both ship with the Notes client.

I originally thought I'd stick with Netscape Navigator after I had moved to the Notes R4.5 client. However, I've now grown quite fond of the Lotus Web Navigator with an InterNotes server. I have the "Web Navigator" database from the InterNotes server sitting on my Notes desktop, and that makes navigating into and out of Web browsing from my Notes desktop very easy. I'm able to use the "right double-click" to navigate backwards when I'm surfing the Web, just as I do with Notes. To set up your Notes desktop to use the Lotus Web Navigator, first edit your location record. That's done with a File|Mobile|Edit Current Location. You'll then get a screen that looks like Figure 8.2. Under Internet Browser, I've chosen Notes as my browser and to retrieve Web pages from the InterNotes server. My Advantis InterNotes server is named ADVI0001. The pull-downs by the Internet browser and Retrieve/open pages areas let me select other browsers (Netscape Navigator or Internet Explorer) or to bypass the InterNotes server and go to Web pages directly from my Notes workstation.

When I open the Web Navigator database on my InterNotes server I get the display shown in Figure 8.3.

If I open (left double-click) the "Computing and Internet" Sampler shown in Figure 8.3, a screen similar to Figure 8.4 is displayed.

If I "right double-click" from the screen in Figure 8.4 to navigate back to the screen in Figure 8.3 and then select (left double-click) Database Views, I get a screen similar to that shown in Figure 8.5.

Figure 8.2
Editing your location
record for Web
browsing.

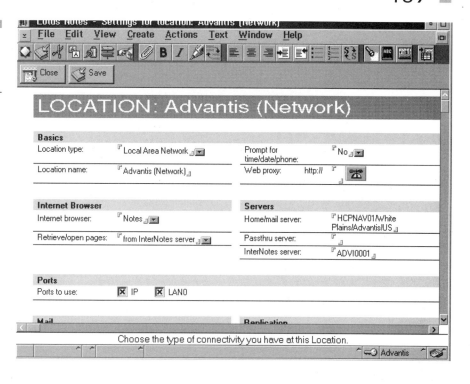

Figure 8.2
Editing your location
record for Web
browsing.

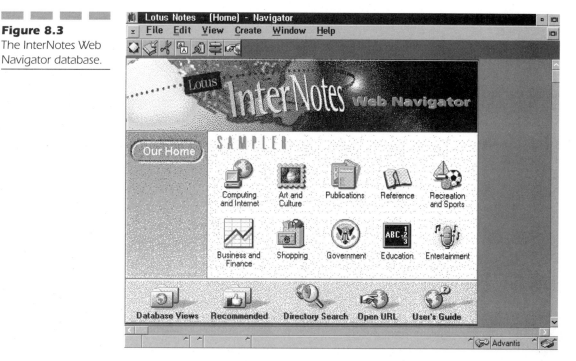

Figure 8.3
The InterNotes Web
Navigator database.

Figure 8.4
The "Computing and Internet" Sampler under InterNotes Web Navigator.

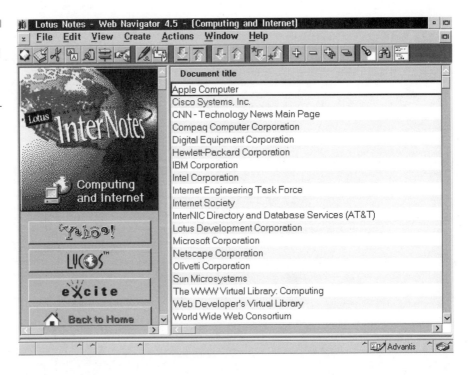

Figure 8.4
The "Computing and Internet" Sampler under InterNotes Web Navigator.

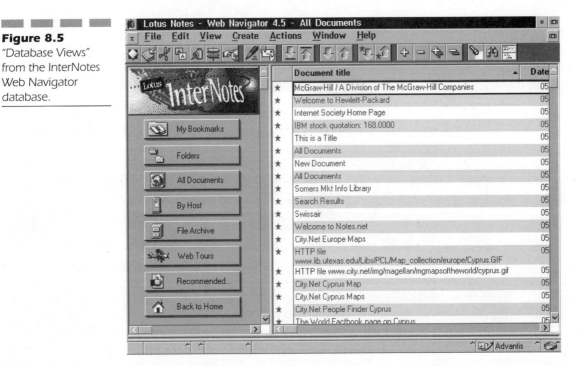

Figure 8.5
"Database Views" from the InterNotes Web Navigator database.

This "Database Views" screen is the ultimate in bookmarks. It shows me all of the Web pages everyone has been surfing via the InterNotes server. By the way, there still is security in the design. Even though you can see all the places everyone surfed, there is no way to tell who surfed what.

Replication vs. Cache

Web browsers do not replicate information, but they synchronize information by preemptively distributing frequently used Web pages to server caches. For many interactive requirements on a user's workstation, caching and replication give a user the same benefits. A user is able to work with the latest database information on his workstation. This limits the impact of network bandwidth on the user's ability to quickly browse through information.

Caching and replication both bring the latest information to the hard drive on the user's workstation and allow the user to work with the information on his local drive. Replication, however, has the clear advantage when a user wants to work with data while disconnected from the network and then later connect to the network and synchronize his work with the data on the server.

Push vs. Pull

A Notes or Domino server can replicate with another server by using both a "push" and a "pull." Originally, Lotus Notes and Web technology were "pull" oriented. The end user has to know where to get the desired information and "pull" the information from the server. For Lotus Notes, you need to know which server to go to, then which directory on that server, then which database in that directory, and then which document in that database.

Once you have used the database and added the icon for the database to your Notes workspace things get easier. You just click on the icon to open the database, and you can scan the databases for new (unread) documents. You can even set up Notes to automatically scan all selected databases each time you log onto Notes. This provides you with notification that there are new documents, and then you can read through each one. Notes keeps track of which documents you have read, so you only have to read new or modified documents. This has been a valuable feature of Notes from the start.

Web browsing in the early days (around 1993!) required that you know which server to go to, and then you had to navigate down the different links of the Web site until you found your information. You could add an entry to your text-based bookmark list to save the link so that you could return to the same Web page with a single click. During a single session, Web browsers kept track of which links you have already seen. However, if a Web page changed after you looked at it, you would not know. You had to go back to the Web page and read it over carefully to see if there were any changes. Furthermore, once you closed your Web browser and brought it up again, all links were reset to indicate that you had not read the Web page yet. This was not convenient. You had to know where the information is, "pull" the information from the server, and look for anything that is new or modified. Search engines, such as Alta Vista or InfoSeek, could help you find where the information was, but you still had to sift through the search results to find what you actually are interested in.

"Push" technology sends information from the server to you so that you do not have to go get ("pull") the specific item. For example, you can configure your client once to have the server send you the local weather map and weather forecast whenever there is an update. You can ask for stock price updates for your stock portfolio (an application can plot the price versus time). You can get the latest press releases and corporate announcements from companies you select. These are all examples of time-sensitive information for which you want the latest information. You do not want to constantly check databases or Web sites to see if there is new or modified information.

Using "push" technology, you let the "push" server keep track of any new or modified information on your selected databases or Web pages and send you the information. PointCast, Inc. is a company that provides this type of a "push" server. PointCast gives away a free screensaver that receives information from the PointCast server over the Internet. Headlines are delivered with colorful, animated graphics and more detail is available with a click.

"Push" technology is useful for Internet information and also for certain intranet information. Corporate Communications often needs to let all employees know about new corporate policies, re-organizations, financial reports, sales reports, benefits changes, and so on. The Lotus Domino.broadcast product together with the PointCast I-Server product are examples of tools for doing this. Any new document placed in a Notes database will be broadcast to all PointCast users. This "push" method complements the traditional "pull" method of delivering information using Lotus Notes.

"Push" technology is one method for tracking new and updated information. Lotus Notes/Domino also uses intelligent agents to accomplish the same goal. Using the Lotus Notes Web Navigator (the "browser" that comes with Notes 4.5), Web pages are stored on disk in a Notes database. Intelligent agents are scheduled to check the Web pages for any changes, update the Web page in Notes database, and send an e-mail summary to the end user with a live link to the updated Web page. This can be used with the Web Navigator proxy on the Domino server, with the stand-alone Web Navigator on the Notes client, or with the Weblicator add-on to a Netscape, Microsoft, or other Web browser. This is a very convenient feature Lotus provides that Netscape and Microsoft are beginning to add to their products. "Push" technology is an important trend that is becoming more popular with Internet and intranet users.

Using Intelligent Agents

Agent technology first appeared in Notes with Release 4. Agents, which are similar to macros, are created to automate time- or event-driven processes. You identify the name, event trigger, documents to act on, and the action itself to create a simple agent.

For example, suppose you are waiting for Prince Charming to accept your wedding proposal. You know how he always procrastinates and has trouble making quick decisions. You just do not have time to continually refresh your mail box and look for his e-mail reply. Set up an agent to do this for you! Have the agent trigger on the event of receiving any mail from Prince Charming (test the "From" field for his name). If there is mail from him, send an e-mail to your pager service provider to page you with the message "Mail from Prince Charming!" (Most pager service providers offer this type of service. Send e-mail with your pager identification number and your message, and the pager service provider will send the message to your pager.)

Of course, in the business world, you might be waiting for urgent messages from your customer, your lawyer, and your boss! Set up an agent to check for mail from each of these people and page you with customized messages. In this situation, an event-driven agent is more efficient than a time-driven agent. A time-driven agent would have to be scheduled for every 5 minutes to check the mail box. This would consume too much system resource.

Agents can perform convenient actions on your databases. I am often asked if Notes mail can be presorted. Some people get so much mail that they would like to have all mail from their customer in one folder, all mail from their lawyer in another, all mail from the boss in a third folder, and so on. This can be performed automatically when mail is delivered using agents. The agent can check who the mail is from and sort it into the respective folder. You can then read mail from your customer first, from your lawyer next, and then from your boss. (You can read the mail from the office pest last or with whatever priority you decide!)

The Lotus Domino Server

The Lotus Domino Server is a combined Web server/Notes server. The data is stored in a Notes database and can be accessed by either a Web browser or a Notes client. Likewise, content can be added to the Web site using any Web browser or a Notes client.

Lotus Domino is a very significant step in the rapid evolution of Lotus Notes. Lotus Domino is a Web server integrated with a Lotus Notes server, running on the same physical box. In fact, the Web server is a Notes server task (load web at the server console, or add to ServerTasks= in the NOTES.INI file), configured and managed within the Notes environment using Notes documents in a Notes database. Domino converts Notes documents to Web pages, and vice versa, on demand. As a consequence, Domino does not store HTML files on disk and just keeps the HTML files in memory.

Lotus Domino has many functions similar to Lotus InterNotes Web Publisher but uses a different approach. In contrast with Lotus Domino, Lotus InterNotes Web Publisher does the conversion from Notes documents to Web pages (and vice versa) on a scheduled basis and stores HTML files on disk. Lotus InterNotes Web Publisher itself is a Notes server task, but it must be used with a separate Web server that runs outside of the Lotus Notes environment. This means that the Web server must be configured and managed using tools separate from Notes.

Some of the benefits and features common to Lotus Domino and Lotus InterNotes Web Publisher include:

- Bringing Notes functions to the World Wide Web (a better Web server)—Workflow, replication, and security.
- Bringing Web access to Notes databases (a better Notes network).
- Distributing the writing of Web pages.
- Improving management of the Web site.

The Importance of Full Interactivity

The close interaction with the customer is what distinguishes Lotus Domino and Lotus InterNotes Web Publisher from other Web servers. Initially, Web servers were used primarily as a publishing channel. Information was presented for users to look at. Increasingly, businesses, governments, and institutions want to conduct services electronically. This means Web servers must now provide two-way interaction with the user. Lotus Domino and Lotus InterNotes Web Publisher provide this interactivity by hooking into the workflow capabilities of Lotus Notes. Information can be returned to the user immediately from a Notes database or any other database linked up using Notes.

Ages ago (around 1993!), Web sites were rather static, and content did not change rapidly. Lotus InterNotes Web Publisher was designed for this situation, and converting Notes documents to HTML on a scheduled basis was appropriate. Web sites became more dynamic, and content did change rapidly. Lotus InterNotes Web Publisher can be configured to publish on a short time interval (e.g., 5 seconds) to simulate conversion on demand. (In fact, we demonstrated Lotus InterNotes Web Publisher in this manner at numerous trade shows to contribute to a Notes discussion database with a Web browser.) However, Lotus Domino is designed to provide HTML on demand.

One important benefit of providing HTML on demand is the capability of expanding and collapsing lists on the Web, just like you can in a Notes database. Any Web surfer will appreciate this feature when confronted with long lists of links. It is much clearer to navigate when the list is collapsed to just show the category titles.

Why Bring Web Access to a Notes Database?

The benefit of bringing Web access to a Notes database might not appear immediately obvious. Consider, however, that many organizations operate in a mixed environment. This might be due to a corporate merger and acquisition. One organization uses Lotus Notes, one uses Microsoft Exchange, and one uses Novell Groupwise. Most users have a Web browser and this becomes the common denominator. You can convert Notes, Exchange, and Groupwise documents to Web pages so that everybody has access to the information. This is a temporary workaround solution until you migrate to one groupware platform (which, in our "humble, unbiased" opinion, should be Lotus Notes, right?). Even without mergers and acquisitions, large organizations often have different groupware platforms sprouting up independently.

Details on Domino 4.5 Server and Notes 4.5 Client

Domino 4.5 server (Domino 4.5) and the Notes 4.5 client (Notes 4.5) were first shipped on December 12, 1996. They brought to the market the leading solution for messaging, groupware, and collaborative Web applications. Domino 4.5 provides a new level of functionality for creating groupware applications for the Internet and corporate intranets that are accessible via Web browsers or Notes 4.5 clients. Domino 4.5 is the only solution that delivers to Web browser users powerful groupware applications for business activities, including human resources, customer support, and sales support. The Notes 4.5 client further improves productivity by taking maximum advantage of the Domino 4.5 groupware, messaging, and security features.

Domino 4.5 extends the ability of users to conduct business on the Web through enhancements in key areas such as messaging, scalability, manageability, and security, in addition to support of Internet standards, such as HTTP, POP3, SMTP, and SSL 2.0. Domino 4.5 also enables developers to rapidly design business applications for both Web browsers and Notes clients, and seamlessly integrate those applications with existing corporate systems. Some of these enhancements include:

■ Domino.Action—A powerful set of easy-to-use tools to allow users to create, populate, and manage a Web site. Domino.Action leverages the collaborative functionality of Domino to enable and manage multi-site distributed authoring.

■ POP3 mail support—Enables e-mail access and storage for any Internet POP3 mail client, such as QualComm's Eudora, Netscape Navigator, or Lotus cc:Mail.

■ SMTP MTA—A reliable, standards-based messaging backbone for SMTP networks by extending native protocol and directory support to the Lotus Domino 4.5 server.

■ Domino Advanced Services—Clustering and partitioning as well as usage tracking and billing capabilities can be added to the single processor or the SMP version of the Domino 4.5 server.

■ Notes 4.5 Client—Combines enhanced Web support with powerful Notes functionality.

In conjunction with the Domino 4.5 server, Lotus also shipped the Notes 4.5 client to create the only integrated full-service intranet client, providing support for the full range of activities in which users engage daily. Enhancements to the Notes 4.5 client include:

■ *Web browsing*—Direct access to the Web, and the ability to retrieve and cache Web pages for offline access. The Notes Web browser supports HTML 3.2, Java applet execution, Netscape plug-ins, SSL, and other key Internet standards.

■ *Personal Web Navigator Database, including Web Agents*—With the ability to customize agents for the individual Web user who needs to retrieve and effectively manage information retrieved from the Internet, Page Minder is a background agent for monitoring selected Web pages, with e-mail notification. Agents can also refresh and reduce Web documents to maintain an optimal Web database size for mobile use.

■ *Group Calendaring and Scheduling*—Calendaring and scheduling functionality that is powerful and intuitive for the end user, flexible for application developers, and scalable to even the largest enterprise. Enhancements include calendar views, easy appointment and meeting creation, free-time search, and the ability to view other people's calendars.

A complete list of enhancements to the Notes 4.5 client can be found in "What New in Notes and Domino 4.5: An Overview" located at http://domino.lotus.com.

The Domino 4.5 server is available on the Windows 95, Windows NT (Intel and Alpha), OS/2, NetWare, AIX, HP-UX, and Solaris (SPARC and Intel Edition) platforms. The Notes 4.5 client is available on the Windows 3.1, Windows 95, Windows NT (Intel and Alpha), OS/2, Macintosh (68K and PowerPC), AIX, HP-UX, and Solaris (SPARC and Intel Edition) platforms.

Lotus Notes 4.5 client is an integrated working environment that combines groupware and messaging capabilities with Internet access functionality. It allows individuals and organizations to leverage rich, intuitive, and integrated e-mail, workflow, collaboration, calendaring and scheduling; to support mobile users; to support client side processing for powerful business applications; and to browse the Internet to access and manage Web information.

Lotus Domino 4.5 is an applications and messaging server with an integrated set of services that make possible a broad range of secure, interactive business solutions for the Internet and intranets. With Domino, businesses can rapidly build, deploy, and manage applications that engage co-workers, partners, and customers in online collaboration and coordination of critical business activities.

Direct Use of Web Technology

The Lotus Domino server lets you directly access a Notes database with a Web browser. The Notes database information is converted to HTML "on the fly" as opposed to a translation process where a Notes database is converted to an HTML database on a scheduled basis.

The Pros and Cons

In certain cases, it might be better to convert a Notes database completely to HTML. In this case, InterNotes WebPublisher can be used. For example, if you wanted to convert static documentation that is in Notes database format and host that documentation on a conventional Web Server, then the use of InterNotes WebPublisher to convert the entire database to HTML makes sense.

Weblicator

Lotus Weblicator adds Web information management capabilities to your favorite Web browser, even while you're not connected to the Net. The Weblicator lets you selectively pre-fetch Web pages, then categorize and annotate them, even without a connection to the Web. By leveraging Lotus' replication technology, Weblicator provides tremendous flexibility while you're offline. You can use Weblicator to replicate your strategic Web applications (such as customer service, sales force automation, and HR tracking), work offline, and synchronize your changes once you're reconnected to the Net.

Domino.Action

Domino.Action is a Lotus product (included free with Domino 4.5) that provides templates to quickly set up and manage a typical Web site. By filling in a form in a Notes database, Domino.Action helps forward mail from "webmaster" to the webmaster's personal Notes mail database, add the same footer to each page (e.g., a copyright statement required by the legal department), and automate other features that otherwise require manual maintenance.

I used Domino.Action to set up a basic Web site in one day. It is not immediately obvious how to start Domino.Action. You use File|Database Open to open the database Domino Site Creator. After this, the rest is straight-forward. You just follow the three steps that are laid out for you.

The first step is to select which areas you want for your Web site. I suggest you begin with the minimum number of areas you can get along with. This will get your Web site up and running as soon as possible. In addition, you get a feeling of what's required to configure each area. Each area is very similar to configure, so start with just a few areas. You can go back and add more areas later. (Furthermore, you might want to take a lunch break during the initial creation of your Web site. There is a lot of processing up front in order to make your life easier later on.) Areas you can choose from include the familiar "About the Company," "Frequently Asked Questions (FAQs)," "Products and Services," and "Jobs."

One consideration when using Domino.Action is that, when you add a new Notes database to your Domino server, it will no longer immediately appear when you access the Domino server with a Web browser. The new database has to be explicitly added to appear on your Domino.Action Web page. That's both good and bad. The good part is that you can add test or development databases and not have them appear to Web browsers. You can still access your test or development database by specifying the exact path and database when you open your URL.

For example, in Advantis, we have a Domino server with the URL advi0001.ims.advantis.com. If I add a test Notes database to the Domino server in the SOMERS subdirectory, with a database name of MIC-SOM.NSF, then I can access this database with my web browser using the URL http://advi0001.ims.advantis.com/somers/micsom.nsf/?OpenDatabase. The URL of http://advi0001.ims.advantis.com will show me our Domino.Action home page, which does not include a hypertext link to my development database.

A Domino Web site, using Domino.Action, can be set up in one day by just one person. Then, many people can add content using any Web browser or a Notes client without knowing any HTML (the Hypertext Markup Language). This is a great convenience and speeds development and deployment.

How the Notes Functions Make Domino a Better Web Server

Domino is a unique Web server because it takes advantage of the powerful integrated Notes functions, including security, replication, and workflow. These Notes functions provide for secure, distributed content authoring with approval routing. Each area of the Web site corresponds to a Notes database.

The Communications department authors the content for the "About the Company" and "Frequently Asked Questions (FAQs)" areas, the Marketing department authors the content for the "Products and Services" area, and the Human Resources department authors the content for the "Jobs" area. The Human Resources department members cannot edit Web pages in the "Products and Services" area because their access is denied using the Access Control List (ACL) feature of Lotus Notes.

Because authors can use any Web browser or a Notes client, an entry in the Notes Name and Address Book directory is made for the Web browser users. Each Web browser user is given a user ID and password and should be added to the appropriate groups. The Notes users get their user ID, encryption keys, and password assigned when their user IDs are created. Each Notes user should be assigned to the appropriate group also. Not everybody should be allowed to edit the Web pages, so the Marketing group in the ACL would contain a subset of the entire Marketing department. Contributors to the Web site areas are often located in distant areas, even around the world. They can work on their Notes database on their local Notes server and, using Notes replication, synchronize the changes with the Domino server hosting the official Web site.

Authoring authority is delegated and distributed readily using Domino. Responsibility for the content must rest with a small group of individuals, and Domino offers approval routing, based on Notes workflow. I suggest that you require that all new Web pages be approved before they are published on the Web site. Depending on the type of document, approval might be needed from the Communications, Marketing, Human Resources, Legal, or Development departments. Once approved, Web pages can be held until they are ready to be published.

For example, a Press Release or Product Announcement can be prepared, approved, and held until the desired release/announcement date. In fact, using Notes scheduling and agents, Web pages can be automatically published and deleted according to a preset schedule (e.g., you can advertise a beta code download that is available for a limited time only). The ability to hold Web pages simplifies development. On traditional Web servers, you often need to maintain a production Web site and a development Web site. Using Domino, production and development Web pages are stored in the same Notes database, with the development Web pages held for publication. The development pages are available for developers to see and use but are not available for end users.

Domino.Merchant

Domino.Merchant is a Lotus product used for quickly and inexpensively setting up a merchant Web site. The user's Web experience at a Domino.Merchant site is analogous to shopping in a physical retail store.

You go shopping for your product, in this case by browsing through the product catalog. Once you've found the item you want, you select the options, depending on the product: blue, red, or green; small, medium, or large; spicy, medium, or mild; etc. You then place your items into your "shopping cart" and continue shopping. Of course, you can always "look" into your shopping cart to "see" what you have selected. You can take items out of your "shopping cart," exchange items, or add items. When you are done shopping, you go check out. By clicking on buttons, your subtotal is calculated, the tax is calculated (appropriate for your address), shipping and handling is calculated (depending on the item's weight, dimensions, and possibly shipping distance), and the total is calculated. You can now make a secure payment utilizing SSL encrypted credit card numbers.

After you finish shopping and have completed payment, the merchant sends you confirmation and will process your order. If you paid for a software download, you will be presented with an enabled link. Let's say you paid for an annual subscription for an online magazine. Using Notes programmable macros and agents, your ID will be added to a group in the Access Control List (ACL) for the Notes database containing the magazine content. Before your subscription expires a year from now, you will be automatically e-mailed a reminder to renew your subscription. If you are a preferred customer, you get a discount. Depending on your preferred customer level, you get a different discount. A special sale can be offered for a limited time only by using a Notes agent to automatically post a sale on the Web site and take it off after a predetermined time.

The best known merchant site, perhaps, is the virtual bookstore at www.amazon.com. Although it is not powered by Domino, it is a great example of shopping on the Web.

Domino.Service

Domino.Service is a Lotus product that is similar to Domino.Merchant. Whereas, Domino.Merchant is oriented towards merchandise (physical and "soft" merchandise), Domino.Service is oriented toward services.

Broadcasting to the Desktop: Domino.Broadcast and PointCast

Lotus Notes was designed for information to be primarily "pulled" by the end user. Documents are put in a database, and it is the user's responsibility to "pull" the information out of the database. Discussion data-bases and other forums are examples of "pull" type of databases. In some cases, it is desirable to "push" information to users. Corporate news bul-letins, competitive news flashes, and network connectivity alerts are examples of more urgent information that needs to be "pushed" to the user without waiting for the user to go "pull" it when the user is available.

Domino.Broadcast, a Lotus product, can be used in conjunction with products like PointCast to broadcast information from a Notes database to the users desktop. The PointCast client is free from the PointCast Web site and was originally a screensaver with a personalized newsfeed from multiple public news sources. A private corporate channel can now be set up on the intranet to alert employees of important information. Users can respond to the alert by filling out a Web form and submitting it to a Notes database using the features of Domino.

Broadcasting to the desktop is an important complement to the "pull" type of databases offered by Lotus Notes. You will want to evaluate the bandwidth used by Domino.Broadcast and PointCast so as not to over-burden your network.

An Advertiser's Dream

Combining Domino.Merchant, Domino.Broadcast, and PointCast can ful-fill an advertiser's dream. Domino.Broadcast and PointCast can provide continuous advertising to the consumer. A coupon for a sale or discount can (literally) flash on the consumer's screen. The consumer simply clicks on the coupon and is brought to a Domino.Merchant site where the con-sumer clicks to pay for the item. Impulse buying has never been quicker or easier. The idea for impulse buying is to have the consumers buy before they have a chance to think and change their mind. This combina-tion of Domino products can quickly implement such an approach.

Java

Java is an object-oriented programming language developed by Sun Microsystems. It holds the promise of fulfilling the long-sought dream of creating applications that work the same on any hardware on any operating system. Java also supports the networking computing concept whereby applications are taken off the network (anywhere in the world, perhaps) instead of off the local hard disk or off the Local Area Network (LAN) file server hard disk.

Java compared with C and C++

My first reaction to hearing about Java was to wonder why a new programming language was developed. What is wrong with C and C++? C was supposed to be the language for developing applications that could run on any hardware on any operating system. C++ is the object-oriented version of C that allows the re-use of code. One important difference between Java and C/C++ is that Java is interpreted whereas C/C++ is compiled. Java code is directly executed at the time the application is used. C and C++ code must be compiled for each hardware/operating system, and then the compiled code is executed when the application is used. In general, compiled code runs faster than interpreted code.

The Impact of Java on the Future of Computing

John Gage, of Sun Microsystems, has been an articulate spokesman for Java. He notes that there are two or three things that are very different from the past, in that everybody's agreed upon a common environment and Web-based software and TCP/IP networking software that just runs on any computer, and that includes IBM and Apple and Sun and AT&T and Alcatel and British Telecom—all the telephone companies—and all the chip companies and all of the cellular companies.

That's a major shift; it never happened before. This shift brings old-style computers up-to-date, because they instantly run the same software as the hand-held computers. It changes everyone's business in the software markets enormously. Your market's no longer a Windows 95 machine or a Macintosh or a UNIX workstation, but every computing device.

Using the Internet, you're able to build a new hierarchy of computing. Any computer linked into the network is part of a huge parallel computer. You can distribute some parts, and you can put intermediate memory a little closer. Today, processors, storage, and display have to be pretty close. The processor and memory speak at very high speed, so you're not going to pull them apart. However, disks are very slow—10,000 times slower than hardware memory. You don't even notice if you take a disk and move it from New York to Chicago. Because connecting things up costs very little, we can create a new form of computer architecture.

There are three or four factors converging that make for watershed change. One element is the hardware side—the development of processors, memory, storage, and display—with all these things now getting cheaper. They no longer belong only to the rich but can move out to hundreds of millions of people. That's the inevitable progress of hardware.

Couple that with the enormous development in networking. You can move enormous volumes of data—622 Mbps—across cheap, phone wire that costs 5¢ a foot. Fiber's being put in everywhere, and the wires are going everywhere, so there's ubiquity coupled with speed. We have these developments that allow all the machines to be connected to each other.

A third development is software that runs on all computers. You no longer ask, "Is it a Mac or a PC?" Everything from a wristwatch to computing elements in automobiles to the ATM to the telephone to the television can run common software. The change has come with Java. Suddenly there's a language that runs on everything. People can pick any computer in the world from the Nokia cell phone to the desktop computer, and write all the applications in Java. You write applications one time, and they just run.

The conjunction of these three forces—cheap platforms, ubiquitous and very fast network connections and communications software running on all computers—allows an explosion of development that until recently has been fragmented.

A lot of people view the world of computer-based devices as being divided. In fact, however, they all are identical. They all have little display screens; they all have memory. They're all becoming smart devices. Those old industry divisions are evaporating. The "Java stuff" will remove the barriers. Companies will rewrite the basic parts of software you use every day so that it will run on every computer.

How Lotus Notes and Domino use Java Technology

Lotus Notes clients had the ability to execute Java applets beginning with Domino 4.5. The server serves Web pages that support Java applets and Internet scripting languages, such as VBScript (for Visual Basic) and JavaScript. Domino 5.0 will offer full support of Java in Lotus APIs (Application Programming Interfaces for developing applications) and Java Beans. Beyond release 4.5, Domino will support Java classes via IIOP (Internet Inter-ORB Protocol, where ORB stands for Object Request Broker) to access Domino application services, such as the object store, security, and messaging.

Domino and Lotus Notes also support Microsoft's ActiveX (OCX) applets. ActiveX can be thought of as Microsoft's "Java-like" software. The Java classes for Notes provide a high-level object library that can be used either on Domino server machines from standalone Java programs or from any Java-enabled Web browser via Java applets. Notes can be used to open the power of Java for distributed application development.

Notes and Java Beans

Lotus Components running within Lotus Notes were demonstrated at the January 1996 Lotusphere trade exposition. These Components are small applications (applets) that can be embedded in a Notes document and function independently within the document! These applets are "Java Beans."

For example, I can send you Notes mail that contains a spreadsheet for calculating your monthly car loan payments for the next three years. I have preprogrammed the spreadsheet with the formulas to calculate amortization. All you have to do is enter the car price, and how many years over which the loan will be paid off. I could also send you a spreadsheet that calculates mortgage payments for houses based on the house price, down payment, interest rate, points, and length of loan. These spreadsheet components work right inside of your mail and are not linked to any other external program. I do not have to worry about which spreadsheet you have on your hard disk, which might be incompatible with the spreadsheet I am using. You do not even need to have any spreadsheet program on your hard disk! This can save you valuable disk space.

The Starter Pack of Lotus Components includes six applets that perform common office functions: Chart, Comment, Draw/Diagram, File Viewer, Project Scheduler, and Spreadsheet. Lotus Components have been downloadable from the Lotus Web site since January of 1996. The initial Components were based on Microsoft's ActiveX controls, which is the latest name for what was called OCX (OLE Controls eXtensions?), which is an extension of OLE (Object Linking and Embedding). Is that clear?

ActiveX only works with Microsoft Windows operating systems. Java Beans is analogous to ActiveX for applications written with the Java programming language. Java Beans works across multiple platforms, and support for Java Beans has been announced by many companies, including Sun, IBM, Lotus, Netscape, Borland, Symantec, and Oracle. Lotus is applying the experience it has gained with ActiveX Components to developing Java Beans Components. Kona is the code name for the Lotus Java Beans Components. (Kona is easier to say than Lotus Java Beans Components, don't you think?)

Java and OpenDoc

ActiveX is the components architecture for applications written for the Microsoft Windows operating systems. OpenDoc is the components architecture for applications written for IBM's OS/2 operating system and for Apple operating systems. CORBA is the architecture for UNIX-based operating systems. Java Beans is the architecture for applications written with the Java programming language. Given the widespread popularity of Java, Java Beans holds the potential for becoming the standard cross-platform architecture.

Integrating Notes and the Web

The Lotus Domino server name replaced the Lotus Notes server name partly to emphasize the integration of the Notes server with the Web. The Lotus Domino server can now act as a full-functioned Web server. This section describes some of the Lotus Notes and Domino forays into the realm of projects that would previously have been the realm of traditional Web servers.

Distance Learning over the Web

Lotus Notes- and Domino-based technologies can be used with the Web for the support and development of rich, multimedia training and interpersonal, collaborative learning online or offline. The following sections discuss some of the projects that use Lotus Notes and Domino to support distance learning over the Web.

Pace University Project

Pace University, with its main campus located in New York City and with other campuses nearby, including White Plains, embarked on a distance learning project using Lotus Notes and the Internet during the first part of 1996. I participated in the pilot class to gain a student's perspective on distance learning. The class I took was Telecommunications

Essentials, taught by Pete Stair (coincidentally, a fellow Caltech alumnus) and based on a classroom course that Pete has taught for several years. This course is one of a series of telecommunications courses offered to IBM's networking staff to provide a common foundation in telecommunications including voice and data communications.

The "class" began in my own house by watching a video tape introduction by Pete Stair. I then completed the reading assignment in the traditional textbook and read Pete Stair's lecture material online using my Notes client to access the Notes document library database. For "homework," I had to answer questions in the Notes discussion database to demonstrate my understanding of the material (or at least to demonstrate that I could find the answer in the textbook or in the lecture material!).

Finally, I was given an assignment to surf a wide variety of Web sites with material related to telecommunications and to comment (in the Notes discussion database) on what I observed there. These Web sites included phone companies, government regulators, and cellular telephone equipment makers. The sites provide good overviews of the organizations, and many have interesting tutorials. I saved the Web sites' addresses as bookmarks so that I can reference them in the future. Every week, new material would be available in the databases. I like the introduction to these Web sites because I would not have found them so readily on my own.

I liked taking this course with the distance learning method because I was able to "attend" class when my schedule permitted. I was not traveling on business during the class; however, if I were, I could still participate. Using Notes replication, I could even "attend" class while in an airplane! Distance learning is usually thought of as a means for students to learn anytime, anyplace. The instructor can also teach anytime, anyplace. (We could call this distance teaching instead of distance learning!) In fact, Pete Stair was traveling in Asia while teaching part of this course, so he took advantage of this feature to present new lecture material and answer questions.

The interactions between the teacher and students, and among students themselves, are analogous to regular classroom discussions. The discussion database is like an open classroom discussion with teacher and student participation. Because each student has Notes e-mail, the teacher can communicate with any student in private "outside of the classroom." Students chat with each other spontaneously within the discussion database and with e-mail, just as you would during a coffee break or after class. Two students discovered they had a friend in common and began a conversation. Another student asked me via e-mail about a project I am working on.

Electronic communications are similar, but not entirely equivalent, to talking in person. Conversing in person is more immediate and involves body language. Electronic communications used in distance learning is not necessarily immediate and does not involve body language unless video conferencing is used. The benefit of distance learning is to bring together people distributed in time and location. The class I took had people from Illinois, New York, and Florida, spread over two time zones. Although we could not gather in one classroom, we saved a lot of travel expenses by staying at our own location. This benefit is even greater when students are literally around the world in up to 24 time zones.

One challenge for Notes-based distance learning is distributing, installing, and configuring Notes client code to each student. At the time, Notes did not come on a CD-ROM, so sets of about 20 floppy disks had to be created and sent to each of the 30 students, along with a few pages of instructions. Separate instructions were required for OS/2 and Windows operating systems. After the class ended, the user IDs were expired and were reissued for the next class. In this manner, the 30 Notes licenses were recycled, and the old students could not take a class they were not registered for.

Now, nearly all IBM employees have a Notes client, so taking a distance learning class internally is not such a logistical effort. The logistics are still a concern for a university offering distance learning classes to the public at large, although a potential market of 15 million Notes users worldwide is attractive nevertheless.

An alternative to using Notes for distance learning is to use the World Wide Web. Lecture material can be presented, and a discussion database can be set up on a Web site, although mail might have to be set up separately. Most people have a Web browser, but the Web site must be tested to be compatible with all the different Web browsers and release levels. One drawback of using the Web is implementing security measures. It is desirable to verify and authenticate users so that students can only take classes they are registered for (and have paid for!). Using just a user ID and password might not be secure enough, and using SSL could be an administrative burden. Notes has security built in and is quite strong. These are trade-offs to be considered when deploying distance learning services.

Lotus LearningSpace

Today's need is for students to take courses anywhere, anytime. Lotus Notes is a natural platform to meet this demand. In fact, Lotus has packaged a set of five databases to deliver distance learning. This is the Lotus

LearningSpace product that includes the Schedule, the Profiles, the MediaCenter, the CourseRoom, and the Assessment Manager databases.

The Schedule database begins with a course overview and presents the objectives for each section of the course. This database also lets students find out when to complete reading and homework assignments and when examinations will be given.

The Profiles database contains student and teacher profiles, including contact information, education and work experience, interests, and even photographs.

The MediaCenter database contains the actual course material, including text, graphics, images, audio clips, video clips, Computer-Based-Training (CBT) sections, and simulations. Access is also provided to external sources such as the World Wide Web and other content repositories that are increasingly important components of curricula.

The CourseRoom database is a virtual classroom where discussions take place publicly (using a Notes discussion database) or privately with the instructor (using secure Notes e-mail).

The Assessment Manager database is like the grade book and set of tests for the course. It is an evaluation tool for the instructor to privately test and receive feedback on participant performance. Quizzes, exams, and surveys are posted in the Schedule database and, when complete, are sent via e-mail into the Assessment Manager database for private review by the instructors. Instructors can review, grade, and provide feedback to participants privately.

These five Notes databases which comprise the Lotus LearningSpace, provide the tools necessary to teach a class anywhere, anytime.

Distance learning is analogous to taking a course in a classroom. The instructor first introduces the class and the objectives (in the Schedule database, in the distance learning virtual classroom). After presenting the instructor's own education, work experience, and interests, the class members present their own background. When I take a class, I value the introductions that each person gives at the beginning of the class, and I value the ensuing class discussion. I make a mental note of who is an expert in an area I need help in, or who is working on a project similar to mine and where we would benefit from collaborating. I have established many important business contacts from taking classes.

The Profiles database is the distance learning class introductions, and the CourseRoom discussion database is the distance learning class discussion. The class introductions and class discussions are often as valuable as the course material itself. The course material is, nevertheless, the central part of the course. Having the material in Notes database gives the additional benefits of utilizing the Notes full-text search capabilities and replication functions. Students can replicate the course material to their local workstation or laptop and take the class when disconnected from the network. This is vital to allowing students to learn anywhere, anytime.

Web Searching When You Have Domino Servers

Searching your company's Domino-based intranet can be done either by using native Domino search capabilities or by using external search engines. The following method for implementing Domino multisite searches was worked out by Philippe Bondono of IBM Global Services in France.

IMPLEMENTING SEARCHING WITH NATIVE DOMINO FACILITIES.   There are two parts to this section. The first part is written from the Help database and presents the way to implement a search database. The second part gives recommendations on how to use the searching facility on your intranet.

SETTING UP MULTI-DATABASE SEARCH   The principle here is to use a special database that will index the various databases to search. Then users will point to this special database in order to run their search queries. The following paragraphs describe how to implement such a facility. Note that this material has been based on the Domino Help database. You can refer to that database for further details.

To set up a search site database to enable users to search for information across multiple databases, you need to perform the step described in the following sections.

Enabling Databases for Multi-database Searching   You must have at least Manager access to enable a database to be included in a multi-database search. If you don't, ask the database manager or designer to complete these steps for you:

1. Choose *File|Tools|Server Administration*.

2. Click *Database Tools*.

3. In the Tool box, select *Multi Database Indexing*.

4. In the Server box, select a server.

5. In the Databases box, select the databases you want to enable for multi-database indexing.

6. Click *Enable*, and then *Update*.

Creating a Search Site Database   After enabling databases for multi-database searching, create a search site database. Do not choose the option Create full text index for searching as you create the search site database. Create the full text index after you configure a search scope:

1. Choose FileIDatabaseINew.

2. Select the server on which to store the database.

3. Enter a title for the database (for example, Marketing Search Site).

4. Enter a file name for the database (for example, MKTGSRCH.NSF).

5. Select the Search Site template (SRCHSITE.NTF).

6. Click OK.

7. The Search form launches by default. Click FileIClose to close it.

8. Disable automatic launching of the Search form. Choose FileIDatabaseIProperties, and click Launch. In the On Database Open box, select a different option (for example, Restore as last viewed by user).

9. Choose FileIDatabaseIAccess Control, and set the default access to No Access to prevent users from using the database while you configure it.

Configuring a Search Scope for a Search Site Database   After you create the search site database, configure a search scope:

1. Select the search site database you created.

2. Define a search scope. Choose CreateISearch Scope Configuration. Select a scope and an indexing option for the scope, then close and save the configuration document. The Search Scope Configurations views display the new document. Create additional configuration documents as necessary.

3. Switch to the server console, and type the following command to display in the search site database a database entry for each database included in the search scope: Load updall SEARCHSITE arguments, where SEARCHSITE is the filename of the search site database and arguments are these optional arguments: -A and -B.

4. Select the search site database, choose ViewIGo To, choose one of the database views, and click OK to see the databases you've included in the scope.

5. (Optional) To change the view used to display the information users see in a list of search results for a specific database, open a database entry and select a different view in the Search Results Should Use box. The default database view is used otherwise.

6. (Optional) To refine the search scope, do one or more of the following, then repeat steps 3 through 5.

In one of the database views, open a database entry, and select a different indexing option for a specific database.

Repeat step 2 to create additional Search Scope Configuration documents.

Creating a Multi-database Full-text Index   After you configure a search scope, you create a multi-database full-text index for the search site database. This requires at least Designer access to the database:

1. Choose FilelToolslServer Administration, **and click** Database Tools.

2. In the Server box, select the server that stores the search site database.

3. In the Databases box, select the search site database.

4. In the Tool box, select Full Text Index.

5. (Optional) Select Case sensitive index **and/or** Exclude words in stop word file.

6. Below Index breaks, select Word breaks only **or** Word, sentence, and paragraph.

7. Click Create.

Notifying Users of a Search Site Database   After you create a multi-database index for a search site database, notify users:

1. Change the default access to Reader.

2. If you disabled automatic launching of the Search form, enable it again. Choose FilelDatabaselProperties. **Click** Launch. **In the On Database Open box, select** Launch 1st doclink in About database.

3. Send an e-mail or otherwise notify users of the new search site database.

SEARCHING YOUR INTRANET USING MULTI-DATABASE SEARCHES   In order to be able to search your Intranet, it is possible to use the previous techniques to implement multi-databases searches. However, it is recommended for most installations that you have a dedicated server that will be the main point where searches will be redirected. This server should be located near your main Notes hub in order to minimize network traffic. The search server would be used to replicate databases to be searched, and then index them as shown previously. This server would be dedicated to searches, and it would be accessible by all your users in reader mode. Depending on the size and make-up of your installation, you can:

1. Add more servers for searching, in order to maximize network utilization.

2. Index directly the databases to be searched on their original server. However, this is usually not recommended for both network resource optimization (assuming the search servers are located on the same LAN as your hub servers) and ease of administration.

USING SEARCH ENGINES.   Due to the way Domino handles documents (especially with the syntax used for URLs, which uses characters such as the question mark), it is not easy right now to index Domino-based web sites with conventional Web search engines (such as Alta Vista).

The solution is to use special dedicated servers that are knowledgeable of Notes databases format. Two companies have announced support for such Domino-based Intranet search servers: Fulcrum Technologies and Verity Inc.

FULCRUM KNOWLEDGE NETWORK   The Fulcrum Knowledge Network is an integrated product that connects diverse information sources across the enterprise. At the client level, the Knowledge Desktop can be any Web browser, a Web browser enhanced with ActiveX controls for enhanced viewing, or a Microsoft Exchange or Microsoft Outlook client. Using the Knowledge Builder Toolkit, which contains a rich Object Library and the Fulcrum WebFIND! query application, developers can create custom Microsoft Windows or Web-based clients for specific business applications.

At the application layer, the Knowledge Server is the heart of the Fulcrum Knowledge Network. It provides the searching and indexing services that integrate the multiple, distributed information sources. The searching service dynamically constructs a personalized Knowledge Map on a per-user basis according to the information sources available for searching and the user's access rights.

Finally, Fulcrum provides Knowledge Activators to allow access to corporate knowledge repositories on multiple servers and running on multiple platforms, and in sources as diverse as Microsoft Exchange Server, Lotus Notes databases, Web sites, file systems, existing Fulcrum SearchServer applications, and more. Knowledge Activators embrace diverse information sources and leverage their security models, compound document architectures, and administration profiles.

You should note the following:

- Fulcrum Knowledge Network is fully synchronized with Lotus Notes databases so that users can search on and retrieve the most up-to-date information stored in these databases.

- Lotus Notes database and document security is preserved.

- It allows searching on compound information in Notes databases.

- Multi-platform support: The Knowledge Activators bridge from the Knowledge Server on Windows NT into the data repositories on multiple UNIX and NT servers.

- Fielded-searching: The Knowledge Activator for Lotus Notes ensures that the Fulcrum index creates an attribute store that inherits the fields created in Notes databases.

- It is fully integrated with the Domino Server to preserve the Web user interface provided by Lotus' product.

- Fulcrum needs either Netscape servers or a Microsoft Internet Information Server in order to make information available to web browsers.

For more information on Fulcrum and Fulcrum products, visit Fulcrum's Web site at http://www.fulcrum.com.

VERITY INFORMATION SERVER Verity is a set of tools built upon what is called an *information server*. This server is a central server that will be the main point of contact for searches. The server uses spiders in order to collect information on the intranet and then indexes them. It must be complemented with a *Notes Connector,* in order to be able to retrieve information located inside Lotus Notes databases.

Verity copes with dynamic HTML and thus allows you to combine both standard HTML files and Domino pages in one place. Another benefit of using Verity is that it is also able to index attached files—such as Acrobat, Word, etc.—either attached to Notes documents or to standard HTML pages.

You should note the following:

- The Notes Connector is scheduled for shipment in September of 1997.

- Verity has provided Lotus with the technology, which is currently implemented in the Domino search engine.

For more information on Verity and Verity products, visit Verity's Web site at http://www.verity.com.

Frequently Asked Questions on Lotus Notes and the Internet

Iris Associates, the creator of Lotus Notes and a wholly owned subsidiary of the Lotus Development Corporation, has developed a very useful online document that contains Frequently Asked Questions (FAQ) on Lotus Notes and the Internet. Most of this document, with permission of Iris and Lotus, has been included in appendix C. These FAQs should answer many of your detailed questions on the actual steps to take to use the Internet for your Lotus Notes wide area network. The database is also included on the CD-ROM that comes with this book.

CHAPTER **9**

Directory Services

The emergence of LDAP as the Internet standard protocol for directory services is a watershed event. It will enable interoperability between clients and servers, and between servers, on the Internet. In addition, it will be the catalyst for directory interoperability between the Internet, intranet products based on Internet technologies, and existing NOS-based and application-specific directories used in today's organizations. The degree to which that interoperability is achieved, however, will be determined by how many vendors support version 3 of the protocol—which goes beyond the functionality of the current standard in RFC 1777—and Netscape's extensions beyond version 3.

The Burton Group
Network Services/Directory Services, May, 1996

Most network products—such as Lotus Notes, calendar systems, e-mail programs, and gateways—have directories. And that's the problem. Often the network administrator must deal with many different directories and sometimes manually update these separate directories. The ideal would be to have one directory that meets the needs of all the different systems. The standard directory of the future will be one that follows international standards for directories. That standard is the X.500 standard. This chapter discusses this important topic.

Models of Directory Services

Much of the material in this section is taken from the Internet's "Request for Comment" RFC 1309 on "Technical Overview of Directory Services Using the X.500 Protocol." This RFC discusses various models of directory services (with an Internet slant, of course), the limitations of some current models, and some solutions provided by the X.500 standard to these limitations.

The Telephone Company's Directory Services

A model many people think of when they hear the words "Directory Services" is the directory service provided by the local telephone company. A local telephone company keeps an online list of the names of people with phone service, along with their phone numbers and their address. This information is available by calling up Directory Assistance, giving the name and address of the party whose number you are seeking and waiting for the operator to search his or her database. It is additionally available by looking in a phone book published yearly on paper.

The phone companies are able to offer this invaluable service because they administer the pool of phone numbers. However, this service has some limitations. For example, you can find someone's number only if you know their name and the city or location in which they live. If two or more people have listings for the same name in the same locality, there is no additional information with which to select the correct number. In addition, the printed phone book can have information that is as much as a year out of date, and the phone company's internal directory can be as much as two weeks out of date. A third problem is that one actually has to call Directory Assistance in a given area code to get information for that area; one cannot call a single number consistently.

For businesses that advertise in the Yellow Pages, there is some additional information stored for each business. Unfortunately, that information is unavailable through Directory Assistance and must be gleaned from the phone book.

Some Currently Available Directory Services on the Internet

Since the Internet is comprised of a vast conglomeration of different people, computers, and computer networks, with none of the hierarchy imposed by the phone system on the area codes and exchange prefixes, any directory service must be able to deal with the fact that the Internet is not structured. For example, the hosts foo.com and v2.foo.com might be on opposite sides of the world, the .edu domain maps onto an enormous number of organizations, etc. Let's look at a few of the services currently available on the Internet for directory type services.

THE FINGER PROTOCOL. The finger protocol, which has been implemented for UNIX systems and a small number of other machines, allows one to "finger" a specific person or user name to a host running the protocol. This is invoked by typing, for example, finger clw@mazatzal.merit.edu. A certain set of information is returned, as this example from a UNIX system finger operation shows, although the output format is not specified by the protocol:

```
Login name: clw                    In real life: Chris Weider
Directory: /usr/clw                Shell: /bin/csh
On since Jul 25 09:43:42           4 hours 52 minutes Idle Time
Plan:
Home: 971-5581
```

where the first three lines of information are taken from the UNIX operating systems information and the lines of information following the "Plan:" line are taken from a file named .plan, which each user modifies. Limitations of the fingerd (the finger daemon) program include:

■ One must already know which host to finger to find a specific person.

■ Because primarily UNIX machines run the "fingerd" application, people who reside on other types of operating systems are not locatable by this method.

■ Fingerd is often disabled on UNIX systems for security purposes.

■ If one wants to be found on more than one system, one must make sure that all the .plan files are consistent.

■ There is no way to search the .plan files on a given host to (for example) find everyone on mazatzal.merit.edu who works on X.500.

Thus, fingerd has a limited usefulness as a piece of the Internet Directory.

WHOIS. The whois utility, which is available on a wide of variety of systems, works by querying a centralized database maintained at the DDN NIC, which was for many years located at SRI (originally Stanford Research Institute) International in Menlo Park, California, and is now located at GSI. This database contains a large amount of information that primarily deals with people and equipment that is used to build the Internet. SRI (and now GSI) has been able to collect the information in the WHOIS database as part of its role as the Network Information Center for the TCP/IP portion of the Internet.

The whois utility is ubiquitous and has a very simple interface. A typical whois query look like:

```
whois Reynolds
```

and returns information like:

```
Reynolds, John F. (JFR22) 532JFR@DOM1.NWAC.SEA06.NAVY.MIL
        (702) 426-2604 (DSN) 830-2604
Reynolds, John J. (JJR40) amsel-lg-pl-a@MONMOUTH-EMH3.ARMY.MIL
        (908) 532-3817 (DSN) 992-3817
Reynolds, John W. (JWR46) EAAV-AP@SEOUL-EMH1.ARMY.MIL
        (DSN) 723-3358
Reynolds, Joseph T. (JTR10)  JREYNOLDS@PAXRV-NES.NAVY.MIL
        011-63-47-885-3194 (DSN) 885-3194
Reynolds, Joyce K. (JKR1) JKREY@ISI.EDU
        (213) 822-1511
Reynolds, Keith (KR35)      keithr@SCO.CO
        (408) 425-7222
Reynolds, Kenneth (KR94)
        (502) 454-2950
```

```
Reynolds, Kevin A. (KR39)     REYNOLDS@DUGWAY-EMH1.ARMY.MIL
        (801) 831-5441 (DSN) 789-5441
Reynolds, Lee B. (LBR9)   reynolds@TECHNET.NM.ORG
        (505) 345-6555
```

A further look-up on Joyce Reynolds with this command line:

```
whois JKR1
```

returns:

```
Reynolds, Joyce K. (JKR1)           JKREY@ISI.EDU
    University of Southern California
    Information Sciences Institute
    4676 Admiralty Way
    Marina del Rey, CA 90292
    (310) 822-1511
    Record last updated on 07-Jan-91.
```

The whois database also contains information about the Domain Name System (DNS) and has some information about hosts, major regional networks, and large parts of the U.S. Military's "MILNET" system.

The WHOIS database is large enough and comprehensive enough to exhibit many of the flaws of a large centralized database:

- As the database is maintained on one machine, a processor bottleneck forces slow response during times of peak querying activity, even if many of these queries are unrelated.

- As the database is maintained on one machine, a storage bottleneck forces the database administrators to severely limit the amount of information that can be kept on each entry in the database.

- All changes to the database have to be mailed to a "hostmaster" and then physically reentered into the database, increasing both the turnaround time and the likelihood for a mistake in transcription.

THE DOMAIN NAME SYSTEM (DNS). The Domain Name System (DNS) is used in the Internet to keep track of host to IP address mapping. The basic mechanism is that each domain, such as merit.edu or k-12.edu, is registered with the NIC and that, at the time of registration, a primary and (perhaps) some secondary nameservers are identified for that domain. Each of these nameservers must provide host name to IP address mapping for each host in the domain. Thus, the nameservice is supplied in a distributed fashion. It is also possible to split a domain into subdomains, with a different nameserver for each subdomain.

Although, in many cases, one uses the DNS without being aware of it, because humans prefer to remember names and not IP addresses, it is possible to interactively query the DNS with the nslookup utility. The following is a sample session using the nslookup utility:

```
home.merit.edu(1): nslookup
Default Server:  merit.edu
Address:   35.42.1.42
> scanf.merit.edu
Server:  merit.edu
Address:   35.42.1.42
Name:    scanf.merit.edu
Address: 35.42.1.92
> 35.42.1.92
Server:  merit.edu
Address: 35.42.1.42
Name:   [35.42.1.92]
Address: 35.42.1.92
```

Thus, we can explicitly determine the address associated with a given host. Reverse name mapping is also possible with the DNS, as in this example:

```
home.merit.edu(2): traceroute ans.net
traceroute to ans.net (147.225.1.2), 30 hops max, 40 byte packets
1 t3peer (35.1.1.33) 11 ms 5 ms 5 ms
2 enss (35.1.1.1) 6 ms 6 ms 6 ms
   ................
9 192.77.154.1 (192.77.154.1) 51 ms 43 ms 49 ms
10 nis.ans.net (147.225.1.2) 53 ms 53 ms 46 ms
```

At each hop of the traceroute, the program attempts to do a reverse look-up through the DNS and displays the results when successful.

Although the DNS has served superlatively for the purpose it was developed (i.e., to allow maintenance of the namespace in a distributed fashion and to provide very rapid look-ups in the namespace), there are, of course, some limitations. Although there has been some discussion of including other types of information in the DNS, to find a given person at this time, assuming you know where she works, you have to use a combination of the DNS and finger to even make a stab at finding her. Also, the DNS has very limited search capabilities right now. The lack of search capabilities alone shows that we cannot provide a rich Directory Service through the DNS.

The X.500 Model of Directory Service

X.500 is a CCITT protocol that is designed to build a distributed, global directory. It offers the following features:

■ *Decentralized maintenance*—Each site running X.500 is responsible *only* for its local part of the directory, so updates and maintenance can be done instantly.

- *Powerful searching capabilities*—X.500 provides powerful searching facilities that allow users to construct arbitrarily complex queries.

- *Single global namespace*—Much like the DNS, X.500 provides a single homogeneous namespace to users. The X.500 namespace is more flexible and expandable than the DNS.

- *Structured information framework*—X.500 defines the information framework used in the directory, allowing local extensions.

- *Standards-based directory*—As X.500 can be used to build a standards-based directory, applications that require directory information (e-mail, automated resource locators, special-purpose directory tools, etc.) can access a planet's worth of information in a uniform manner, no matter where they are based or currently running.

How X.500 Works

The '88 version of the X.500 standard talks about three models required to build the X.500 Directory Service: the Directory Model, the Information Model, and the Security Model. In this section, we will provide a brief overview of the Directory and Information Models sufficient to explain the vast functionality of X.500.

THE INFORMATION MODEL. To illustrate the Information Model, we will first show how information is held in the Directory, then we will show what types of information can be held in the Directory, and then we will see how the information is arranged so that we can retrieve the desired pieces from the Directory.

Entries The primary construct holding information in the Directory is the *entry*. Each Directory entry contains information about one object; for example, a person, a computer network, or an organization. Each entry is built from a collection of *attributes*, each of which holds a single piece of information about the object. Some attributes that might be used to build an entry for a person would be "surname," "telephonenumber," "postaladdress," etc. Each attribute has an associated *attribute syntax*, which describes the type of data that attribute contains; for example, photo data, a time code, or a string of letters and numbers.

The attribute syntax for the surname attribute would be CaseIgnoreString, which would tell X.500 that surname could contain any string, and case would not matter; the attribute syntax for the telephonenumber attribute would be TelephoneNumber, which would specify that telephonenumber could contain a string composed of digits, dashes, parenthesis, and a plus sign. The attribute syntax for the title attribute would also be CaseIgnoreString. A good analogy in database terms for what we've seen so far might be to think of a Directory entry as a database record, an attribute as a field in that record, and an attribute syntax as a field type (decimal number, string) for a field in a record.

Object Classes   At this point in our description of the information model, we have no way of knowing what type of object a given entry represents. X.500 uses the concept of an *object class* to specify that information and an attribute named "objectClass," which each entry contains to specify to which object class(es) the entry belongs.

Each object class in X.500 has a definition that lists the set of mandatory attributes, which must be present, and a set of optional attributes, which might be present, in an entry of that class. A given object class A might be a subclass of another class B, in which case object class A inherits all the mandatory and optional attributes of B in addition to its own.

The object classes in X.500 are arranged in a hierarchical manner according to class inheritance. One major benefit of the object class concept is that it is, in many cases, very easy to create a new object class that is only a slight modification or extension of a previous class. For example, if I have already defined an object class for "person" that contains a person's name, phone number, address, and fax number, I can easily define an "Internet person" object class by defining "Internet person" as a subclass of "person," with the additional optional attribute of "e-mail address." Thus, in my definition of the "Internet Person" object class, all my "person" type attributes are inherited from "person."

X.500's Namespace   X.500 hierarchically organizes the namespace in the Directory Information Base (DIB); recall that this hierarchical organization is called the Directory Information Tree (DIT). Each entry in the DIB occupies a certain location in the DIT. An entry that has no children is called a *leaf entry*, and an entry that has children is called a *non-leaf node*. Each entry in the DIT contains one or more attributes that together comprise the Relative Distinguished Name (RDN) of that entry. There is a "root" entry (which has no attributes, a special case) that forms the base node of the DIT. The Distinguished Name of a specific entry is the sequence of RDNs of the entries on the path from the root entry to the entry in question.

Each entry in the RDN tree contains many attributes. As noted previously, any entry in the tree could use more than one attribute to build its RDN. X.500 also allows the use of alias names, so that the entry {C=US, o=Merit, cn=Chris Weider} could be also found through an alias entry such as {c=us, o=SRI, ou=FOX Project, cn=Drone 1} , which would point to the first entry. Note that for X.500 (and Lotus Notes). c = Country, o = Organization, ou = Organizational unit, and cn = Common name.

THE DIRECTORY MODEL. Now that we've seen what kinds of information can be kept in the Directory, we should look at how the Directory stores this information and how a Directory user accesses the information. There are two components of this model: a Directory User Agent (DUA), which accesses the Directory on behalf of a user, and the Directory System Agent (DSA), which can be viewed as holding a particular subset of the DIB and can also provide an access point to the Directory for a DUA.

Now, the entire DIB is distributed through the worldwide collection of DSAs that form the Directory, and the DSAs employ two techniques to allow this distribution to be transparent to the user: *chaining* and *referral.* The details of these two techniques would take up another page, so it suffices to say that, to each user, it appears that the entire global directory is on his or her desktop. (Of course, if the information requested is on the other side of the world, it might seem that the desktop directory is a bit slow for that request....)

The Functionality of X.500

To describe the functionality of X.500, we will need to separate three stages in the evolution of X.500:

1. The 1988 standard
2. X.500 as implemented in QUIPU
3. The 1992 standard

We will list some of the features described in the 1988 standard, show how they were implemented in QUIPU, and discuss the 1992 standard. The QUIPU implementation was chosen because it is widely used in the U.S. and European Directory Services Pilot projects and it works well.

FUNCTIONALITY IN X.500 (88). There are a number of advantages that the X.500 Directory accrues simply by virtue of the fact that it is distributed, not limited to a single machine. Among these are:

■ *An enormously large potential namespace*—Because the Directory is not limited to a single machine, many hundreds of machines can be used to store Directory entries.

■ *The ability to allow local administration of local data*—An organization or group can run a local DSA to master their information, facilitating much more accurate data throughout the Directory.

The functionality built into the X.500(88) standard includes:

■ *Advanced searching capabilities*—The Directory supports arbitrarily complex searches at an attribute level. As the object classes that a specific entry belongs to are maintained in the objectClass attribute, this also allows Directory searches for specific types of objects. Thus, one could search the c=US subtree for anyone with a last name beginning with S, who also has either a fax number in the (313) area code or an e-mail address ending in umich.edu. This feature of X.500 also helps to provide the basic functionality for a Yellow Pages service.

■ *A uniform namespace with local extensibility*—The Directory provides a uniform namespace, but local specialized directories can also be implemented. Locally defined extensions can include new object classes, new attributes, and new attribute types.

■ *Security issues*—The X.500 (88) standards define two types of security for Directory data: simple authentication (which uses passwords) and strong authentication (which uses cryptographic keys). Simple authentication has been widely implemented, but strong authentication has been less widely implemented. Each of these authentication techniques are invoked when a user or process attempts a Directory operation through a DUA.

In addition to the global benefits of the X.500 standard, there are many local benefits. One can use their local DSA for company or campus wide directory services; for example, the University of Michigan is providing all the campus directory services through X.500. The DUAs are available for a wide range of platforms, including X-Windows systems and Macintoshes.

FUNCTIONALITY ADDED BY QUIPU. Functionality beyond the X.500 (88) standard implemented by QUIPU includes:

■ Access control lists
■ Replication

Access Control Lists   An access control list is a way to provide security for each attribute of an entry. For example, each attribute in a given entry can be permitted for detect, compare, read, and modify permissions based on the reader's membership in various groups. For example, one can specify that some information in a given entry is public, some can be read only by members of the organization, and some can only be modified by the owner of the entry.

Replication   Replication provides a method whereby frequently accessed information in a DSA other than the local one can be kept by the local DSA on a "slave" basis, with updates of the "slave" data provided automatically by QUIPU from the "master" data residing on the foreign DSA. This provides alternate access points to that data, and can make searches and retrievals more rapid as there is much less overhead in the form or network transport.

Current Limitations of the X.500 Standard and Implementations

As flexible and forward looking as X.500 is, it certainly was not designed to solve everyone's needs for all time to come. X.500 is not a general-purpose database, nor is it a Data Base Management System (DBMS). X.500 defines no standards for output formats, and it certainly doesn't have a report-generation capability. The technical mechanisms are not yet in place for the Directory to contain information about itself, thus new attributes and new attribute types are rather slowly distributed (by hand).

Searches can be slow, for two reasons:

- Searches across a widely distributed portion of the namespace (c=US, for example) have a delay that is partially caused by network transmission times and can be compounded by implementations that cache the partial search returns until everyone has reported back.
- Some implementations are slow at searching anyway, and this is very sensitive to such things as processor speed and available swap space.

Another implementation "problem" is a trade-off with security for the Directory: Most implementations have an administrative limit on the amount of information that can be returned for a specific search. For example, if a search returns 1000 hits, 20 of those might be displayed, with the rest lost. Thus a person performing a large search might have to perform a number of small searches. This was implemented because an organization might want to make it hard to "troll" for the organization's entire database.

Also, there is at the moment no clear consensus on the ideal shape of the DIT, or on the idea structure of the object tree. This can make it hard to add to the current corpus of X.500 work, and the number of RFCs on various aspects of the X.500 deployment is growing monthly.

Despite this, however, X.500 is very good at what it was designed to do (i.e., to provide primary directory services and "resource location" for a wide band of types of information).

Things to be Added in X.500 (92)

The 1988 version of the X.500 standard proved to be quite sufficient to start building a Directory Service. However, many of the new functions implemented in QUIPU were necessary if the Directory were to function in a reasonable manner. X.500 (92) will include formalized and standardized versions of those advances, including:

■ A formalized replication procedure

■ Enhanced searching capacities

■ Formalization of access control mechanisms, including access control lists

Each of these will provide a richer Directory, but you don't have to wait for them! You can become part of the Directory today!

Some Current Applications of X.500

X.500 is filling Directory Services needs in a large number of countries. As a directory to locate people, it is provided in the U.S. as the White Pages Pilot Project, run by PSI, and in Europe under the PARADISE Project as a series of nation-wide pilots. It is also being used by the FOX Project in the United States to provide WHOIS services for people and networks and to provide directories of objects as disparate as NIC Profiles and a pilot K-12 Educators directory. It is also being investigated for its ability to provide resource location facilities and to provide source location for WAIS servers. In fact, in almost every area where one could imagine needing a directory service (particularly for distributed directory services), X.500 is either providing those services or being expanded to provide those services.

In particular, X.500 was envisioned by its creators as providing directory services for electronic mail, specifically for X.400. It is being used in this fashion today at the University of Michigan: Everyone at the University has a unified mail address (e.g., Chris.Weider@umich.edu). An X.500 server then reroutes that mail to the appropriate user's real mail address in a transparent fashion. Similarly, Sprint is using X.500 to administrate the address space for its internal X.400 mail systems.

Those of us working on X.500 feel that X.500's strengths lie in providing directory services for people and objects and in providing primary resource location for a large number of online services. We think that X.500 is a major component (though not the only one) of a global Yellow Pages service. We would also like to encourage each of you to join your national pilot projects; the more coverage we can get, the easier you will be able to find the people you need to contact.

LDAP: A Fast Way to Get Started on X.500

At the Electronic Messaging Association (EMA) '96 Conference, the X.500 talk was all about the LDAP (Lightweight Directory Access Protocol) direction, which is a simpler subset of the very complex X.500 protocol standard, DAP. Marc Andreessen, who gave one of the keynote addresses indicated that LDAP was the only way to go. Of course, Andreessen is the technology leader at Netscape. Lotus and IBM are also endorsing LDAP as the practical step towards X.500. LDAP does not require the full 7-layer OSI stack, as does X.500.

The Burton Group, in a May 1996 analysis report, stated: "The emergence of LDAP as the Internet standard protocol for directory services as a watershed event. LDAP will enable interoperability between clients and servers, and between servers, on the Internet. In addition, it will be the catalyst for directory interoperability between the Internet, intranet products based on Internet technologies, and existing NOS-based and application-specific directories used in today's organizations. The degree to which that interoperability is achieved, however, will be determined by how many vendors support version 3 of the protocol—which goes beyond the functionality of the current standard in RFC 1777—and Netscape's extensions beyond version 3."

The real momentum for LDAP started in April of 1996, when Netscape, Novell, Banyan, and more than 40 other companies and organizations announced support for LDAP as the open standard for directory services on the Internet. The proponents stated that the use of the protocol would allow customers to mix and match clients (directory user agents) and services (directory services agents) from multiple vendors.

The companies and organizations announcing support for LDAP at that time included AT&T, Computer Associates, Control Data Systems, Digital Equipment Corp., Hewlett-Packard, The Internet Factory, Macromedia, NCR, Network Applications Consortium, Software.com, Starfish Software, University of Michigan, VeriSign, and Yahoo!. Since then, many other companies have endorsed support for LDAP. Both IBM and Lotus have been active in their support for LDAP. The LDAP specification, as well as information on extensions to the current standard, can be obtained at the IETF Access, Searching, and Indexing of Directories (ASID) working group home page, http://www.ietf.cnri.reston.va.us/html.charters/asid-charter.html.

LDAP Defined

The University of Michigan developed LDAP in conjunction with the Internet Engineering Task Froce (IETF). LDAP 2 is a current Internet standard. Further extensions to the protocol are being formulated and will appear as LDAP 3.

LDAP is a simplification of the X.500 Directory Access Protocol (DAP). DAP was originally designed for use between clients (directory user agents) and servers (directory services agents). DAP creates so much overhead that it's not practical for use in the DOS and Windows client environments.

Thus, the University of Michigan developed LDAP as a streamlined way to access and update directory information in a client/server model. Version 1 of LDAP, defined in RFC 1777, is an Internet standard. LDAP v2 is an Internet Draft Standard, and further extensions to the protocol have been made in version 3, which is specified in an Internet Draft.

NOTE: *The progression for Internet standards is Internet Draft, Proposed Standard, Draft Standard, and then Internet Standard.*

The latest status on LDAP 3 can be found via the URL http://www.ietf.cnri.reston.va.us/html.charters/asid-charter.html, which is the same URL given above for the ASID home page.

With LDAP, applications can add, delete, and modify objects and their attributes in a directory database. One or more LDAP servers contain the data comprising the directory tree, and LDAP clients connect to an LDAP server to query or modify the contents of the tree.

LDAP does not require an X.500-compliant directory. The protocol can communicate with any hierarchical, attribute-based directory. For interoperability, LDAP assumes support for the X.500 naming model. For example, object classes include "country" and "organization" and generally follow the hierarchy defined by X.500. An X.500-compliant attribute syntax for LDAP is defined in RFC 1778.

LDAP also includes support for authentication. In RFC 1777, simple authentication (use of a clear text password) and Kerberos version 4 are supported. LDAP 3 will take advantage of X.509 strong authentication, which uses public-key security certificates. However, LDAP does not provide for standard access-control mechanisms. During mid-1997 there was much concern in the technical press about LDAP 3's lack of replication and synchronization capabilities. Replication would take changes to directory-entry attributes and propagate them to another directory, while synchronization would monitor for attribute changes and provide for automatic replication to keep all directories up to date (sounds like Lotus Notes functions!). These LDAP capabilities will be handled as separate drafts and would not be available until 1998 or later.

Under LDAP's client-server model, one or more LDAP servers contain the data comprising the directory tree. LDAP clients connect to an LDAP server to query or modify the contents of the tree. LDAP uses a single-master topology in that one single server is the master of the database; only the master can make changes in the directory. Multiple "slave" servers provide replicas of the entire directory database designed to balance the load of searches and access.

To handle failed queries, version 3 employs a referral capability that allows an LDAP server to refer a query to another server. The referral capability is also used to refer all write operations to the master server. In such cases, the referring server passes the name of the master server to the LDAP client (transparently to the user), which then connects to the master server. LDAP does not, however, allow multiple trees (in other words, multiple master servers) to learn about each other and their contents automatically. Referrals are static, based on entries for other servers manually made in the directory database.

The University of Michigan has been working on extensions to LDAP that would allow master servers to create indices of their contents and pass them on to other master servers. These so-called *forwarding indexes* would allow referrals to be made more dynamically, based on the nature of the client query and knowledge of the content of other trees.

There are several Internet drafts related to LDAP that will be implemented by some vendors that have decided to use the protocol. For example, the Internet draft "An LDAP URL Format" defines a URL syntax for issuing LDAP searches via an HTML browser. (See ftp://ds.internic.net/internet-drafts/draft-ietf-asid-ldap-format-03.txt.) The Internet draft "A MIME Content-Type for Directory Information" defines a MIME-compliant method of performing bulk import/export of directory information.

Novell Use of LDAP

Novell announced that it will support the LDAP specifications detailed in RFCs 1777 and 1778 in Novell Directory Services (NDS). Novell is developing a NetWare Loadable Module (NLM) that will allow any LDAP client to access and browse NDS. As announced, the NLM will allow the equivalent of a "guest" login to NDS, letting NetWare administrators make subsets of NDS publicly available over the Internet and available to anonymous users over intranets. The company made some LDAP support available in 1996.

Microsoft Use of LDAP

Microsoft has announced that it will support the version of LDAP specified in RFC 1777, and it outlined a road map for supporting LDAP in its products. For example, Microsoft pointed to third parties that provide MAPI drivers that allow the Exchange client to act as a DAP client to X.500 directories. An add-on to Exchange to support the Defense Messaging System (DMS) will include support for DAP, which will allow X.500 DAP clients to access the Exchange directory. The DMS add-on was shipped in 1996. Microsoft now supports LDAP in Exchange. Microsoft also added support for LDAP to Internet Explorer in 1996.

In addition, Microsoft said that it is building the service providers necessary to support LDAP under its Open Directory Services Interface (ODSI), a set of APIs designed to allow applications to work with multiple directories while using a consistent set of Windows interfaces. ODSI includes OLE DB and OLE DS and integrated login APIs that are part of Win32. ODSI service providers will allow the LDAP protocol to run under the ODSI APIs. For example, Microsoft is developing providers that will allow applications using both the OLE DB and OLE DS interfaces to access directories via LDAP.

Finally, Microsoft said that Windows NT Server Cairo will include LDAP support, allowing LDAP clients to access the NT directory via LDAP. A beta of Windows NT Server Cairo is due in 1997.

Lotus Use of LDAP

Lotus, also has announced support for LDAP. LDAP support was added to the SoftSwitch Directory product and the Lotus Messaging Switch in 1996. Domino 4.6 has both IMAP4 and LDAP. LotusMail also supports

IMAP4 and LDAP with Notes 4.6. Domino 5.0 and Lotus Notes Client 5.0, which will be formally available in late 1997, will support the full Internet version of mail and directory, adding HTML, IIOP, MIME, NNTP, and ICAP support in addition to IMAP4 and LDAP.

LDAP and the Future of Directory Services

LDAP is significant for both the Internet, where public directory access and interoperability are a pressing issue, and the intranet, where directory integration is a significant customer problem. There will never be a single Internet directory, for example. On the contrary, there will be many directories for the Internet, including both "white" and "yellow" page listings hosted by communications companies, e-mail address directories, and directories published by individual corporations.

Public directories on the Internet will also vary in how they're accessed. Some will be completely open to anonymous browsing, just as you anonymously browse many Web sites today. Others will be available only via secure login. Moreover, public directories must be interoperable, capable of exchanging information on an as-needed basis, both with each other and with private intranet directories. For example, an organization might want to pull a subset of a public directory (say entries for its business partners) down from a public directory and into its private intranet directory. In all of these cases, a standard directory access protocol is a baseline requirement.

Similar levels of interoperability are already necessary on corporate intranets. As we've already discussed, the integration of application-specific directories, Network Operating System (NOS)-based directories, and now intranet directories continues to be one of the biggest problems facing corporate customers. To be effective, intranet products must become interoperable with existing NOS-based and application-specific directories.

The emergence of LDAP as the Internet directory standard, is good news on all fronts. Internet standards are driving the development of enterprise networking products, so the same directory standards will work in both the Internet and intranet environments. LDAP is playing that role, providing a simple access protocol that can work on many levels, both for the Internet and for LAN-based directories. LDAP is well-suited to that role because it delivers the benefits of X.500—interoperability being the most significant—while eliminating the inefficiencies associated with full implementations of the standard.

Through the efforts of the University of Michigan and the Network Applications Consortium (of which the university is a member), LDAP emerged over the last few years as the leading directory standard. With

AT&T, Banyan, Novell, Sun, and other vendors joining Netscape to announce support for it, LDAP's position as the Internet/intranet directory standard is firmly established. In particular, Netscape's current clout has given LDAP the credibility it needed. Netscape's relationship with the University of Michigan ensures that LDAP will remain an important part of the company's enterprise intranet strategy. That relationship also ensures that Netscape will drive the LDAP standard in a market-driven fashion as its product line matures. The entire industry will benefit because directory development can now begin in earnest, based on an interoperable standard.

However, it's important to understand the differences in the LDAP protocol and the LDAP API.

NOTE: *For clarity, we mean the protocol when we use the term "LDAP"; we will use the term "LDAP API" when referring to the API.*

As a protocol, LDAP will see widespread use. On the Internet, for example, LDAP will allow users to browse public directories, both anonymously and through secure logins. Through the efforts to integrate LDAP with DNS and Web indexing engines—and the development of the URL LDAP syntax—browser users will be able to transparently access both directory and Web site content.

On corporate intranets, LDAP will be an increasingly important standard, driven primarily by Netscape's use of the protocol in several critical roles. For example, Netscape will support LDAP as the basic Internet directory browsing protocol, just as Novell has done. However, Netscape has gone much further, using the protocol to enable a simple yet effective replication model between directory servers.

Although the use of LDAP as a replication protocol isn't explicitly specified in the LDAP standard, the basic operations LDAP enables (such as reading, comparing, writing, modifying, and deleting data) are the basic operations required for the creation and maintenance of replicas. Netscape's use of LDAP for replication, then, is a new use of the protocol rather than a function explicitly specified in the standard.

NOTE: *That might change in future versions of the specification as the need for more functionality causes the addition of replication-specific capabilities.*

Today, most of the vendors that have announced support for LDAP are using it as a browsing protocol, not as a replication protocol. However, once other vendors support LDAP as a replication protocol, directories from multiple vendors will be able to interoperate. Netscape is also supporting and pushing the development of standards-based import/export methods that will further ease the directory integration problems that customers face.

In addition, Netscape is using LDAP in a way that could well encourage a more widespread use of existing directories by applications, particularly server applications that require a directory to operate. E-mail systems, for example, need a directory in which to store user names and e-mail addresses.

Developers of such systems have long been forced to build their own directories for two basic reasons. First is the lack of a standard directory API, which requires developers to either become dependent on a particular vendor's directory product or write multiple versions of their applications, each for a different directory. Second and less obvious is the lack of market penetration. Developers simply can't take it for granted that customers already have a general-purpose directory service they could use instead. Less than 20% of all installed network operating systems use a general-purpose directory, for example. Developers simply have to create a directory in order to ensure that their products will run, right out of the box.

Netscape, of course, faces exactly the same problem with its intranet server products; the Netscape Enterprise, Fast Track, Mail, and News Servers must be capable of working out of the box. However, Netscape also wants to provide directory interoperability. The company is using both LDAP and the LDAP API to address these problems in an elegant fashion, and it's an example other developers should follow.

Future versions of Netscape's Web, mail and other server products will come with an embedded LDAP server, usable only by each of those servers. For example, a version of the Enterprise Server will use an embedded LDAP server to build its own directory, allowing it be installed and used without requiring the purchase and installation of a separate directory. However, if a general-purpose directory that fully supports LDAP is present, any of Netscape's servers can work with that directory. The application-specific LDAP implementation can replicate its contents to the more general-purpose directory. As a result, administrators will be able to manage the server from the general-purpose directory instead of an application-specific directory. LDAP, then, gives developers a completely vendor-neutral way to solve their immediate directory problems, while ensuring that their products can be integrated with more general-purpose directory products if and when necessary.

Netscape has certainly been the leader in support of LDAP. Although Novell, Microsoft, Lotus, and other vendors have announced support for the protocol, initially that support was largely limited to client browsing access. However, Netscape's efforts have set an example for the industry, and the growing momentum behind Internet/intranet technologies will make such support an imperative for other vendors.

The LDAP API also stands a significant chance of gaining widespread use. Though simple, the LDAP API supports many of the functions that applications need on a day-to-day basis. It also has the luxury of being usable. Key components of Microsoft's ODSI initiative (OLE DS and OLE DB in particular) are still works in progress, but the LDAP API is usable today. Because of its cross-platform nature, the LDAP API is a better fit with the "universal client" nature of the HTML browser. Netscape plans to have a beta software development kit out in the same time frame as the release of the Netscape Directory Server, and once LDAP API is exposed via Java class libraries and JavaScript, it will be a valuable tool.

However, it's important to understand that, while the LDAP API is a competitor to portions of ODSI—OLE DS in particular—LDAP the protocol is a complement to ODSI. Microsoft's efforts to create drivers that allow LDAP to run under the ODSI APIs, including OLE DS, will ensure that applications written to ODSI will be able to run over LDAP and work with any LDAP-compliant directory. The choice between APIs, then, will not force a choice in protocols. The choice between APIs will largely be made based on how Windows-specific or how cross-platform developers want their applications to be. As a Windows-specific API set, for example, ODSI will be tightly integrated with Windows, while the LDAP API will be more cross-platform in nature.

The choice between APIs will also be made on the basis of functionality. OLE DS is designed to enable two kinds of basic functions. First, it allows applications to browse and modify directory objects. Second, OLE DS is also designed to allow applications to operate on the resources that are represented by directory objects. In addition to adding, deleting, or modifying a printer object in a directory, for example, OLE DS is also designed to allow an application to perform an operation on the actual printer that the object in the directory represents. The LDAP API is specifically designed to enable the former, but not the latter, functionality. Developers creating applications designed for resource-specific administrative functions will likely find OLE DS attractive, then, while developers needing basic look-up, login, and authentication functions will find the LDAP API useful.

Of course there are downsides to LDAP, since LDAP does have its limitations. RFC 1777, for example, doesn't support strong authentication, and some developers find the search model limited. Also, LDAP doesn't support multimaster replication or the ability for multiple masters to learn about and communicate with each other automatically. These limitations could affect any implementation of the protocol. The lack of automatic communication between multiple masters is a significant weakness of the standard. Thus, we can expect vendors to extend the standard in order to build competitive products. One of the biggest issues with LDAP, then, is the level of interoperability vendors will be able to achieve. With too many additions, LDAP might become HDAP, the "heavyweight protocol"! Not adding features to the core protocol is important to keep directories simple.

Netscape is basing much of its work on LDAP version 3 and extensions that go beyond the current draft of version 3. LDAP's referral capabilities are not part of RFC 1777, for example, but are in version 3. Access controls aren't supported in any current draft of the standard, but they are an extension Netscape is making in the Netscape Directory Server.

The issue of standards extension underscores these points: The market almost always moves faster than standards bodies, and vendors are forced to extend standards to remain competitive. Netscape has taken precisely that approach with HTTP and HTML, and it is taking the same approach to LDAP. For example, the Netscape Directory Server would be severely handicapped without access controls, so Netscape is adding that functionality to its implementation of the product in order to produce a competitive product.

Interoperability will depend on both the degree to which Netscape publishes its extensions and on the degree to which Microsoft, Novell, and other vendors support version 3 of the protocol and track Netscape's extensions to it. Although none of the vendors announcing support for LDAP have explicitly stated that they will support version 3 and Netscape's extensions to the standard, the Netscape's market clout will create pressure to do just that. Netscape has also made it clear that it intends to publish its extensions, and it will, in fact, work to enable most of the extensions through the Internet standards process.

Overall, LDAP will guarantee at least basic levels of interoperability, which is a major step forward in directory services. The Internet, by providing both a market imperative and an economic incentive, is driving progress in network computing, and that's good news.

User Impact

It appears that the majority of the network marketplace has yet to be convinced that directory services are strategic products. Few customers view directory services as substantially more than either an electronic mail directory or a file and print administrative utility.

This low penetration is due to two basic factors: Few applications exploit directory services today, and few directories interoperate to any significant degree. These factors can in turn be attributed to the simple fact that no directory standards (at either the API or protocol level) have emerged. As a result, the market for directory services has remained relatively small when compared with the network operating system market.

However, the emergence of LDAP as a standard (and Netscape's entry into the market with Netscape Directory Server) should change things. As companies build intranets and link those intranets with the Internet, the need for interoperable directory services and directory/application integration will increase exponentially. By providing both a standard protocol and a standard API, LDAP should jump-start things in a big way, providing the foundation on which directory interoperability and integration can be built.

From a server aspect, previously there was neither an economic incentive nor a market imperative for vendors to solve the interoperability problems customers have faced for so long. However, the Internet now provides that incentive, and the vendors that don't effectively leverage the Internet will lose. The incredible pace of the market has eliminated the ability to stall, making it impossible for vendors to avoid addressing the LDAP issue in short order. As a result of these factors, LDAP, driven largely by Netscape, has emerged as the directory standard for which the market has been looking.

It will still take some time before products and applications show up in full force, and it will take another year beyond that for large shifts in directory usage patterns to show up in quantitative market studies. However, the shift to directory-enabled computing has occurred, catalyzed by Netscape and LDAP.

The Future

Directory integration is a critical problem for many customers, and LDAP has emerged as the most promising standard solution to that problem. In short, LDAP is a standard for which you should be demanding support from your vendors, no matter who they are. NOS vendors and application-server vendors alike should view LDAP as the common ground on which they can come together, eliminating the directory conflicts that cause so much duplication of effort on today's networks. LDAP is emerging as the Internet standard on which many directories are based, and support for it is quickly becoming an imperative.

It's important, however, for customers to understand the reality of what LDAP will bring and their role in making the promise of the standard an implementable reality. The LDAP API, for example, is relatively simple, so the first generation of applications based on it won't make extensive use of directories.

Support for LDAP as a replication protocol between servers, version 3 of the protocol, and Netscape's extensions beyond version 3 will be important baselines for interoperability, and all product vendors should be encouraged to commit to that baseline.

SoftSwitch Directory Publisher

SoftSwitch Directory Publisher is a suite of directory products that provide flexibility and modularity to Lotus customers seeking to establish an enterprise directory based on the X.500 standard. It provides synchronization of data from the Lotus Messaging Switch Names Directory to an X.500-compliant directory server.

SoftSwitch Directory Publisher replaces the "Lotus Pages" directory effort that was also to be an X.500 offering. However, Lotus Pages was to be limited to the LMS platform. SoftSwitch Directory Publisher will run on most of the popular platforms.

Novell Directory Services (NDS)

Novell Directory Services (NDS), which is the new name for NetWare Directory Services, has become a very popular directory service. However, it is not X.500, so that's a problem. Novell would, of course, like to have NDS as the standard for the Internet. Because it is popular, SoftSwitch Directory Publisher will interoperate with NDS.

X.500 Directory Technology for Lotus Messaging Switch

Lotus Development Corporation has licensed X.500 directory technology for use with the Lotus Messaging Switch. Formerly Soft •Switch EMX, the Lotus Messaging Switch is the industry's leading enterprise network backbone switch used to link different electronic messaging systems. LMS has employed an X.500-architected directory for several years but has not used the actual X.500 protocols, called Directory System Protocol (DSP) and Directory Access Protocol (DAP). Lotus is now adding support for these protocols to LMS.

X.500 defines common protocols for accessing globally distributed directories containing information such as electronic mail addresses and other data that can be accessed by electronic messaging systems and other networking applications. The X.500 standard for distributed directories is published by the International Telecommunications Union, a worldwide standards body.

X.500 protocols allow the Lotus X.500 directory service to interoperate with other X.500 directories and will allow the Lotus X.500 directory to be accessed by X.500-based Directory User Agents (DUAs). Lotus will work with suppliers of X.500 DUAs to assure DUAs used by customers operate properly with the Lotus X.500 implementation.

Lotus licensed the X.500 directory services software from Unisys Corp., which has been developing it since 1987. The Unisys software has proven its interoperability in extensive conformance testing as well as an ongoing pilot project among five large utility companies managed by the Corporation for Open Systems.

The Unisys software—called TransIT 500 Directory Services—features technology from the latest version of the standard, X.500(93), including directory replication. In addition, it provides the ability to access a relational database for storing directory information and includes a Simple Network Management Protocol interface, which allows the software to interoperate with the popular network management standard.

Company-Wide Name and Address Book

IBM has long had a company-wide directory on their mainframes that is accessed by the IBM CallUp program. This directory provides very quick access to any IBM employee's telephone number, e-mail address, internal and external mailing address, manager's name, and other important information. The directory offers "fuzzy" search capability and allows you to search by area, state, country, or the whole world. There is also the ability to generate an e-mail message after finding the right name.

All of this functionality is not available with a lot of separate Lotus Notes Name and Address Books. The office support section of IBM's ISSC division (now IBM Global Services) came up with a company-wide Lotus Notes N&A Book that gave many of the features of the directory under CallUp. This consists of a "consolidated" N&A Book called the "IBM Name and Address Book," which is often referred to by its Lotus

Notes file name IBM.NSF. The single N&A Book consists of approximately 70,000 names.

There are performance questions with such a consolidated N&A Book. The information in each entry for this Name and Address Book was cut to the minimum (i.e., full name, short name, and Notes address). Its main purpose is to allow Notes domains to concatenate this Name and Address Book with the domain N&A Book. Then users in the domain can use the "address" feature in Lotus Notes to find almost anyone with a Notes address in IBM.

However, things change! With IBM's new Notes architecture, the IBM.NSF N&A Book was renamed to "Legacy Notes Domains." As of June of 1997, there are still about 70,000 names in IBM.NSF, but that number will steadily go down with the migration to the new architecture. Currently I still have an ADVANTIS user ID, so I'm listed in the IBM.NSF N&A Book with that "Legacy" entry, but I also have an IBMUS user ID, so I'm also listed in that new domain. By the end of 1997, all ADVANTIS domain users will be migrated to IBMUS users. At that time, I'll have given up my user ID in the ADVANTIS domain.

In IBM's new Notes architecture, the Public Address Books (such as IBMUS) only contain the information that Notes requires to route mail and some basic employee information, such as phone number, fax number, etc. The Public Address Books are maintained with a minimum of employee information in order to provide reasonable database performance. The IBMUS Public Address Book is over one gigabyte in size with the minimum employee information! In order to provide employees with the information they were used to under the OV/VM CallUp program, a Notes application called Directory-on-Notes (DoN) was developed. DoN duplicates much of the directory function and data that had been available when users had mail on the VM platform.

The Directory-on-Notes data about people and departments is much more extensive than that in the Public Address Book. While it uses standard Notes database technology, its function as a directory access tool has been optimized so that it does not exhibit the performance problems of the Public Address Book when scaled to over 300,000 entries. The DoN database is built through the use of a feed from the master CallUp repository on VM. This process can be controlled much more tightly than the one that produces Notes Public Address Books because there is no technical need for many administrators to have update authority. Therefore, the DoN database is considered to be a secure source of employee and department data and, through the use of standard Notes APIs, can be accessed by local or corporate-wide applications.

Although DoN provides rich employee directory data, its current function does not provide seamless integration between the search results and the Notes mail facility. Users cannot perform on-the-fly queries against the DoN database to populate the address fields in a mail message. Because IBM employees had this ability with OV/VM, it was imperative that the Lotus Notes solution provide similar access to enterprise mail addresses from within Notes, and this required that the Public Address Books from all mail domains be accessible by the mail servers.

To simplify management and to enable their specification in the NOTES.INI file, IBM devised a plan to consolidate the NAMES.NSF files from the 131 mail domains (one for each country in which IBM does business) into four "super books." Geographic alignment was selected because IBMers send most mail to others in their country or region. The resulting databases were AP.NSF (Asia/Pacific), EMEA.NSF (Europe/Middle East/Africa), LA.NSF (Latin America), and NA.NSF (North America).

To improve the performance of these multi-domain Public Address Books, all nonperson related views and documents were removed, and the indexing parameters were tuned. In addition, the IBM.NSF Legacy Domain Address Book is used by IBMers in the strategic domains (e.g., the IBMUS domain) to send mail to users in legacy domains (e.g., the ADVANTIS domain). By 1998, the legacy domains within IBM will have very few mail users; however, in the meantime, much effort is being spent in the migration effort!

Address Book Synchronization

Once you've brought an address book to Lotus Notes from another system (e.g., the mainframe), how do you keep these address books synchronized? One answer to this question is to use an Address Book Synchronization (ABS) product. Here's what IBM does.

Every night, the IBM Address Book Synchronization/2 product captures all of the mainframe directory change records and downloads them to a directory manager server. There, those records that also belong to Notes users are extracted and imported into the IBM.NSF database using the Lotus Notes API.

All of this processing is centralized (in Atlanta, GA) to ensure integrity and to minimize risk. ABS/2 is also the tool used to keep the four geographic super books, mentioned in the previous section, in synch with the CallUp directory files on VM. ABS/2 supports both the CallUp and Lotus Notes directory formats by default, so little customization was done other than to configure the source and target directory names.

10

Developing Corporate Standards for Lotus Notes

Standards: Everyone needs these just to compete.
Gartner Group, December, 1996.

This section discusses Local Area Network Standards, Standards for the Wide area Network, Lotus Notes Naming Standards, and Administrative Standards.

Local Area Network Design Standards

In February of 1990, an IBM LAN Council was formed to define LAN standards, implementation guidelines, and product requirements for the internal IBM community. These guidelines and standards were discussed and developed at on-going meetings since that time. In addition to the internal IBM community, the documentation resulting from the IBM LAN Council has been distributed to many IBM customers.

Two important areas where the LAN Council developed standards and guidelines are LAN topology and naming and addressing. These two areas are described in this section.

LAN Topology Recommendations

LAN topology addresses the designs needed to meet LAN requirements. These LAN requirements might range from a small departmental LAN to a Wide Area Network providing LAN connectivity to widely separated users.

LAN SEGMENT CLASSES. The LAN Council defined three classes of LAN segments (A segment is a single token-ring, FDDI ring, or Ethernet. Segments are interconnected via bridges or routers.) These classes are from the perspective of a central Information Systems (I/S) support group. Note that these classes have nothing to do with TCP/IP classes of address spaces. The three classes defined by the LAN Council are:

■ *Class A segment*—I/S has complete responsibility and control over these LAN segments. No customer devices are allowed to be attached to a class A segment. The following are possible examples of these types of segments:

     Campus and backbone segments
     Gateway segments
     Isolation segments (for intersite communications)

- *Class B segment*—I/S has responsibility for installation and support of these segments. Customer devices are attached to these segments. The following are possible examples of these types of segments:
  Local segments (for office connections)
  Manufacturing floor segments

- *Class C segment*—The independent department is responsible for installation and support of these segments. I/S is responsible for the management of any connections between these segments and class A or B segments. The independent department is responsible for funding the equipment for the segments and connections. The following are possible examples of these types of segments:
  LAB segments (for development or testing)
  Manufacturing line Segments
  Other nonsupported LANs

CLASS A AND B TOPOLOGIES. For LANs of greater than three segments, a hierarchic backbone structure is recommended. This type of structure provides the best configuration in terms of flexibility, growth, network response time, capacity management, and routing control.

SINGLE-LEVEL BACKBONE. The simplest of these structures is shown in Figure 10.1. This structure would be used for a small campus consisting of less than six buildings. This design consists of multiple local segments that are bridged to a pair of backbone segments. It is strongly recommended that paired backbone segments be installed to provide for high availability and fault tolerance.

This parallel backbone segment structure serves two purposes. First, if there is a single failure of a bridge or a backbone segment, there will be an alternate path through the network. In addition, if source routing (Token-Ring is the primary LAN that uses source routing) is being used, the parallel segments will provide load distribution. The traffic through the network will tend to be distributed over the two parallel backbones.

The local segments can be designed to cover a specific area, such as floor, or can be spread between floors. It is recommended that local segments be designed geographically instead of departmentally. In general, departments tend to be located together; therefore, most of the traffic between requester and server will be on the same local segment.

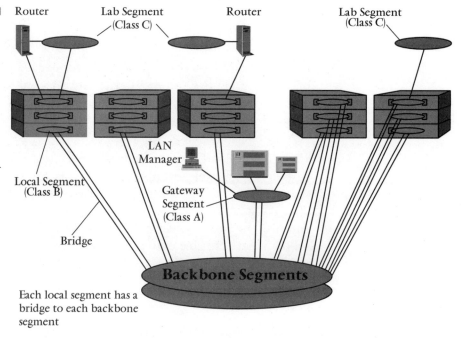

Figure 10.1
Single-level backbone
design. Parallel
backbone segments
connect the local
segments together.
The multiple layer
design provides high
availability and
throughput.

TWO-LEVEL BACKBONE. For a campus with six or more buildings, a two-level backbone structure is recommended. This design is shown in Figure 10.2.

The local segments in multiple buildings are connected to parallel campus segments via two bridges. The campus segments are connected via bridges to parallel backbone segments. Notice that the top campus segment is only connected to the top backbone segment and likewise for the bottom layer. This provides for two independent paths through the network. One might be tempted to connect the campus segments to both backbones or to connect the campus and backbone segments together. This is not recommended because it will result in additional broadcast traffic with no gain in availability. (Dual redundancy will handle any single failure. Triple redundancy will handle any two failures and so on. Other variations on the dual redundancy design simply change which two items must fail to cause an outage.)

The main reason for the campus segment is for problem isolation and backbone traffic reduction. The backbone segment only terminates in a few buildings. Campus segments in other buildings are extended to the closest building that has the backbone. If the campus segments were eliminated, the backbone segments would have more repeaters and terminations. This would tend to decrease reliability. Also, in many buildings, communication will tend to be primarily within the building or

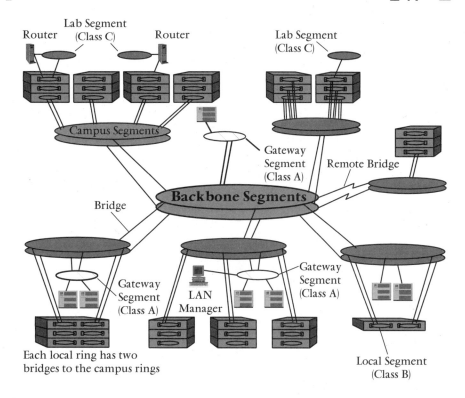

Router Lab Segment (Class C) Router Lab Segment (Class C)

Campus Segments

Gateway Segment (Class A)

Remote Bridge

Backbone Segments

Bridge

Gateway Segment (Class A)

LAN Manager

Gateway Segment (Class A)

Each local ring has two bridges to the campus rings

Local Segment (Class B)

set of buildings. The campus segments support this traffic without using the backbone segments.

GATEWAY SEGMENTS.   Gateway segments are used to connect mainframe gateways (e.g., IBM 3745, 3174, and 3172) and terminal servers (e.g., IBM 3174s for coax-attached terminals). These segments are bridged either to the nearest campus or backbone segment.

The purpose of these segments is twofold. First, they allow isolation of these connections for security and reliability. Second, they provide alternate routing though the network because they are bridged to both pairs of campus/backbone segments. Because these are class A segments, they only have devices on them that are supported by the I/S group. Placing gateways for the same protocols on the same LAN segment also allows for better traffic control and management through the use of protocol sensitive bridge filters. Experience has shown that the separation of protocols on these gateway rings (segments) greatly reduces gateway failures and enhances through-put and availability of these devices.

CONNECTING CLASS C SEGMENTS.   In a development environment, it will become necessary to connect lab segments to the production LAN. These segments are called *class C segments*. In many cases, the development groups want complete flexibility in configuring their lab environments and might require connectivity to the rest of the network. This requirement raises the issue of how to provide this connectivity with minimal cost and minimal risk to the production network.

There are two choices of how to connect to the network: bridge or router. If the lab only requires a single protocol like TCP/IP or OSI, a router should be used to connect to the local network. The lab network should have a different IP network address from the main production network. If a separate IP network address is not required, a bridge should be used. In some cases, multiple protocols are required on the lab segment, thus a bridge (with filters) and a router should be used.

There are two choices of where to connect a class C segment to the network: local segments or campus/backbone segments. The recommendation is that connections from the production network to the class C segments be via local segments. These routers or bridges should be managed by the I/S group. Connections to other segments in the lab would be managed by the independent department. If a bridge connection is made to a lab, the hop count settings of the network might have to be changed.

INCREASED BANDWIDTH FLEXIBILITY. As faster speed LANs become available, these recommended structures allow easy increases in the capacity of the network. As network usage increases and the utilization of the backbone and campus segments increases, the backbone and campus segments can be upgraded to a higher speed. This can be done without impacting the use of the network. Users will perceive no change to the environment. To upgrade a local segment, however, all of the users must have the adapters needed for the faster speed. This change requires coordination, but only has to be done once per segment.

INTERSITE CONNECTIVITY. There are a number of different methodologies that can be implemented for intersite LAN connectivity, each having its own set of limitations and protocols supported.

The LAN protocols most commonly in use within IBM are: SNA/SNI, LEN (PU 2.1/LU 6.2), APPN, NetBIOS, IPX, and TCP/IP. The recommendation is that the use of multi-protocol routers be considered for interconnection of a large number of LANs. Many multi- protocol routers (e.g., IBM 6611, Cisco, and Bay Networks) provide for the transportation of all of the protocols previously listed. Remote-bridge technology will remain the simplest and most economical way to interconnect a small number of LANs at different sites.

Standards for the Wide Area Network

In the United States, IBM uses two major internal wide -area networks. One is based around the Systems Network Architecture (SNA) protocol. This network links both LAN attached workstations and non-programmable terminals to IBM's central applications for order entry, accounts receivable, configuration aids, electronic mail, bulk data, and file transfer.

A second internal network, implemented in 1991, is the Multi-Protocol Network (MPN). The main requirement for this network is for TCP/IP transport, which is in wide use by IBM's scientific and engineering personnel and increasingly by employees who access the Internet and Lotus Notes. MPN uses multi-protocol routers connected by T1 circuits. There are currently six MPN backbone nodes in the U.S. Each node has routers with token ring connectivity, which provides an interface to host computers and bridges. User sites in both U.S. and international locations ordinarily connect directly to backbone routers via T1 circuits, although connectivity via a remote bridge is an option.

The SNA network and MPN provide the major wide-area connectivity between LANs in the IBM Corporation and the LAN Council naming standards are a requirement before LANs are allowed to connect to these wide area networks. We have found these standards and guidelines and the registration process based on the standards to be an absolute necessity for the smooth operation of our internal networks.

The SNA network and MPN are now operated and managed by the Advantis Company, which was formed at the end of 1992 by IBM and the Sears Corporation. The IBM LAN Council has provided guidelines for all aspects of LANs, including central LAN Management, measurements, software distribution and security.

When Your Wide Area Network is TCP/IP Router Based

Many corporate networks are router based with TCP/IP as the major protocol supported. IBM's internal MPN (Multi-Protocol Network) is this type of network, and we have found that it lends itself very well to supporting a corporate- wide Lotus Notes system. Because TCP/IP has become the most popular protocol over wide area data networks, there are many hardware and software products available to support the administration and management of such networks. IBM uses a variety of routers in its networks (e.g., IBM, Cisco, and Bay Networks), which are all managed by NetView/6000 systems using the Simple Network Management Protocol (SNMP). The OSI standards based management protocol CMIP (Common Management Information Protocol) is becoming more popula, and IBM's network management systems are evolving in this direction.

Some of the naming/addressing standards used by IBM for this type of network are discussed later in this chapter in the section "Local Area Network Naming Standards."

When the Internet is Your Wide Area Network

The Internet can be thought of as a public wide area network that is TCP/IP router based. TCP/IP is the only protocol used on the Internet. IBM's OpenNet is used for providing IBM Internet access to our commercial customers, and the OpenNet can be thought as IBM's part of the Internet.

The OpenNet is designed around routers interconnected with T1 circuits. SNMP (Simple Network Management Protocol) is used for the management of the OpenNet. NetSP firewalls provide security between the OpenNet and private networks (including IBM's internal networks).

When XPC Dial Connects Your Notes Servers over the Wide Area

The use of XPC dial is a very common way of quickly setting up a Lotus Notes network over the wide area. IBM Consulting, for example, very quickly set up a worldwide Lotus Notes system based around an "Intellectual Capital" application.

XPC dial was used to reach many of the locations outside the United States. As the Lotus Notes database grew, replication over XPC dial grew to be a problem because of the increasing connect time required. The IBM Consulting Lotus Notes system has now migrated mostly to using TCP/IP over IBM's internal network for it's worldwide reach. However, XPC dial always remains a way to get to those remote worldwide locations. As long as you can contact your customer or colleague via the telephone, you can also use XPC dial to let that same customer or colleague have access to your Lotus Notes application.

The use of XPC dial does not create any specific requirements for Lotus Notes standards over the wide area network. It can always be thought as a backup for other wide area access methods or as the one way to reach any Notes user anywhere. With the increasing use of wireless communications, including cellular modems, XPC dial access literally allows Notes access to anyone, anywhere.

Local Area Network Naming Standards

This section first describes general LAN naming standards and guidelines used by IBM and then discusses some of the specific standards that apply to Lotus Notes. The idea behind this is to describe what IBM has done with naming standards as input into what you might find useful for your own networks and Lotus Notes system. It might be that only a part of these standards prove useful to you. In any event, it should prove useful to find out what IBM has found useful in developing naming standards for our internal and commercial networks.

LAN Naming/Addressing Standards

The general idea behind the IBM LAN naming/addressing standards is to keep the corporate standards down to the minimum necessary to prevent duplicates. The different corporate business areas then can expand on these minimum requirements as they see fit.

When possible, the process has been to register a part of the name (usually the prefix) rather than try to define the whole name as a standard format. The different business areas can then define a rigid format if they want, as long as they use the registered prefix. Examples of how some of the IBM business areas have expanded on the basic standards are given in this book.

The following Corporate Wide Standards have been established by the IBM LAN Council and must be followed by all users who intend to communicate with other networks using TCP/IP, SNA (APPN, LEN), NetBIOS, or IPX protocols over the IBM internal networks. These standards are controlled and administered by Advantis. Registry requests are sent to a registry ID via IBM internal e-Mail. Registration for the NetBIOS (or site) and SNA prefixes are handled on a "first-come, first-served" basis. The recommendation is to make the prefix meaningful if possible (e.g., RAL is the site prefix registered for the IBM Raleigh site).

NETBIOS: THREE-CHARACTER SITE PREFIX. Used by Domain and Server Names that will be used for inter-site communications.

NetBIOS is a well-accepted, proprietary IBM protocol that operates at the OSI transport and session layer on top of the IEEE 802.2 Logical Link Control.

NetBIOS uses names to identify users/applications in the network. NetBIOS names are 16-byte (unstructured) names that need to be unique across an entire LAN implementation. To allow delegation of administrative functions to local organizations, a 3- character prefix is used.

```
NetBIOS-name = LLLXXXXXXXXXXXXX
```

where LLL is the prefix assigned to the IBM organization responsible for defining (local) NetBIOS names and X...X is a 13-byte string.

NOTE: *Most products only allow the use of 8-character NetBIOS-names. The names are padded by those products with default character strings.*

The NetBIOS name prefixes are managed centrally by Advantis.

SNA/APPN: TWO-CHARACTER PREFIX FOR LUS AND NETID REGISTRY. In an SNA environment, end users and programs communicate via Logical Units (LUs). An LU is identified by a network qualified name (NETID, LU-name). Because some products do not support network-qualified LU names, both NETID and LU-name need to be unique. Naming standards for NETIDs and LU-names have already been defined and agreed to, separately, and are summarized here:

```
NETID = CCEEEESS
```

where CC is the ISO country code, EEEE is the identification of the enterprise, and SS is the suffix.

```
LU-name = AANNNNNN
```

where AA is the unique two-character prefix assigned to the IBM organization responsible for the resource and NNNNN is an alphanumeric suffix.

USIBMCD0, is an example of a NETID used for the backbone IBM SNA network.

NETIDs are registered in the IBM SNA NETID Registry (IBM Networking Systems, Raleigh). LU-name prefixes are managed centrally by Advantis.

Names for other SNA network resources (e.g., PU-name, Exchange Identification (XID), line names, etc.). need only to be unique within the resource owning the network.

> ■■ ■■ ■■ ■■ ■■ ■■ ■■ ■■ ■■ ■■ ■■ ■■ ■■ ■■ ■■
>
> **NOTE:**   *Particular applications exploit SNA LU 6.2 sessions across a LAN. An example of such an application is Remote Data Base Manager. The administrative organization that assigns LU names for these applications can be different from the organization that is responsible for SNA subarea networking at that location. All departments that can assign LU names need to be aware of the LU naming standard and need to know which LU-prefix they should use.*

TCP/IP: INTERNET ADDRESSES AND DOMAIN NAMES.   These names and addresses must be unique, must follow standard form, and must be registered centrally.

IBM has an IP Class A address of 9.0.0.0. Advantis registers and manages the Class B and Class C subnets created off of the "9." address.

Subnet names must be unique within a network. For example, within the network level of IBM.COM, there must be unique subnet names, such as RAL.IBM.COM. This requirement creates a registration requirement for subnet names within IBM. These subnet names should reflect some geographic reference. This is the only consideration given to subnet name content.

Within a subnet, there should also be a central registration facility for Internet names. This can be provided on a Domain Name Server (DNS) that is accessible by the users.

Name servers can be made to "talk" to each other across subnets, simplifying the registration process between subnets. Routing statements and a routing function should also be provided to the users as required for connectivity outside the subnet.

RING SEGMENT NUMBER REGISTRY.   Source routing, the protocol used by Token Ring bridges, does not tolerate duplicate segment numbers. For this reason, groups of segment numbers are registered for backbone and isolation segment numbers and are assigned uniquely by Advantis. They are assigned from the range 100 to 7FF. Other segment numbers are assigned locally from the ranges 001 to 0FF and 800 to FFF.

LOCALLY ADMINISTERED ADDRESS REGISTRY. Communication between adapters on a local area network requires that each adapter be recognized by a unique 12-digit hexadecimal address. These adapters come with a Universally Administered Address (UAA) that is permanently encoded in the microcode on the adapter. Blocks of these addresses are assigned to each manufacturer by the IEEE, which ensures uniqueness.

IBM local area networks allow the UAA to be overridden by a 12-digit hexadecimal Locally Administered Address (LAA). The LAN Council recommends the use of UAAs wherever possible, but certain products require the use of LAAs, and in some cases, LAAs are much more convenient. Most IBM products require the LAA to be in the hexadecimal form "4000 XXXX XXXX."

To prevent LAA conflicts for devices on backbone or isolation rings, a registry is provided for ranges of LAAs. These ranges are of the following form:

nnnnyxxxzzzz

where *yxxx is* the only portion of the LAA registered.

In the previous address, *y* designates the type of LAN device, and it is recommended that the *xxx* be the tie line for IBM sites or the area code for other sites.

LOTUS NOTES, AS/400, AND WORDPERFECT OFFICE. Use a 3-character site prefix for all of these names. The other characters are up to the business area. See the section "LAN Naming/Addressing Guidelines" later in this chapter for examples of how some IBM sites are doing this.

NOVELL NETWARE/IPX NAMES. Use a 3-character NetBIOS site prefix for generating NetWare server names (i.e., the names are of the form "LLLxxxxx" where "LLL" is the registered 3-character prefix). The *xxxxx* is up to the area. See the section "LAN Naming/Addressing Guidelines" later in this chapter for examples of how some sites are using this.

External addresses for the Novell servers will follow a convention based on the IP address for each site. The convention is to use each octet of the IP address and express it in hexadecimal notation. For example, the IP address 255.255.255.0 would be expressed by the 8 hexadecimal digits FFFFFF00.

IBM has an IP Class A address of 9.0.0.0, and groups within IBM, such as the Detroit marketing area, are given a Class B subnet off of this Class A address. The Detroit marketing area was given 9.27 as its Class B address. The external network address will be the 9.27.0.0 expressed in hexadecimal, or 091B0000, which each server should specify within the Detroit area.

If routers are used, each site within the Detroit area would implement a Class C address with a subnet mask of 255.255.252.0. Then the Novell external address would be changed to reflect the third octet as well. For example, a Class C range of 9.27.32.1 to 9.27.35.255 for a site would dictate an external address for the server of 091B2000.

Internal addresses should follow a scheme where the Class C-based external address is used, adding one to the last octet for each server's internal address. In the previous example for the Detroit area, 091B2001 would be used for the first internal address, 091B2002 for the second, etc.

LAN Naming/Addressing Guidelines

Guidelines have been established to provide a model for the definition of LAN names and addresses. They are *not* intended to be required formats by the IBM LAN Council. They are to serve as a format guide for users who are looking to set up naming schemes for their LAN environment. The guidelines are, for the most part, examples of how some of the business areas have extended the LAN Council naming standards to fill in the unspecified parts of a name.

NOVELL NETWARE, AS/400, AND WORDPERFECT OFFICE NAMING GUIDELINES. The LAN Council standard is to use the 3-character NetBIOS site prefix for generating NetWare "internal" names. As a guideline, IBM Dallas uses "NN" following the prefix to indicate Novell NetWare.

IBM Branch Offices follow the convention:

```
preSnNOV
```

where pre is the NetBIOS prefix, S denotes server, n denotes the number of the server, and NOV denotes Novell.

The standard for AS/400 and WordPerfect Office names is to use the 3-character site prefix. As a guideline, IBM Dallas follows the prefix with "AS" for the AS/400, and "WPO" for WordPerfect Office.

LOTUS NOTES NAMING GUIDELINES. These guidelines are intended for use by LAN and Lotus Notes administrators for IBM Branch Offices. For LAN administrators, it provides guidelines on naming. For Notes administrators, it provides guidelines on naming as well as the use of Notes' security and replication features.

LAN SERVER NAME The recommendation is to use the convention:

LLLABCCC

where:

■ LLL is the 3-character NetBIOS prefix per the IBM LAN Council Standard. The NetBIOS name prefixes are administered centrally by Advantis.

■ A is the resource type (S=file server, N=Lotus Notes Server, G=gateway, D=domain controller).

■ B is the sequence number (0 to 9) for each resource of the same type at the same location.

■ CCC is the 3-character branch or department number.

LAN DOMAIN NAME The recommendation is the same as for the server name, except that, for a domain controller, the resource type is D.

USER NAMES Users are free to choose names using clear English that are descriptive and meaningful to other users. It is suggested that users choose their VM name as their LAN and Notes (if applicable) user names.

LAN user name conflicts will be found and corrected by LAN administrators as new users are added to each LAN. Notes users at sites that are networked together will share a common replicated address book. Notes ensures that each name is unique as it is added to this address book.

Recommendations

Well-documented standards and guidelines for LANs are an absolute must when LANs are interconnected on a large-scale basis, as is the general rule in Corporate America. Without these standards, the interconnection of LANs becomes a problem.

We have found that it is better to provide a registration process for part of the LAN name (e.g., the prefix) rather than try to define the whole name in a standard format.

A separate corporate resource, such as the IBM LAN Council, is an effective way to:

- Provide uniform design recommendations.
- Establish standards and guidelines.
- Provide centralized administration for naming and addressing.
- Act as a "clearing house" for implementers to compare, share, and develop solutions for LAN problems.

Using the Prefix of the Name

The prefix of the NetBIOS name can be used for routing. Over IBM's internal T1 router-based network (MPN), for example, routers will only bridge NetBIOS traffic over the wide area network when the NetBIOS names have valid prefixes. This means that, from White Plains, only NetBIOS names with a prefix of SCH will be allowed when NetBIOS traffic is to be bridged to Schaumburg, IL; only NetBIOS names with a prefix of TAP or TPA will allow traffic to be bridged to Tampa, FL, etc.

TCP/IP Addresses and Domains

TCP/IP addresses and domains within IBM and Advantis use standard conventions based on site. For the ADVANTIS.COM TCP/IP domain, for example, users at different sites have the following TCP/IP addresses:

- WHP.ADVANTIS.COM—White Plains
- SCH.ADVANTIS.COM—Schaumburg
- BLD.ADVANTIS.COM—Boulder
- TAP.ADVANTIS.COM—Tampa

Notes Domain Names

Notes domain names *must* be unique. That doesn't mean, of course, that it's impossible to have two domains with the same name. It's only when those domains are to be interconnected that problems occur. For example, early on in the growth of IBM's internal Notes system, one of the IBM Marketing Areas chose to call their domain "IBM." This domain was connected to the IBM_INTERNAL hub domain. Later on, two other IBM domains with the name "IBM" requested connection to

IBM_INTERNAL. Of course that was not possible. Mail routing is based on domain name. With two "IBM" domains connected to IBM_INTERNAL, mail sent to "John Doe@IBM" would always be routed to the first "IBM" encountered during a search of the connection records. This is one of the types of problems encountered by "bottoms-up" Notes design in a company. Several small groups thought they would use the domain name "IBM."

The LAN Council recommendation is to use "IBM" as the prefix for the Lotus Notes domain name. For example, one of the domain names for IBM in the United Kingdom is IBMUK. Examples of other Lotus Notes domain names currently used within the IBM internal network are: IBMJAPAN, IBMNORDIC for the IBM Nordic countries, IBMISSC for the IBM Integrated Systems Services Corporation, and IBM_M&S for the IBM Marketing and Services group in the U.S.

NOTE: *As part of IBM's corporate move to completely move to Lotus Notes, the Lotus Notes domain name standard was defined as an expanded version of the LAN Council recommendation. The standard, developed in the fall of 1995, specifies a domain name in the form "IBMxxn," where "xx" is to be the 2-character CCITT country designator and "n" is a sequence number for those countries where IBM will require multiple domain names. For example, IBMNZ was designated as the Notes domain name for IBM New Zealand (NZ is the CCITT designator for New Zealand). Notes domain names for IBM in the United States would have the names IBMUS1, IBMUS2, etc., although it appears that a single domain name IBMUS, will be sufficient.*

Notes Server Names

The recommendation is to use the convention:

`LLLABCCC`

where:

- LLL is the 3-character NetBIOS prefix per the IBM LAN Council Standard. The NetBIOS name prefixes are administered centrally by Advantis.

- A is the resource type (S=file server, N=Lotus Notes Server, G=gateway, D=domain controller).

- B is the designator for each resource of the same type at the same location.

- CCC is the 3-character branch or department number.

Because the server name is only a recommendation, many IBM sites use the site prefix for the first three characters of a server name and then modify the recommendation for the remaining characters. The registered 3-character prefix will assure that the server name does not conflict with a server name at another site, and this is the important thing. Some examples of Notes server names used within IBM that follow the complete recommendation are in the IBMISSC Notes domain:

- STFNMI01 and STFNMI02 are Lotus Notes mail servers at the IBM ISSC Sterling Forest, NY, site (STF is the site designator, N is for Notes, M is for mail).

- ALANHI00 is a Lotus Notes hub server at IBM ISSC in Atlanta, GA.

NOTE: *As part of IBM's corporate move to completely move to Lotus Notes, the Lotus Notes server name standard was designated as a modified form of the LAN Council recommendation. The Notes server name is defined with the site prefix followed by characters that designate the server. Because new Notes domains defined within IBM will have hierarchical certificates, the characters designating the server name will also be hierarchical and have Organizational Unit and Organizational elements.*

Administrative Standards

Because Notes might be just one of several network applications that are administered for your network, many of the administrative standards for Notes should follow the general standards set up for the rest of your network. If Notes is the first (and only) application being used on your network, then other network applications should follow the general standards set up for Notes. Of course, each specific network application will have its specific administrative standards. However, there is no reason, other than scale, why the Notes administrator cannot also be the administrator for other network applications (such as cc:Mail, LAN Server, etc.).

Centralized Administration

The two main factors that will ultimately affect server placement (centralized versus decentralized servers) are administrative costs to support decentralized servers and network traffic costs created by centralizing servers. Centralized server placement minimizes support resource costs.

If all servers were to be placed in a central location, a high level of dedicated server support could be expected. Also, it costs less to administer multiple servers in one location than in multiple locations. Additionally, server maintenance and troubleshooting is centralized and can be more rigorously controlled.

User Names and IDs

One major consideration with Lotus Notes user names is whether or not to use a middle initial. The general practice for IBM Notes domains has been not to use middle initials. That will probably change as more users are migrated to Notes and the need to differentiate users with very similar names within a Notes domains becomes a factor. Of course, even with your middle initial defined, you will also be recognized by Notes without your middle initial. For example, even with the middle initial, I would still be listed in the ADVANTIS N&A book as "John Lamb," "John P. Lamb," and "John P. Lamb/White Plains/Advantis/US."

The advantage of not using middle initials is simplicity. It's easier to type "John Lamb" rather than "John P. Lamb." The disadvantage is less differentiation between users. Actually, on the side of not using middle initials is the fact that some Notes add-on products get confused with the "." after the initial. Our SMTP gateway would not route mail to a "david l. reich@watson.ibm.com" because of the period after the initial.

The Notes "Shortname"

The Lotus Notes "shortname" can be very useful. The obvious use is to make it very easy to send mail to a user within your domain. It's only necessary to type his short name in the "To:" area of your message. The default shortname will be the user's first initial plus his or her last name, where the last name will be truncated after the shortname reaches eight characters.

Because all IBM users have (or had) mainframe mail, the Lotus Notes shortname was often selected by IBM Notes domain administrators as the users mainframe mail ID, which is limited to eight characters and usually is selected by the user. Often a user selected his mainframe mail ID so that it contained his last name (e.g., my mail ID is LAMB) and sometimes as a nickname (e.g., RANDY). In any event, it was a mail ID that the user might have had for many years. This was important because the Gateway between the mainframe and Notes used by IBM (the IMLG/2 product) uses the Notes shortname to route mail from the mainframe to Notes. Because the directory on the mainframe makes it easy to find the mainframe mail ID (or User ID) for anyone in IBM, it is also easy to know how to send mail to that user on Notes.

Password Standards

There are two aspects to password standards. One is defining what they should be (e.g., minimum length, how often they must be changed, etc.). The other aspect is how to enforce the standards you have decided on. The Notes product itself has limited function for enforcement. Version 4.1 (and higher) allow you to limit the number of failed attempts to enter a password.

Recovering Lost IDs or Passwords

Lotus Notes is different than other LAN systems in its concept of user ID and password recovery. In most other LAN systems, passwords are stored centrally on a LAN server, and the LAN administrator can "reset" a lost password. With Notes, the password is stored with the USER.ID, which a Notes user ordinarily keeps as a file on his workstation or as a file on a floppy disk. Thus the USER.ID file is not available to a central LAN administrator, and there is no concept of a LAN administrator "resetting" a lost password.

There are two common options for the LAN Administrator to deal with the lost password. One is to delete the user's entry in the Name and Address Book and to reregister the user, after having made a copy of the user's mail file so that no mail will be lost. A second option is to keep a central file of all the original USER.IDs registered for your domain. These USER.IDs would all have the original password and can be given to the user with the last password. That's the method we use for the ADVANTIS Lotus Notes domain.

Passwords for a Domino Web Server

A Domino Server requires a user ID and password when a user accesses a Notes database from a Web browser if the default ACL for that database is "no access." To change, or initially "set" this password, go to the N&A book on your Notes/Domino server. Edit (e.g., Ctrl—E) your entry in that N&A book. You'll notice there is an entry "HTTP password" right below your shortname. This entry might only be in the copy of the N&A book on your Domino server because the template is unique for Domino servers.

If you already have a password, the entry will appear as "gibberish" because it's encoded. If you don't have a password, it will be blank. If you don't have a password, you won't have access to a Notes database, which requires one (i.e., no password, no access). To change your password, "swipe" the gibberish with your mouse pointer, and then type in your new password. The password is case sensitive. There is one advantage to this password over a Notes password: If you forget it, you can always go and reset it. However, nobody can see it, because it gets encoded as soon as you save it.

There is one other security consideration when dealing with database access from a Web Browser. The user ID that you enter must match the user name in the ACL for the Notes database. Most of our Notes databases in our ADVANTIS domain had our fully distinguished names listed in the ACL (or group in the ACL). Without changes, we'd have to enter that fully distinguished name (e.g., Terry Steilen/Schaumburg/ Advantis/US) for a user ID. The Domino documentation recommends adding the common names to the ACL.

In the ADVANTIS domain we added common names to the ACL so that the ACL had both our fully distinguished and common names (just like our regular entries in the N&A book). As you start to use your Domino server more, you might decide to start using just common names in your ACL lists. However, remember that the fully distinguished names are more secure, so there is a possible trade-off between convenience and security.

Recent Updates to IBM's Standards for Lotus Notes used Internally

Standards are a very important component of a successful Notes implementation. Standardization is a valuable tool for preventing problems or inconsistencies as organizational use of Notes grows. The following section reviews corporate naming standards for use within IBM's implementation of Lotus Notes.

By establishing rules on how to implement changes to key system components, administrators can eliminate or minimize service interruptions. This is particularly important in a distributed architecture where a number of different individuals will provide a consistent system management service.

Every effort has been made to incorporate existing IBM standards that might have application within Notes.

Domain Names

A Notes domain is a group of servers that share the same public Name and Address book. A domain can span multiple geographies and LAN types. For administrators, the domain concept provides an easy, centralized method for management of servers within the domain. Users are assigned to a domain based upon the location of their mail file.

To facilitate a consistent and global method for addressing mail, IBM's internal Notes domain name strategy will be based on "IBM" as leading characters, the standard CCITT country code, and an optional sequence number. The sequence number will be used only when there is more than one mail domain within a country. For example:

    IBMGB2

where IBM are the lead characters, GB is the CCITT for the UK, and *2* is the optional sequencer (second of two domains).

Network Names

Lotus Notes uses the Network Name in the Name & Address book for two purposes:

- To group servers together for presentation to clients in the File|Open dialog box. Only servers that have the same network name as the workstation's home server are shown in this dialog box.

- To route mail directly to other servers on the same named network without using connection records.

Servers on the same network name should be physically available to each other through the network and share a common protocol. Servers that access each other via remote connection (e.g., dial-up) should use different network names.

In a large network with multiple domains and/or Network names, it is important to standardize on a descriptive network naming convention. Doing so will assist administrative personnel with the overall management of the system. It will also benefit programs or automated tools that might need to specify these names.

Connection Records

Server connections for replication and remote mail routing are defined by connection records in the Name and Address book.

The naming standard used in Name and Address book connection records will be a combination of the conventions used for:

- Domain naming
- Network naming
- Server naming

User Group Names

PUBLIC GROUP NAMES. Public group names are group names that reside in a public N&A book. They are accessible to both servers and clients.

Group names are used in three ways:

- As e-mail distribution lists
- For access control
- Dual-purpose

Group names that are used for ACL purposes must reside in the first Name and Address book in the daisy chain list on the server.

To facilitate system management, application development, and client ease of use, it is important to establish a public group naming convention.

RESERVED PUBLIC GROUP NAMES. "Reserved" groups are special group names set aside for system or administration use only. The following reserved group names are recommended for use within IBM:

- AdminGlobal—All administrators with global administrator rights.
- Administrators—List of all administrators within IBM. The "Administrators" group is assumed to include administrator groups for the various sites: Admin{Site Name 1}, Admin{Site Name 2}, through Admin{Site Name 24}.
- Contractors—List of all members of IBM's complementary work force. The "Contractors" group is assumed to include groups for all domains: Contr{Domain Name 1}, Contr{Domain Name 2}, through Contr{Domain Name 131}.

- LocalDomainServers—Names of all servers in a given domain.
- OtherDomainServers—Names of all servers in other domains.
- Terminations—Individuals that have left the company. The "Terminations" groups are assumed to include groups for all other domains: D01 Terminations, D02 Terminations, through D27 Terminations.

NOTE: *Termination maintenance occurs at the domain/site level. Because individuals are certified to access servers cross-domain, termination names need to be present in all domain deny access lists.*

STANDARD PUBLIC GROUP NAMES. Most individuals who use group names do so based on three pieces of information: name, organization, and geography. To assist users in locating and using group names, it is proposed that the following syntax be used when creating group names: Group name-Organization-Region.

Group name will usually be comprised of an application indicator and will often include further delineation of roles within the application. For example, "Registry Authors" would easily convey the application and the intended ACL authority of the group. Application specific "roles" can also be used. For example, "MktAutomotive" or "MktManufacturing" can be used to indicate sectors of users of a Market Management application.

Interpretation of the Organization component is flexible, depending on the nature of the user group being defined. It is expected that this will most often be a divisional indicator; for example, "SNS" for Sales and Services. It can also be used to indicate external company names. Organization code should be included in the user group name if the group is intended to apply to a single organization. If the particular group being defined is cross organization, then specifying this component would only add to the number of overall groups being defined. Given that the number of groups allowable is limited within Notes, the overall emphasis is on limiting the number of groups defined, and therefore this component should be excluded.

Likewise, the Region component is useful in defining the geographic scope user group but should be used judiciously. For example, if the geographic scope is limited only to the European countries, then include this component. However, if the user group being defined spans Europe and North America, then omit this component, as it would require an additional group to make the distinction. For user groups that are worldwide in scope, then including "WW" is useful as it will help administrators easily identifying the need for the group.

The hierarchy makes it easy to search for entire groups or organizations of people. For example:

- ICAP Best Practices Editor-ICG-WW
- ICAP Best Practices Reader-ICG-WW

STANDARD PUBLIC GROUP NAMES RULES. The following rules should be adhered to when dealing with group names:

- Do not use server names as group names.
- Do not include the characters underscore ("_"), ampersand ("&"), or tilde ("~"). The dash ("-") should also not be used as it is reserved for delimiting the component parts.
- Can vary in length. While brevity is important, recognition of meaning is more critical.
- Do not use "reserved" group names as "standard" group names
- "Standard" group names should be unique across domains. This will help facilitate consolidation of N&A books.

To ensure adherence to standards and to avoid security problems, ISSC will create group name shells and the owner will modify its members.

User Names

User names in the IBM environment will be hierarchical and follow one of two formats, depending on whether the user is a regular employee or a member of the complementary work force:

- Regular: User name/location/IBM@ Domain Name
- Contractor: User name/location/Contr/IBM@ Domain Name

Briefly, a hierarchical name is composed of a set of discrete naming elements that, together, unambiguously identify each person and resource in the Lotus Notes network. The rightmost portion of the name is the domain, which is separated from the hierarchical name by an "at" symbol ("@"). To the left of the "@" are naming elements that become more granular until the actual user name is reached. Traditionally, a hierarchical name can contain six elements: the Organization (O), up to four Organizational Units (OUs), and the User Name or Common Name. IBM has determined that no more than two Organizational Units will be used for brevity and simplicity. Thus, a hierarchical Notes name in the IBM environment will never be any longer than:

```
User name/OU1/OU2/O @ Domain
```

Following are the detailed specifications of user names within IBM.

The "user name" portion of the hierarchical name will be comprised of the individual's full first name and last name:

```
FirstName LastName
```

Optionally, the client can elect to include a middle initial. Note that a period should *not* be used after the middle initial:

```
FirstName M LastName
```

Use of a period might yield unexpected results when running applications that work with the name field.

The first and last name should comply with the current corporate data standard for these fields. Compliance with this standard will ensure consistency from an applications perspective.

The overall name should be the same as the Internet user name conforming to the new IBM Internet naming standards for mail (User Name@*cc*.ibm.com). The "*cc*" being the same country code as the IBM GNA Notes mail domain.

Other reference names, such as nicknames, will be applied as aliases in the Notes N&A book person document.

The "location" portion of the hierarchical name, otherwise known as the Organizational Unit, should reflect the physical work location of the user. This information will be used to create a meaningful e-mail address. Moreover, it will often be the only element differentiating two people with the same user name, so accuracy is important.

The Organization element of the IBM hierarchical user name will always be "IBM."

The following is an example of a fully qualified regular employee user name in the IBM environment:

```
Susan User/Dallas/IBM @ IBMUS
```

For contractors and other complementary personnel, a second Organizational Unit, Contr, is added to the hierarchical name:

```
Tom R User/Dallas/Contr/IBM @ IBMUS
```

Server Names

Server names are used with connection records, server records, access control lists, user group and desktop icons, and as components in the hierarchical naming scheme. Due to the size of IBM's installation, special thought needs to be given to how server names will be used so that the naming standard built will enhance this environment.

The data access strategy calls for clients to access data primarily though a navigational tool. The need for direct server access through the Notes browser screen should diminish over time. As a result, the requirement for server naming will focus on easing administration and overall system organization.

The key points used for developing a server naming standard were that it be:

- Descriptive to allow quick identification of server location.

- Unique across IBM domains worldwide.

- Consistent to facilitate application development.

The common name portion of the hierarchical server name should follow the format:

```
Daabbbbb
```

where D is the lead character, aa is the site code, and bbbbb is the server name.

Like user names, server names will be hierarchical and have Organizational Unit and Organization elements. The Organizational Unit elements for server names are of the form:

```
Daabbbbb/aa/xx/IBM
```

where aa is the site code, bbbbb is the server name, and xx is one of the following:=A (for application), M (for Mail), H (for Hub), or G (for Gateway).

For example, my mail server in the IBMUS domain on the SP2 in Poughkeepsie, NY, has the name D01ML007/01/M/IBM.

This naming standard allows generic ACL entries. For example the entry "*/H/IBM" in the server "can access" field of the N&A Book will allow all Hub servers to access all other Hub servers.

CHAPTER 11

Migration to Lotus Notes from Existing Mail Systems

IBM faced a significant challenge in migrating 230,000 users
from Office Vision to Lotus Notes.
                    Albert Schneider, Lotusphere '97, January 28, 1997

This chapter will discuss some of the experiences IBM has had in migrating from existing mail systems and some of the recommendations that come from those experiences. One issue that must be considered in a migration is the different types of platforms involved. For example, you might be running a mix of hardware platforms such as IBM-compatible PCs, Macintosh computers, and RISC-based workstations. The operating systems might include Windows, OS/2, and Unix.

Recommendations on How to Do It

When migrating from an existing mail system to Lotus Notes, two areas that need to be addressed are user education in the use of Notes and the design of a Notes support system.

Education

From our experiences, formal education in the use of Lotus Notes is an important requirement for a successful migration to Lotus Notes. Some of the elements of a successful education environment are:

■ Basic Lotus Notes end user education. 1 day. For all Notes users.

■ Lotus Notes Implementation/Administration. 2 to 3 days. Includes Network aspects of Lotus Notes.

■ Lotus Notes Application Development. 4 to 5 days. This is both for the programmers concentrating on writing Lotus Notes applications and for end users interested in how it all works.

■ Appropriate hardware for clients.

Setting up a Notes Support Organization

When you purchase Notes, you will understandably want users installed and running as quickly as possible. However, before you begin, you should have an internal user support organization trained to implement and maintain Notes. This might take some extra time initially, but it will help ensure that your Notes implementation will be efficiently installed and supported.

Most sites already have user support organizations in place. If this is the case with your site, this chapter will help your group prepare for Notes. If your site does not have an existing support group, you could outsource it or have an organization, such as Lotus Consulting Group or IBM Global Services, help you organize and staff one.

If planned properly, Notes does not require a great deal of special training or expertise to implement and support. Notes maintenance consists largely of dealing with typical network issues, rather than specific Notes concerns. An experienced support staff might need little additional training or personnel to maintain Notes, particularly if the site already uses an existing LAN-based application.

The responsibilities of a typical Notes support organization can include:

- Testing and troubleshooting the installed Notes system.
- Administering Notes (creating new user IDs, maintaining the Name and Address book, backing up servers, and so on).
- Responding to user problems.
- Training new users.
- Defining, developing, and testing new Notes applications.
- Coordinating tasks with remote sites.

The number of individuals responsible for these tasks depends largely on the size and complexity of your Notes implementation. A large Notes user community might require several people dedicated to each task. Smaller Notes systems might assign a single individual per task, or even multiple tasks to one person. Also, Notes implementations with geographically remote sites might require local administration and training.

The suggestions and options we discuss in this chapter are intended as guidelines. Some of these might apply directly to your Notes implementation. Others you can modify and adapt to your particular needs.

A typical Notes support organization can include the positions listed in Table 11.1.

NOTES SUPPORT ENGINEER. The Notes support engineer is your primary authority on Notes installation and troubleshooting. The support engineer is responsible for installing and testing Notes hardware and software. This includes the initial Notes implementation and subsequent additions to accommodate a growing Notes community. In corpo-

TABLE 11.1

Positions in a
typical Notes
support
organization

Position	Primary responsibilities
Notes Support Engineer	Install Notes hardware and software. Test and troubleshoot Notes. Address user problems. Contact Lotus phone support for assistance, as required.
Notes Administrator	Perform Notes server maintenance tasks, including running Database Fixup, managing disk space, and backing up files. Assign user Ids to new Notes users. Maintain public Name and Address book.
Application Designer	Help define, create, and test Notes applications. These can be databases, or developed with the API.
Notes Trainer	Train user on Notes. Teach basic application design. Train users on internally created Notes applications.
Remote Site Coordinator	Coordinate implementation/administration/training efforts of remote Notes sites.

rations that include remote sites, this might often require travel across the country or abroad.

The support engineer is also responsible for troubleshooting Notes problems encountered by Notes administrators and users. These problems can be reported directly to the Notes support engineer. The support engineer will contact Lotus if further information is needed. We recommend the following professional experience for a Notes support engineer:

■ Two to five years administering network hardware and software, especially LANs. Ideally, this experience would include networks on which Notes runs (Novell, NETBIOS, Banyan VINES, and so on).

■ Modems. These are among the most common causes of reported user problems and can be particularly difficult to identify and fix.

■ IBM PC, PS/2, or compatible systems.

■ Windows or Presentation Manager.

As you can see, this experience is not Notes-specific. Most support organizations will already include individuals with these qualifications. In many cases, these individuals need not be dedicated to Notes. Instead, they will support Notes in addition to other systems and applications.

In addition, the ideal support engineer will demonstrate the following personal traits when dealing with reported problems:

■ Good listener (understands what is meant, not just what is said)

■ Good communicator

- Tactful (even seemingly self-evident solutions must be explained with diplomacy—users must never feel reluctant or embarrassed to report problems)
- Patient
- Flexible

The number of support engineers you need to support your Notes community depends on several variables. These include the number and experience of Notes users, complexity of network topology, geographic diversity of sites, and number and complexity of Notes applications.

Lotus offers a number of training courses to help your Notes engineers become expert with Notes. These include the Certified Lotus Notes Engineer (CLNE) program.

NOTES ADMINISTRATOR. The Notes administrator is responsible for maintaining Notes users and servers. These tasks can include loading Notes and creating accounts for new users and servers. They also include server maintenance, such as using the Notes server log to monitor server activity, managing calling schedules, backing up servers, and managing disk space.

The Notes administrator might also be responsible for creating replicas of databases on servers. (Some sites might allow users to do this. You should decide your policy on this as part of the site-planning process.)

In most cases, the Notes administrator is not a dedicated full-time position. Instead, the Notes administrator is an experienced end user who has assumed these responsibilities. In addition, Notes administrators should be familiar with Windows and OS/2. The Lotus Notes Administrator's Guide describes all responsibilities of the Notes administrator in detail. Lotus also offers the Introductory Systems Administration (3-day) training course for Notes administrators.

The Notes administrator will often respond to user problems. The administrator can troubleshoot some problems or work with the support engineer when assistance is needed. In some cases, the Notes administrator will refer a user directly to the support engineer.

As with other positions within your support organization, the number of Notes administrators required to support your Notes community depends on several factors. These include the number and experience of users, number and location of servers, and similar issues. A typical Notes administrator might be responsible for up to five servers, each serving 35 to 100+ users.

If your Notes community includes several remote sites, each site regardless of size should include one person designated as Notes administrator.

We highly recommend that you assign one Notes administrator responsibility for creating Notes user IDs and making the corresponding entry in the public Name and Address book. This will help ensure that new users are properly registered and will minimize the chance of incorrect or duplicate entries.

APPLICATION DESIGNER.   The application designer is responsible for developing Notes applications. These applications include Notes databases. They might also include programs that use the Notes API.

The *Site and Systems Planning Guide* (Part No. 12348) describes the Notes applications development and roll-out processes. The application designer's role in these processes can vary. For example, users might develop their own simple applications but request assistance from the application designer for more complex ones, or application designers might be responsible for developing all applications.

Application designers should have previous experience with applications programming. If the application designer will use the Notes API, experience with C language is required. (See the *Site and Systems Planning Guide*, which discusses the API.) Lotus offers the following training for application designers:

- Introductory Applications Development (3 days, no prerequisites)
- Applications Development II (3 days, attendees must have attended Introductory Applications Development or have equivalent skills)
- API (1 day, attendees must have attended at least one of the previous courses)

NOTES TRAINER.   The Notes trainer is responsible for training new users on Notes. This includes how to use the standard Notes features, such as electronic mail and document editing. It also includes basic database application design. (Lotus offers training in these areas. However, it might not be convenient for all users at a large Notes site to participate in this training.) In addition, the Notes trainer will explain how to use new applications created by your application designers.

Notes trainers should be experienced end user training professionals. Most sites will already have a training staff in place. One or more of these individuals can receive Notes training from Lotus, and then train their own users.

REMOTE-SITE COORDINATOR.   As mentioned previously, each remote Notes site should include at least one Notes administrator. You should assign one or more remote site coordinators to standardize the efforts of these remote Notes administrators.

The remote-site coordinator's responsibilities include coordinating replication strategies, maintaining standard maintenance and training procedures, ensuring distribution of the latest software, and similar responsibilities. Experience should be similar to that required for the Notes administrator.

How Lotus Rolled Out Notes

In August of 1990, senior Lotus management began the drive for corporate-wide Notes roll-out. Their objective was to make Lotus the best example of how to implement Notes worldwide, using it as a platform for better communication among traditional and nontraditional work groups. The goal was to complete Notes roll-out within one year.

One of the first problems Lotus needed to address was hardware. At the start of the process, Lotus was not fully networked. Virtually none of the North American field offices were wired. In addition, computers in Cambridge, U.S. field offices, and International were not sufficiently configured to run Notes. Plus, Lotus had in place an obsolete electronic mail system from which users would be gradually converted to Notes mail.

Of course, as the developer of Notes, Lotus began roll-out with previous Notes experience. There were already a small number of scattered Notes users and servers in the company. Their Internal Support Services (ISS) group maintained three servers, and several others existed in Notes Development itself. There were also a few other isolated Notes servers and users in Lotus. These were subject to varying levels of administration within disparate work groups. Once Lotus began roll-out, these isolated servers proved challenging to security and uniformity of administration.

Organizing a Notes Support Group

The initial focus was to organize and develop a Notes roll-out and support team. When Lotus started, they had in place an existing End User Computing group. This was the central organization within ISS that dealt with PC-related issues. They had a broad assortment of PC hardware and software skills, with a focus on Lotus products. The group consisted of approximately 100 people; their responsibilities included software applications (including Notes), hardware repair, and training on Lotus products. The entire effort was headed by a Director of End User Computing.

In preparation for the Notes roll-out, Lotus reorganized the group into four sections:

- Notes Operations and Administration
- Notes Training
- Notes Applications
- Roll-out

NOTES OPERATIONS AND ADMINISTRATION.   This group was responsible for designing the server architecture and establishing standard Notes administrative procedures. This group needed to build the Notes "roadway" capable of handling over 3000 geographically dispersed, multinational users. Initially, this group numbered five people.

NOTES TRAINING.   Prior to Notes roll-out, this group had trained Lotus users on a broad assortment of Lotus products. In preparation for Notes roll-out, they were refocused on Notes. They developed a series of courses on Notes for users in Cambridge and North America sales offices. This group consisted of four people.

NOTES APPLICATIONS.   This group was created specifically for Notes roll-out. It was staffed by individuals from the existing applications group. They initially focused on Notes applications that would either be widely used by many users or were otherwise of critical importance to Lotus. They also began developing an in-house Notes/Vax mail gateway to allow new Notes users to communicate with existing Vax mail users waiting to be converted to Notes. This ensured that corporate-wide communication was not disrupted during the Notes roll-out process.

ROLL-OUT.   This was the last group formed. The group consisted of two managers. These individuals worked closely with business group managers in Lotus to understand the hardware, software, and training requirements of the various groups. They worked closely with Purchasing to acquire equipment and software for these organizations and with Finance on budget issues. The Roll-out group also worked closely with the outside hardware vendor who performed the actual hardware and software upgrades at client sites. (Given the limited time frame involved in the roll-out, it was more efficient to outsource this task, as is explained later in this chapter.)

Developing a Plan

As is appropriate with any large long-term project, the first goal was to develop a comprehensive plan of execution. This plan included executive summary, specific deliverables, and projected time frames. Group managers throughout Lotus reviewed and had input into this plan. When this review was complete, senior level ISS, Finance and Operations, and Sales management reviewed and approved the plan.

The Notes Operations and Administration group did much of the initial effort. The two immediate objectives were to define a comprehensive architecture that was scalable worldwide and to begin moving the existing user base into this new architecture.

DOMAINS.　Lotus decided to place Notes users into three domains. These domains were called Lotus, Notes Development, and International. The Lotus domain was the largest, consisting of all production Notes operations in North America. The Notes Development group needed their own domain due to their unique development environment. (They often run pre-release software, perform tests, and so on.) International, headquartered in Staines, UK, managed the third domain. (Lotus felt administration for Europe and parts of Asia would best be done in Staines to accommodate time differences, localization issues, etc.) All three domains were integrated at key points. Name and Address books for all three would reside on each production server worldwide, providing seamless mail routing.

SERVER TYPES.　Next Lotus defined server types. These included Hubs, Mail, Database, and Gateway. Hub servers were centrally positioned either on a WAN or LAN to communicate with other servers. Each would serve two functions: a bridge for inter-LAN/WAN mail routing and replication and a central point where changes to the Name and Address book could be made and then propagated outward. These machines would be very busy with mail and replication, and many files on these servers would be replicas of files on others servers. Lotus therefore decided to configure these servers to deny access to the general user community.

Dedicated mail servers yielded several benefits. Lotus needed to pinpoint their capital requirements for the coming year. Because they knew all users of Notes within Lotus would convert to Notes mail and that mail usage among users is somewhat linear, they developed estimates on capacity given a certain hardware configuration. For example, they rated a particular 486 server configuration at 100 mail files. Because the number of people in Cambridge was 3000, they knew 30 mail servers were required. From such figures, they derived their capital requirements.

Database server requirements were more difficult to project. Lotus anticipated database usage to vary greatly from workgroup to workgroup. They decided ISS would purchase a limited number of servers for company use. Lotus encouraged individual workgroups to purchase their own database servers, once a "critical mass" of their applications were in place. These servers would still be administered by ISS. However, access and use of these servers was by the business group, through the user of "group" names.

Gateway servers included machines used for outgoing fax, Notes-to-Vax mail routing, and production control functions. The Vax mail gateway was a particularly critical component. As people migrated to Notes mail from Vax mail, this gateway was heavily used to send mail between new Notes users and others still using Vax mail.

As their Notes Applications group developed an increasing number of applications, Lotus dedicated a Production Control server to run production-grade applications and APIs.

For their North American field offices (serving small user communities of 25 to 100), they combined mail and databases on the same server.

FIELD AND INTERNATIONAL COORDINATION.   To ensure successful roll-out in the field offices, it soon became clear that the Information Systems group in Cambridge needed experienced people onsite. Per the roll-out team's suggestions, the Sales organization agreed to fund six Regional Technical Coordinator (RTC) positions. The RTCs were responsible for maintaining the Notes servers as well as print and file servers in the office. They also acted as first-line support for field office employees.

The roll-out team also worked closely with their International counterparts in Staines. The Staines group was simultaneously planning the details of the roll-out in Europe and Asia. The Cambridge roll-out team informed them of their activities and worked to understand the unique needs of the Staines group. Soon both groups agreed on the overall system architecture and certain key administrative policies and procedures, such as certification and mail/replication paths. Throughout the roll-out, both groups maintained close contact. This built a level of mutual trust and cooperation that contributed greatly to the global success of the Notes roll-out.

OTHER SERVER AND WORKSTATION ISSUES. After they defined the architecture, but prior to formal roll-out, Lotus needed to incorporate existing servers and users into the new Notes environment. This required three steps:

1. *Institute the most secure system possible.* Lotus created a set of new master certificates. These certificates are very tightly controlled by a few trusted individuals. All servers and existing Notes users in the company were recertified via a formal recertification program. This process took about three months to complete.

2. *Redefine ISS's existing servers to fit the new dedicated mail and database format.* Lotus' existing servers had both mail and databases residing on them. Therefore, they developed a plan to move mail files to new dedicated mail and database servers. This required careful coordination with users.

3. *Collect existing company servers not administered by ISS.* This sometimes required some negotiation with the people involved to reassure them the servers would still be maintained and available for their use.

The Notes Operations and Administration group prepared for roll-out by ordering 10 new servers. After they had a committed order, all 10 servers were operational within three days.

The Roll-out group focused on user requirements. They defined standard Lotus hardware requirements for both OS/2 and Windows workstations. The workstation had to be of reasonable capacity to support Notes and the growing number of Lotus GUI (Windows or OS/2) products. Each business group within Cambridge was surveyed for their hardware and software needs. If their user did not have sufficient capacity to convert to a GUI environment, the group purchased equipment to upgrade the workstation.

In the field the process worked very similar to Cambridge. Once a RTC was hired for the region, each office was surveyed and appropriate equipment ordered. The field and Cambridge roll-outs were handled in parallel.

Rolling out Notes

Lotus negotiated with IBM to supply them with engineers to assist roll-out. IBM helped survey and upgrade hardware. In Cambridge, these engineers were supplied directly from IBM. In the field, IBM subcontracted local technicians.

Careful coordination was required between ISS organizations as a business group was brought up on Notes. As the roll-out team surveyed work groups, Purchasing ordered and delivered equipment at the point technicians were available to install it. The Roll-out team had a limited capacity to create servers IDs, so the ID run rate was carefully monitored and controlled. The Roll-out team also needed sufficient server capacity to support mail files for new Notes users. Training needed to be available in a reasonable amount of time after completion of the hardware/software upgrade to prevent new users from becoming uncertain or anxious about their new environment. Finally, the Roll-out team created a new service called Deskside Support to assist users. Deskside Support worked directly at deskside with users who required special assistance.

To support roll-out, they used a Notes database. This database tracked surveys, costs incurred, and training. The database provided periodic management reports on progress to date, plus future plans.

A note about the roll-out budget: If a workgroup had sufficient capital, the cost was charged to their cost center. If not, there was a reserve fund maintained by ISS. The Roll-out group worked closely with the Plans and Controls organization to implement and monitor this arrangement.

Notes at Lotus after the Roll-out

The worldwide Notes roll-out, which took approximately a year, was completed in late 1991. In that time, the number of Notes servers in Cambridge had grown from 3 to almost 65. There are now 15 North American field offices installed. The International organization ended 1991 with over 75 servers spread through 18 countries worldwide. In total, there are over 4000 people in Lotus using Notes.

Lotus found that Notes changed the way many of their employees do business. There is a tendency toward fewer meetings because information is exchanged electronically. The amount and quality of communications within and across various groups has improved through the innovative use of this technology. Even inter-company communications is becoming more common through Notes.

This has kept their Application group extremely busy. They have developed a number of key applications, including a Notes-based "shotgun" system, an Executive Schedule system, and a Phone Book directory of all company employees. These are just a few of the many applications they have developed. They have also assisted countless other individuals to develop their own applications.

The Notes roll-out at Lotus was extremely successful, finishing ahead of schedule. The Lotus roll-out process is often used as a model for other companies. Now that they have Notes in place, they're continually finding many new and innovative ways that it can help them all do their jobs faster and more efficiently.

IBM's roll-out of Notes R4

In June of 1995, IBM started a "top-down" effort to move to the client/server environment, which included Lotus Notes at the core. This section gives the reasons behind IBM's decision and is based on the information distributed to IBM employees explaining the decision. Overall, the reasons given were to:

- Reduce IT expenses.
- Use what we sell to our customers.
- Increase productivity through "groupware" and information sharing.
- Follow Corporate direction for IBM.

In more detail, these reasons are based on the major driving force behind the move toward client/server computing, which is the need to empower employees and to support teamwork within the organization. In his book, *Liberation Management: Necessary Disorganization for the Nanosecond Nineties,* Tom Peters states that "finding terrific people and then getting out of their way" is a path to business success. This is the essence of empowerment. This was recently underscored by IBM Chairman Lou Gerstner who has identified teamwork as one of our eight operating principles.

Client/Server applications such as Lotus Notes support empowerment and teamwork by enabling employees to make their own decisions. It provides the information they need in the right format and at the right time.

In a client/server environment, any end user can transparently access any authorized application, data, computer service, or other resource within a work group or throughout the enterprise. Client/server spans multiple computing platforms and heterogeneous communications networks. It encompasses the latest advances in computer workstations, file servers, open systems interoperability, midrange and mainframe computers, databases, graphical user interface applications, and more.

Because of the strong productivity and efficiency gains from client/server computing, a consulting study commissioned by IBM estimates that 84% of U.S. businesses polled favor a move to client/server computing within the next few years. A recent study done by IBM indicated that, out of the top Fortune 1000 companies, slightly more than 50% are already using client/server computing and, of those not using it, more than 50% intend to be using it within two years.

Advantages to IBM for Using Lotus Notes as a Groupware Tool

The following points outline the benefits of this environment to IBM employees:

■ Rich development environment, ideal for rapidly building graphical, client-server database applications that collect, track, route, and distribute documents.

■ Easy addition of workflow to automate procedures, processes, and tasks.

■ Significant database access control (seven access levels, can protect all or part of database).

■ Mobile Client support (ability to work on mail, calendar, and other Notes databases while disconnected from Host/Server; update (replication) of changed information upon return).

■ Automatic indicators to identify new or updated information to users.

■ Only "one" physical copy of data. If the data changes, the user automatically sees the updates (speeding communications). Easy one-button access is available to information from "unlimited" places (via Notes Doclinks). Mail and other documents can be filed under multiple categories, with only one copy of the data existing.

■ Mail can be categorized, subcategorized, and "blind copied."

■ Data encryption and signature verification for mail and other documents.

■ Portability of databases allows them to be replicated to other servers and domains.

■ Notes databases can be shared between remote sites, permitting communication across organizations, across platforms, around the world.

■ Ability to attach files of "any format" to documents/memos and mail or store them; ease of combining text, voice, video, audio, graphics, and soon fax.

- User-customizable views of database information.

- Standard GUI interface making navigation and usage intuitive.

- Customizable SmartIcons that can be used for activating "special" or common/repetitive tasks saving valuable keystrokes.

- Capability of having more than one workspace window open concurrently allowing easy switching between multiple databases, views, and documents.

- Rapid prototyping of application databases with a consistent user interface.

- Availability of numerous Lotus-provided "templates" offering working solutions for database designers.

Why IBM is using Lotus Notes

This section is again based on the information distributed to IBM employees to explain why IBM was migrating to Lotus Notes from the legacy mainframe e-mail and application systems. It was explained that Lotus Notes was first made available as a product in 1989. It is a client/server environment that allows users (or clients) to communicate securely over a LAN or a telecommunications link, with data (document databases) residing on a server machine. Everything in Notes is a "database" and can be shared. It is robust and secure for mission-critical applications. A key concept is that of replication. This keeps geographically dispersed Notes databases in line automatically through the routine transmission of changes (only).

Notes is best used as an environment for the development of a new class of "Information Sharing" applications. According to Lotus promotional material, these can be developed as much as 90% faster with Notes than with traditional action-oriented development tools. Teamwork is a key Notes concept. Notes can be viewed primarily as group collaboration software consisting of two main parts:

- A fully integrated e-mail system.

- An environment for the development and deployment of information-sharing applications.

Lotus Notes is recognized worldwide as the premiere workgroup communication tool and is now an IBM product since the merger with Lotus Corporation. Once adopted, this tool can tremendously improve productivity with groups, giving them ways of performing once-complicated communications and coordinations with ease. In order for IBM to succeed, the company as a whole is asking its employees to work smarter and harder, and tools such as Lotus Notes that will help accomplish this goal.

Lotus Notes is the replacement for OfficeVision/VM electronic mail. Everything in Lotus Notes is a "database." Your mail database, which is located on a mail server machine, is a Lotus Notes database that stores incoming and outgoing mail. All Lotus Notes mail users have their own mail database. In addition to mail and distribution list support, Notes also adds the value of other databases that can be easily created to improve communication and processes in your team/department.

In addition, you can use Lotus Notes to receive and send mail to the Internet, easily attach and mail files to Notes and VM customers, and use Notes' other advanced communication features (such as using Doclinks, pop-ups, buttons, etc., rather than just the text-only limits of OfficeVision).

Top-down Deployment

Lotus Notes within IBM first started from the "bottom up." Many groups started Notes domains on a department, site, division, or country level. In 1995, it was decided that a "top-down" approach was needed. At that time, most IBM employees used the mainframe's PROFS and OV/VM for office systems. The knowledgeable and experienced user population at IBM was accustomed to a high level of service, function, and availability.

Besides the basic capabilities of mail, calendar, and document preparation, which PROFS and OV/VM provided very reliably, IBM employees made extensive use of a highly functional personnel directory system, a secure and flexible electronic forms product, ECFORMS, rich publishing tools, corporate-wide information warehouses, and a substantial number of productivity aids designed to facilitate office tasks from vacation planning to personal equipment inventory. The result was that, in 1996, when IBM Global Services (IGS) began the wide-scale deployment (with a top-down approach) of Lotus Notes to replace OV/VM, IBM had one of the lowest cost, most stable (availability was consistently in the 99.96% range), and most sophisticated office systems in the world.

The design problem was to provide a new platform with similar functionality and equivalent reliability, at the lowest possible cost. Naturally, with requirements such as these, the top-down approach with "buy in" from IBM's very top management was an absolute necessity. That executive "buy in" was no problem because IBM's CEO, Lou Gerstner, was behind the effort from the beginning.

The Geoplex Concept

Given the number of users within IBM and their work patterns, a single Lotus Notes domain for the whole company was beyond the capability of existing technology and administrative processes. All users in IBM were to be registered to home domains for mail only. Application databases would reside in a separate domain. In an environment as large as IBM's, this domain division provided the following benefits:

- Keeps the number of database domains low and constant, thereby simplifying replication connections management.
- Aligns user mail addresses with the IBM corporate strategy for Internet domain naming.
- Reduces the number of replications required to make general-purpose applications and databases available to all employees.
- Makes it easier to separate administrative duties by domain function.
- Makes it easier to size hardware requirements for servers within both types of domains.
- Allows the optimization of domain structure and Public Address Books by function.

There is one database domain at each of the 24 Geoplex sites (see chapter 2 for details on the Geoplex locations and architecture). The database domains follow a hub-and-spoke architecture, with hub servers initiating replication with each of the spoke servers. The 24 Geoplex centers provide mail and database service to IBM's user community. The Notes servers at the Geoplex centers are RISC-based Scalable POWERparallel systems (SPs) running AIX 4.1.4. IBM chose to separate replication hubs from the database servers accessed by end users so that each could be optimized. There is one mail domain for each country in which IBM does business (131 mail domains). The servers for these mail domains were distributed geographically among the 24 Geoplex sites.

An upgrade of IBM's internal TCP/IP network was completed in 1996 in anticipation of a significant increase in Notes IP traffic. IBM believes that increasing the bandwidth of the wide area network (so that users can go to applications) is a more cost-effective approach than expanding local server capacity (so that applications can be brought to users). The enhanced network benefits IBM in many ways, such a improving World Wide Web access speeds. Also, without the centralized Geoplex concept, it was felt that adding local server capacity would have a more significant effect on capital, expenses, and administrative resources than enhancing the wide area network.

For most users in the U.S., and many worldwide, participation in a discussion database that is physically located 2000 miles away is virtually as fast as in one attached to the local LAN due to the capacity and availability of the wide area network. In this environment, there is often no benefit, other than perception, to physically replicating a database from one server to another, and there are often drawbacks, including older data and a higher risk of creating replication and save conflicts. IBM ensures that all their users and servers share a common ancestral certificate in their ID files so that a wealth of information located on servers all over the world can be accessed without regard to physical proximity. Moreover, given the decision to standardize on TCP/IP as the communication protocol, the Geoplex application model offers a good approach to minimizing data redundancy resulting from replication by minimizing the need to have more than a single copy.

In determining the best mail domain architecture, IBM considered two main alternatives. One was to organize the domains by business unit, and the other was to organize the domains by geography. After studying user movement patterns and the reorganization history of IBM, it was determined that most employee transfers tend to be between organizations rather than geographies, and physical moves are rarely made outside the home country. Additionally, the users are administered typically by country human resource groups in their native language and by business function on a worldwide level.

This led IBM to the decision to align mail by country and to align the business-related application on a global basis by Geoplex. The plan for mail also coincided with the company strategy for mail addressing with external companies so that users have Internet and Notes addresses that are as similar as possible. The end result was that each IBM user was a member of the mail domain representing his or her country, of which there are 131 to support all the countries in which IBM does business. Similar to the database domain model, mail servers are deployed within mail domains using a hub-and-spoke architecture. Within each mail domain is at least one hub that is responsible for routing all message traffic into and out of the domain. Please see chapter 2 for additional details on mail routing within the IBM Geoplex architecture.

Creation of 130,000 Notes IDs in 12 Months

As part of the roll-out to Lotus Notes during 1996, IBM needed to create over 130,000 Notes user IDs. To do this required an automated process. The Distributed Administrative Automation Tool (DAAT) was selected as IBM's user account management tool because of its rich functionality and support of both Lotus Notes and the LAN environment. Working from standard input files containing minimal user information, DAAT interacted with the corporate directory system to generate and distribute up to 1000 Lotus Notes and LAN user accounts per day.

In addition to the required user ID files and passwords, DAAT sent each user a welcome letter that had been written specifically for his or her location or business unit. The letter contained the information needed for that user to become productive in Lotus Notes. By eliminating most of the manual labor involved in the user account creation and distribution process, IBM achieved significant reductions in both cost and cycle time and a noticeable improvement in quality.

Notes License Types

For a large company, such as IBM, the ability to use many security features of Notes, such as encryption and electronic signature, becomes more complex when used across geographic boundaries. In 1996, the United States export rules dictating the length of the encryption key that can be used in software taken outside of the U.S. were modified to accommodate traveling citizens. However, the documentation and itinerary requirements were considered too rigorous for consistent use, and IBM felt that employees might unwittingly fail to comply. Concurrent with this change was the strengthening of the encryption key used in the International version of Notes. The new longer (64-bit) encryption key could be used in most countries worldwide and allowed IBM to standardize on a single version of the product without compromising security.

The specific International version of Lotus Notes used for deployment within IBM varies on the region in which Notes is being rolled out, because language and time-zone requirements differ. The use of the International client in the roll-out within IBM does bring up some migration considerations. For example, my ADVANTIS Notes user ID is a North American user ID. Therefore, I could not install an International Notes client and switch between my ADVANTIS and IBMUS ID. My solution was to use the North American client during the migration period, because the North American client allows use of both my user IDs.

Use of Customized Mail Templates

Because IBM employees were already heavy users of mail, the migration plan had to provide similar functions with similar ease and also protect the user's ability to access legacy data by moving his or her current and archived mail files to the new Notes environment. IBM developed several tools to migrate legacy data and designed an enhance mail template in order to meet these requirements.

The standard Notes mail template was enhanced to provide additional features and to ease the transition from OV/VM. These included:

- Setup functions for allowing users to mimic OV/VM's prolog and epilog on mail memos.

- An "Out of Office" feature that re-routes incoming mail for a specified period of time (e.g., vacation) and that more closely resembles the tool that users were comfortable with than the built-in function in Notes R4.

- An attachment viewer that allows formatted host file attachments (e.g., legacy AFPDS LIST3820 files) to be viewed easily from a Notes memo.

- A mail archive facility to allow users to select items from the Notes mail database to be moved to another database for archive.

- A "Who am I?" selection in the custom Mail Tools pull-down that presents comprehensive information about the user's current environment, such as operating system, Notes version, mail template version, OV/VM address, Notes address, license type, and more. This is proving invaluable for help desks when they are working with the user to diagnose a problem.

- The "Address Assistant" prompts the user for the specific address components—such as fax number, pager number, or Internet address—and reformats it on-the-fly into the structure required by the LMS gateway.

The Address Assistant feature is a significant part of the effort IBM made to ensure that users correctly address mail to the variety of environments reachable through the LMS gateway. IBM discovered early in the deployment that the most difficult hurdle both new and experienced users had to overcome were the differences in addressing between either the OV/VM or the legacy Notes environment and IBM's strategic Notes domains.

IBM also developed a set of tools to assist the users in moving their OV/VM correspondence to the Notes environment. These include a tool to select and move note logs (mail folders), the in-basket, and OV/VM documents. These tools were designed to be executed by the end user for two reasons:

- To cause the end user to select which files to move, encouraging him to avoid moving many old, unneeded files.

■ To reduce the load on support personnel who must migrate large numbers of users simultaneously.

Each of the tools creates one or more Notes databases on the user's workspace with their OV/VM data stored accordingly.

Conclusions on IBM's Roll-out

In 1996, IBM moved about 130,000 employees to Lotus Notes. Almost all of the remaining 100,000 employees will be migrated by the end of 1997. Of course, giving an employee the Lotus Notes client and a user ID doesn't necessarily mean the employee is then fully on Lotus Notes. Most employees go through a transition period where they are using both their "legacy" office system and the Lotus Notes platform.

In June of 1997, IBM Japan announced they were the first IBM country to completely migrate to Lotus Notes. They could make this statement because PROFS and OV/VM IDs within IBM Japan were eliminated. VM access was still available for legacy applications, but legacy mail was no longer available. For most of the rest of IBM, the plan is to eliminate PROFS and OV/VM by the end of 1997.

With IBM's migration to Lotus Notes, we've found that, when very large enterprises migrate to Lotus Notes, a level of complexity and a set of technical challenges are introduced that are not usually found with smaller organizations. Developing the technical solutions to these unique problems requires a thorough understanding of the existing environment and the Notes technology. One of the tools that has proved very useful within IBM is The Enterprise Application Registry which is used to help advertise, deploy, and control Lotus Notes Applications across the IBM Corporation. The template for this database is available on the book's CD-ROM.

Lotus Notes can improve an organization's communications flow, ultimately making it more effective. However, if issues like scalability, manageability, and coexistence are not addressed during the planning stages of a migration to Lotus Notes, organizations will not be able to fully experience the benefits of this technology.

IBM's Lotus Notes environment will continue to evolve and embrace new technologies that complement large, diverse companies. The Calendaring & Scheduling facility, Master Address Book, server clustering and partitioning, and Domino are all starting to play key roles in 1997 and beyond as our worldwide infrastructure expands and matures.

Some Pitfalls and how to Avoid Them

The following sections list suggestions and cautions you should keep in mind before starting your Notes roll-out.

Organizational Support for Notes Roll-out

- *Be sure to give proper attention to all locations at which you intend to roll-out Notes.* Think of Notes as similar to a chain, which is only as strong as its weakest link. Put as much effort into your most remote installation as you do in the home office. The more you work together to help everyone succeed, the more you succeed.

- *Work together with your counterparts in the field and international.* They understand best the local idiosyncrasies, cultural issues, etc. You can't do the job everywhere. However, you can develop and foster close ties with your counterparts to ensure global success.

- *Get senior management support up front.* This is a corporate-wide effort. Make sure you have the buy-in you need.

User Training and Support

- *Anticipate your training needs, and prepare a training organization and classes to meet this need.* If possible, use existing internal training resources. They have the training skills. Lotus can help them learn Notes.

- *Offer phone support.* Have resources available that users can call. You're migrating users not only to a new software product but, in many instances, to a new environment (GUI). As with any new product, the majority of calls come within the first 90 days. Prepare for it.

Server Operations

- *Establish a specific organization focus designed to roll-out and maintain Notes.* This organization needs to secure the appropriate talent to design server architecture, policies and procedures (security, backup, ID distribution, replication, and so on), expenses, and project management. It is particularly important that this team is properly trained with Notes. Lotus can help train your team members.

- *Control the N&A book.* The N&A book is the single most important control file in the system. Too many inexperienced users making changes to it will cause problems.

- *Standardize server management.* You should establish corporate-wide policies and procedures used to administer Notes servers. If you don't directly control the servers, be sure whoever does follows the standards you set.

- *Document all procedures.* Consistently review them. Discard what does not work and add what does.

- *Migrate users gradually.* This is particularly true if many of your users are already using another e-mail system. Users will resist a sudden forced transition from one system to another.

- *Maintain close ties with Lotus.* Relay problem areas to Lotus, and stay abreast of future features and functions.

End user Roll-out Issues

- *Work closely with management.* Keep them appraised of activities, preferably before they happen. Minimize surprises—change can be disconcerting. Pick up the phone occasionally and ask your management if they are happy with the progress and what you could do to improve it. Then act on their suggestions.

- *Periodically inform your expanding user base of new applications, new procedures, or policies.* You can do this through company-wide e-mail messages. This is not only important to your user base, but it can also act as a morale booster to your Notes roll-out team, illustrating their progress to date.

Details of a Conversion Process

Here is a summary of the Conversion Utility Button developed to migrate a Notes domain from a single-level hierarchy (/ADVANTIS) to an X.500 hierarchy (Location/Advantis/US)

This section describes the Lotus Notes Code and Pascal Code involved in the process to convert existing Notes users (/Advantis) to the new hierarchy (e.g., /Schaumburg/Advantis/US). The entire process can take anywhere from 2 to 10 minutes based on machine, memory, etc. A listing of the code is given in appendix D of this book. David Frank of Advantis developed the procedure and code.

The simplified procedure is as follows:

1. Send a Lotus Notes form to all current Notes users with the old hierarchy.

2. Once the users receive this Notes form in their mail, they simply click on the button in the form and the conversion process begins. The users will receive some prompts to keep them informed as to the status of the procedure.

3. Once completed, a completion message is sent to a Notes database for our records.

It is a very simple process for the end user.

There are three pieces involved in this process. The first piece is a Lotus Notes database called Notes Announcements. This is for use only by the Notes Administrator. This is where they send out the Notes conversion form to existing users with the old hierarchy. This database then tracks who has received the form, what steps the user has completed, any errors the user might have encountered, and when they have completed the process.

The second piece is a Lotus Notes database called Conversion Process. This will reside on the users current mail server. This database contains all newly registered users along with their IDs, the PASCAL programs (NOTESINI.EXE for Windows, NOTESDOS.EXE for DOS) used to update the NOTES.INI, and the cross certificates so that these users can access both /ADVANTIS and /Schaumburg/Advantis/US servers. The cross certificates will not be needed once we have converted all servers and users to the new hierarchy.

The last piece is the Notes conversion form. This form controls all processes relating to converting the user. The form contains the user's current mail server, the user's new mail server, and the button the user clicks on to convert himself to the new hierarchy. Following are details concerning the button's step-by-step processes:

STEP 1. Captures some user information and temporary variables. These are then passed to the users NOTES.INI file for use by later programs.

STEP 2. Checks two major areas for errors. First we ensure that the user has access to the Conversion Process database so that the button can retrieve their new ID, the correct PASCAL program, and the cross certificates. Then we check to make sure the user has his or her own personal Name and Address book to allow us to paste in the cross certificates. If either of these two areas has a problem, the user is prompted with a message that an error has occurred and that a pager message has been sent to support, then an actual page is sent to the support pager listing the user's phone number and error. The program is then halted.

STEP 3. If no error has occurred, then the user is prompted that the process has been started and that it might take up to 10 minutes of their time.

STEP 4. The Conversion Process database is opened and the cross certificates are copied.

STEP 5. The user's personal Name and Address book is opened and the cross certificates are pasted. This gives the users access to /ADVANTIS servers as well as /Schaumburg/Advantis/US servers.

STEP 6. The Conversion Process database is reopened, and the user's new ID is detached to their C: drive as filename NOTES.ID.

STEP 7. The PASCAL program view is opened in the Conversion Process database, and the correct EXE is detached to the users. There are two versions of the EXE that update the NOTES.INI. One version is for Lotus Notes for Windows users (NOTESINI.EXE), and the other is for Lotus Notes for OS/2 users (NOTESDOS.EXE). The button captures the Notes platform of the user, allowing us to detach the correct EXE.

STEP 8. The "PROMPT FOR UNREAD MARKS" macro is started. This macro notifies the user that he or she might receive more prompt messages following this one and to click on YES or OK to continue the step. They are also informed that the following step might take up to 6 minutes. After the user clicks on OK, the user's mail file and address book are redirected from the old mail server to the new mail server.

STEP 9. The EXE that was detached to the C: drive is now launched. This program checks to see if the user's NOTES.INI file exists. If so, it is copied to NOTES.OLD. Then it copies the user's C:\NOTES.ID to the users current Lotus Notes directory. After the previous two steps have been completed, the new NOTES.INI is updated. The KeyFilename variable is pointed to the new NOTES.ID. The MailServer variable is pointed to the new mail server. If the previous steps work, then $INIPROGRAMOK=YES is added as a line to the NOTES.INI. This variable is an environment variable that the Lotus Notes button can now retrieve for error checking. The EXE program then deletes itself from the users C: drive.

STEP 10. The "CHECK TO VERIFY INI FILE UPDATED" macro is started. This macro checks to see if $INIPROGRAMOK is set to YES. If so, then the process worked, and the user is informed that his new password is his PROFS ID all in uppercase, that the process is complete, and that Lotus Notes will now be shutdown. If an error occurs, the user is prompted with a message that an error has occurred and that a pager message has been sent to support, then an actual page is sent to the support pager listing the users phone number and error. The program is then halted.

Again, the Lotus Notes and PASCAL code for this conversion program is listed in appendix D of this book.

Notes Migration Services

If you want to outsource your migration from a mainframe system to Lotus Notes, you can do that too. One such service is described in this section.

IBM LAN Migration for PROFS, LAN Migration for OV/VM, and Migration Services Offerings

IBM has a service offerings for migration from Host Office to Notes and cc:Mail. IBM's Business Solutions Consulting Organization offers this migration to facilitate a migration from PROFS or OV/VM (IBM's Mainframe Office Software) to Lotus Notes, cc:Mail, and Time and Place/2.

Migration Benefits

Using this migration offering, you will:

- Roll out Notes/cc:Mail faster.
- Achieve the benefits quicker.
- Not have to develop any in-house skills.

Migration assistance is provided by experienced consultants and tested software tools. The software conversion tools were used to downsize internal use of IBM PROFS to Lotus Notes and cc:Mail. IBM has been using and testing these tools since 1994.

IBM's LAN Migration

IBM's LAN migration includes the following features:

- Preliminary planning session
- Requirements questionnaire
- Migration strategy study

- Solution definition
- Topology design
- Connectivity strategy
- Implementation plan
- Onsite assistance
- Pilot implementation
- Follow-on assistance

Web Sites on Migration Products and Services

To get further information on Lotus Notes migration products and migration specialists, take a look at the following Web sites.

EMEA IBM/LOTUS INTEGRATION CENTER.   This site (http://w3-emeailic.ae.boeblingen.ibm.com) gives information on the EMEA (Europe, Middle East, and Africa) IBM/Lotus Integration Center and its mission. It also lets you know about the migration contacts for pre-sales, in both Lotus Consulting and in the IBM Notes Service Line. Information is also given on the EMEA events about migration.

U.S. (DALLAS) IBM/LOTUS INTEGRATION CENTER.   This site (http://dsctssrv.sl.dfw.ibm.com/ilic) gives information on the U.S. IBM/Lotus Integration Center and its mission. You can also get information on all Migration and Coexistence products.

Browse the U. S. Internet Web site as well:

http://www.software.ibm.com/is/ibm-lotus

MIGRATION AND COEXISTENCE TOOLS SITE. This site (http://service2.boulder.ibm.com/nov/) gives information on all migration products. You can download beta code and give your feedback to IBM/Lotus development.

CHAPTER 12

Network/System Management and Administration

When Lotus Notes replaced Office Vision as the messaging solution for IBM, we knew Notes had to be managed with the same level of discipline as users were accustomed to with their mainframe solution.

Albert Schneider, Lotusphere '97, January 28, 1997

This chapter deals with overall Notes management, monitoring, and administration. In general, Lotus Notes is not easy to administer. However, Lotus Notes Release 4 has made the job easier. Release 4 allows automatic recertification of Notes IDs, which is an essential part of network security. End users are able to recertify their IDs without intervention by administrators. Other administration features in Release 4 include the ability to set size limits on Notes databases, along with an Administrative Agent that automates server operations such as adding and deleting user names that appear in multiple places. Release 4 also extends the Notes directory to make it more customizable and comprehensive. It has templates for creating public and private Notes Name and Address Books, allowing users to store their private lists on their Notes client and public directories on the server. The Notes Name and Address Book was extended to include user information for cc:Mail, X.400, and SMTP addresses, which are integrated in the Lotus Communications Server messaging platform. Another feature, the Administration Delegate Agent, enables managers to assign and distribute administrative rights to IS professionals and users throughout a Notes network.

Daily Network/System Management for Lotus Notes consists of:

- Backup and restore
- Virus scan
- Alerting
- Problem determination (local and Remote)

System Administration, the process of maintaining all Notes users and servers, can be considered part of System Management. System Administration includes:

- Determining who in your company is authorized to submit change requests and then validating that person's authenticity.
- Building the Lotus Notes servers.
- Establishing all schedules for replication and mail routing for the servers.
- Backing up the servers.
- Monitoring all server performance/activity.

- Managing the disk space on the servers.
- Gathering server statistics.
- Problem determination for reported errors and alarms on the servers.
- Performing server maintenance.
- Creating Notes user IDs.
- Certifying and cross-certifying new and existing users and servers.
- Ensuring that the Name and Address Book is accurate and consistent.

Additional tasks might include:

- Providing database change management schedules for application developers.
- Creating replica copies of Notes databases onto servers.
- Responding to user problems.
- Assisting other Notes staff members with problem determination and resolution.

No enterprise Notes architecture can be successful without appropriate systems management tools to easily monitor network traffic. As part of the Lotus Communications Architecture, Lotus has developed an SNMP-based (Simple Network Management Protocol) management tool for Lotus Notes networks, called NotesView.

The NotesView management tool is based on Hewlett-Packard's OpenView system management technology. NotesView is described in chapter 13 in the section "NotesView."

Remote Monitoring

In the typical corporate Notes installation, there are Notes servers and gateways at many different locations, perhaps throughout the world. In many cases, a single crash of a machine—a Notes hub server, for example—can cause Notes mail to cease to flow across the wide area network to dozens of mail servers, thus impacting hundreds or thousands of Notes users. Thus some Notes machines are crucial and can potentially affect all users in your Notes network, while other machines, such as mail servers, are used only by a fraction of all users. Monitoring should have a priority level.

Also, over the enterprise, there are many components that have nothing to do, per se, with your Notes system, which will go out and, nevertheless, play havoc with your Notes network. These components include routers, bridges, leased lines, etc. This section looks at the monitoring of all of the components that can impact your Notes network.

When TCP/IP is Your Protocol

TCP/IP is the best choice as a protocol for Lotus Notes over the wide area network. This protocol allows a great deal of flexibility in the area of remote monitoring, as is indicated in the following description of remote monitoring for the IBM Gemini system.

The Gemini system is a "high-profile" mail routing system used within IBM. It's basic use is to route mail between the Internet and Lotus Notes. A diagram showing the Gemini mail flow is shown in Figure 12.1. This system was also discussed in chapter 8 under "Using the Lotus SMTP Gateway Product."

Because of the high profile of the Gemini system, this was one of the first Notes systems in IBM where all components, both network and Notes, were highly monitored on a 24 hours a day, 7 days a week basis. Upon the failure of any component, which could not be fixed by first-level support (IBM Network Control), second-level personnel would be paged. The following sections cover the basic remote monitoring procedure.

IBM NETWORK CONTROL. IBM Network Control monitors all TCP/IP network connectivity (Notes servers, Gateways, routers, bridges,

Figure 12.1
The IBM "Gemini"
Mail Routing System.

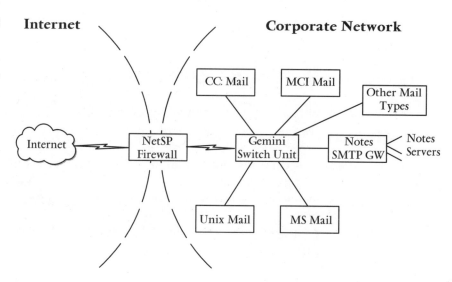

etc.). This is done with a simple TCP/IP "ping," and any alerts are automatically sent to a NetView/6000 screen and marked in red. Because the Gemini system is quite extensive, boxes not under the direct control of IBM's network control in White Plains, NY, are also monitored with a TCP/IP ping, and other IBM Network Control Centers (e.g., Somers, NY) are contacted in case of problems. The other network control centers will page their second-level personnel if the problem is in their area.

WHITE PLAINS NETWORK CONTROL. White Plains Network Control monitors whether Notes is running on all Notes servers in the Gemini path. This is done by a telnet to port 1352, which is the TCP/IP port used by Lotus Notes. If Network Control gets no response to telnet, then there is a problem with Notes. If Notes does respond, then the Notes server is running, but Notes tasks might not be running (e.g., the Notes Router task could be down resulting in no mail flow). We've had the problem on a Notes hub server where the mail.box was corrupted and the Notes server allowed mail to build up in the mail.box but the Notes mail router would not release any mail. The hub server "passed" the TCP/IP ping and telnet to port 1352 checks with flying colors, but no mail was flowing. That's the reason an end-to-end mail flow checking program (discussed in the next item) is required.

In addition to the check on port 1352 to see if Notes is alive, on those boxes running SMTP (e.g., the Gemini box and the Notes SMTP GW), port 25 is tested via telnet. Port 25 is exclusively used by SMTP.

Thus three simple tests indicate whether the box is "alive," running TCP/IP, and is reachable over the network via TCP; whether Notes is responding on port 1352 and hence is "alive"; and whether SMTP, on those boxes where it's supposed to be running, is actually running.

MAILCHECK: END-TO-END MAIL MONITORING. The ultimate, and necessary, check for the Gemini system is with the use of an end-to-end mail monitoring system. A program called MailCheck is used for this. MailCheck consists of a "poller" program at a central Notes client and "echoer" programs that should be resident at each of the mail servers that need to be constantly monitored. If the MailCheck program detects that mail is not getting to a target mail server, alerts are sent via Notes e-mail. A pager add-on program is used to send pager alerts on a 24/7 basis to those personnel responsible for the health of the Gemini mail system.

SNMP MONITORS SNMP. (Simple Network Management Protocol) monitors were also placed on the Gemini boxes. SNMP allows you to monitor specific parameters, and the results of this monitoring are sent to IBM's Network Control. The parameters are monitored via NetView/6000 (more to come on this type of monitoring).

REDUNDANCY. One way to help solve the reliability problems of a Notes mail system is to eliminate as much as possible the single points of failure. One single point of failure that is difficult to eliminate is the user's home mail server. Of course, it is possible to replicate all mail from a mail server to a second mail server. Then, if a mail server fails, a user could manually switch over to a backup mail server. A better way to solve this problem is to build redundancy into the mail server. A Notes mail server such as a SP2 can solve many of these problems for the large enterprise (see appendix A on the SP2).

However, for a mail system such as IBM's Gemini, redundancy in routing is far more important than redundancy in the mail server. If you have an existing wide area network, putting in alternate paths for Lotus Notes is often a "freebie" because it just involves putting in additional connection records or domain records.

Redundancy for the Gemini Mail System consisted of installing a second set of Gemini boxes at Schaumburg, IL (White Plains, NY is the first site). Thus, for this system, there are two firewalls (NetSP RS/6000s), two Gemini servers (RS/6000 box), and two Notes SMTP gateways (PCs). In addition to redundancy in the Gemini boxes, the Notes servers and domains all have dual attachments to other servers and domains in the path. This only requires changes to the Notes Name and Address Book (see chapter 4, the section "Using Alternate Routes for Notes Network Backup").

The redundant sites are primarily used for load balancing; however, in case of a failure, there is an automatic switch-over to the redundant site with no user intervention required. MX records in the IP Domain Name Server (DNS) provide an easy way to set up that automatic switchover. RAID 5 storage installed on the Notes SMTP gateway servers provides additional redundancy for those machines. RAID helps in the overall Gemini goal of making mail routing as reliable as possible and helps keep with the precept that we cannot risk losing any Notes mail. A RAID (Redundant Array of Independent Disks) strategy for Lotus Notes servers is described in appendix A.

E-mail reliability and integrity has always been the highest priority within the IBM Corporation, but historically, the emphasis has been on e-mail via the mainframe. E-mail via Notes has only recently been receiving the same emphasis as e-mail via the mainframe. IBM's purchase of Lotus in mid-1995 naturally gave emphasis to Notes as the primary e-mail system for the IBM Corporation. Migration to Notes as the primary mail system for more than 200,000 employees, most of whom have relied on mainframe mail (i.e., PROFS) for all their time at IBM, will take some time. The GUI and rich text fields is a clear advantage of Notes mail. However, unless the reliability and integrity of Notes e-mail is at the level IBM has historically enjoyed with PROFS, the significant additional functions of Notes mail will not be appreciated to the extent they should be.

When XPC is Your Protocol

XPC, the ubiquitous dial protocol for Lotus Notes, despite its limitations, will continue to be a popular protocol to use. After all, it's available with every copy of Lotus Notes, so all you need is a modem and an analog telephone line, and you can get to any other Notes server.

With XPC, you can use the Remote Console option to issue commands to servers from your workstation or from another server. The Remote Console option gives you access to all the commands you would have if you were sitting right at the server console. To have remote access to the server console, you (or the server, if you're at another server) must be listed in the Server document in the Administrator Access field. From your workstation, issue File|Tools|Server Administration and click the Console button to gain access.

Statistics Gathering

In order to monitor your Notes network and make network changes based on how the network has been performing, it is necessary to have a good history of your network in order to analyze your network. That's one of the reasons for statistics gathering.

Using the Notes Logs

The Lotus Notes log can contain almost all of the information you'll need to analyze your Notes network. The key word here is "can." The amount (and type) of statistics data gathered is very much dependent on what you choose to collect. The Notes defaults for the log will probably not be adequate.

Centralized Statistics Gathering

Along with centralized administration goes centralized statistics gathering. All of the Notes logs should be rolled up into one statistics database at a central location.

For the ADVANTIS domain, Lotus Notes logs and NetFinity information are relayed to a Mail-In database on a Notes statistics server. The Mail-In database name is "Statistics." These statistics are mailed every two hours. Events that have exceeded their thresholds are mailed to each Local Notes Administrator as well as the Central Notes Administrator. These events are also logged in the Statistics database. Examples of thresholds defined are for:

- Hard disk free space for each drive
- Free space on the swap disk
- Dead mail
- Memory swap file size

Central Management of Physical Nodes

Central management of your Notes network assumes that there is also in place central management of the underlying physical notes on which your Notes network rides. These physical nodes include routers, bridges, modems, gateways, and all the links interconnecting these nodes.

Remote Management of Lotus Notes over the Wide Area Network

Remote management of Lotus Notes must include management systems for both the physical and logical parts of your network. You probably already have a management system for the physical part of your network. If the management system (e.g., Network Control Center, Help Desk.) for the physical network is well-established, then it makes sense to have the same network management system involved in the management of your logical network, which includes Lotus Notes. Then the same 24-hours-a-day, 7-days-a-week system can handle all of your Lotus

Notes management requirements. This requires additional training for your network control people, but the results of having one management group for all of your network is well worth the effort.

Bandwidth Requirements Measurement and Analysis

The bandwidth analysis tools used by IBM were discussed in chapter 2 (in the section "Measuring Bandwidth Requirements"). They are based around TCP/IP trace analysis. The bandwidth tools are provided on the CD-ROM at the back of this book. See appendix F and the README file on the CD-ROM for usage instructions. The bandwidth analysis tool used by IBM for OS/2 is based on the IPTRACE utility that comes with TCP/IP for OS/2. Just run IPTRACE in an OS/2 Window, and it will dump all IP activity going through the stack on that machine to a file called IPTRACE.DMP.

The .DMP file can then be formatted using the IPFORMAT command. Do an IPFORMAT -? for details on the parameters. The most useful parameters are -a, which prevents ARP/RARP packets from being shown (there will be an awful lot of these!), and -s, which allows you to filter out all but the specified MAC address. For example, use IPFORMAT -a -s<mac address of Notes server> to format the trace for all non-ARP activity with the Notes server as source or destination.

Obviously, this has its limitations, the most notable being that it will only show the data that has gone through the stack on the machine you're running the trace on. However, if your server is OS/2, run it on there and you should get all the Notes traffic. Look for Notes' well-known port 1352 in the output, and you're away!

A DatagLANce trace is more comprehensive than IPTRACE but is more time-consuming, and you'll need to do a lot more filtering to get the desired information.

The Role for NotesBench

NotesBench is a Lotus Tool that is made available to PC hardware vendors and the 13,000 Lotus Business Partners. NotesBench is an estimating tool, used to help customers determine how many users a particular hardware/software platform will support. IBM used NotesBench in their internal roll-out of Notes, for example, to determine how many users each node on the SP2 would support. So NotesBench, together with tools to determine bandwidth requirements, will help you establish the server size and "pipe" sizes needed for your Notes installation.

Server Backup and Restore over the Network

It will happen that a Notes user will accidentally delete a document from his mail file or a document from a report he is writing. When this happens, it would be nice if the user could call his Notes Administrator and ask to have that document restored. Even better would be to allow the user to restore the document from a backup system himself.

Backup systems often allow a user to restore a backup from a list of backed-up versions of his document. For example, the document might be backed up every night, and backup versions of your document from the previous day, a week ago, and a month ago are available. Usually, the backed up version from the previous day is what you want to restore. However, say you only deleted part of your document and didn't realize it until two days later. Then the backup version from the previous day will also have that part of your document deleted and won't do you any good. In this case, you might want to restore the version from a week ago.

One product designed for backup and restore is IBM's Adstar Distributed Storage Manager (ADSM). ADSM runs on the AIX or MVS platforms and provides hierarchical storage management for a variety of workstations and server platforms (e.g., Windows NT servers and workstations, Silicon Graphics Inc.'s Irix, etc.). ADSM allows general backup and restore for workstations and servers but also features a special backup agent for Lotus Notes. The Lotus Notes backup agent allows for "incremental" backup and restore of Notes databases similar to way Notes uses replication. The Notes servers in the ADVANTIS Lotus Notes domain are backed up via an ADSM system running on MVS at a "hardened" site in Columbus, OH. MVS was used as the platform because the mainframe and DASD were available and also because a high-speed network to Columbus was available from all the Advantis Notes sites.

The Tivoli Management System

IBM uses the Tivoli TME 10 Module for Domino/Notes to manage their internal Notes systems. The TME 10 product allows administrators to use one product to deploy, monitor, and administer an entire Notes environment while also reducing the need for additional Notes Information Technology staff at remote locations.

The TME Module is intended for large Notes installation (1000+ Notes seats). It is an extensive management system meant for customers who need consistent, integrated life-cycle management for their mission-critical applications. It is meant for organizations committed to an enterprise Notes roll-out, either new installations or upgrades. Because TME 10 is an extensive management system, it also means that it is a fairly expensive system, meant for large organizations.

TME 10 system manages both the Notes infrastructure and Notes applications. The management system is designed especially for mission-critical applications, based on the Applications Management Specification (AMS), the Tivoli Developer Kit, and the TME 10 suite of application management products.

Disaster Recovery Considerations

Disaster recovery, as the name implies, refers to getting your Notes system back online when a system, network, or site failure occurs. The "disaster" could be catastrophic (e.g., your office building burned down or was destroyed by an earthquake, and all the Notes servers, workstations, and network components were destroyed). In this case, all hardware and software must be replaced. All Notes databases and operating system software must be replaced from backup copies stored outside the building. This type of Hierarchical Storage Manager disaster recovery has been available for mainframes for many years and only recently has been made available for file servers.

Disaster recovery planning for Notes covers much more than the catastrophic disasters just discussed. The disaster scenario could be the result of a system failure. System failures are conditions resulting from either hardware or software failures. Both hardware and software failures can result in either intermittent or hard-down (unrecoverable) failures. Any number of hardware component failures can result from manufacture flaw, electrical problems, continuous operation, or abuse. Software product failures result from the inability of software to coexist rendering the system unavailable, programming bugs, data loss, data corruption or incorrect software customization.

The installation of Notes services for specific sites is determined in part by the available environmental contingencies (i.e., physical security, power, network route redundancy, on-site personnel). Should a site become unavailable, Notes users' sessions need to be reconnected to Notes systems at alternate sites that have mirror images of that Notes data.

Disaster recovery expectations will vary based on customer requirements. A general list of user requirements that need to be designed as a baseline to satisfy these requirements follow:

- Notes server failure
- Data corruption or loss
- Damage to physical site

Using a Test/Development Domain

In order to test out changes to your Notes network, it is highly desirable to have an isolated ring with test Notes servers. Then test changes to your servers or network can be first run in the test domain/network with no fear of impacting your production Notes system.

Management of both Hierarchical and Nonhierarchical Domains

The following material is from the IBMNOTES instructions and will be useful to you when you deal with the mixed hierarchical and nonhierarchical domains. This information is quite IBM specific, but the concepts and the step-by-step instructions should prove useful to many other Notes procedures and systems.

TCP/IP

■ ■

NOTE: *When installing TCP/IP, make sure that a route to the IBMInternal backbone (MPN) is available. (You might need to contact network support for information on your local router.) If you have any problems connecting with the IBMNotes server, you might need to add the server to your tcpipfile (this is coreon NetDoor). See the HOSTS SAMPLE file included with the IBMNotes package for sample entries.*

Notes Connections

There are two ways to connect with the IBMNOTES server. First, an end-user workstation can connect through a File|Database|Open, then add databases to their folder, accessing the server directly. Second, where there are multiple people accessing IBMNOTES from a site, database replication can be scheduled between IBMLN01 and a local site server.

NOTE: *If you are not the Notes Administrator, you might be able to jump ahead to the section "Opening An IBMNOTES Database."*

Stamping SafeIDs

To enable you to directly access the IBMNOTES server, the following steps are needed:

1. Have your Notes domain administrator cross-certify between your domain and the IBMNOTES domain.

2. Make sure your workstation ID is properly certified.

The details of these two steps will vary with whether your domain is a hierarchical Notes domain or a flat domain. (To determine your domain type, go to FileIToolsIServer AdministrationIID...IAdministration...IID File, select your local ID file—usually in your Notes subdirectory as UserID. The type will be listed as either Hierarchical User or Server or Non-Hierarchical ID) IBMNOTES is a nonhierarchical (or flat) domain.

For more detail about the following steps, consult the Lotus Notes Administrator's Guide.

Nonhierarchical (or Flat) Domains

Have your Notes domain administrator certify the IBMLN01.ID file included in this package and return it to "IBMNOTES Administrator@ IBMNOTES." Then have them send a safe copy of their certifier ID file for cross-certification by IBMNOTES. (This will be stamped and returned to be merged into your domain server's ID.) Once the exchange is complete, anyone in your Notes domain will be certified to access the IBMNOTES server.

Hierarchical Domains

To cross-certify with IBMNOTES, a hierarchical domain must first create a nonhierarchical (or "flat") certifier ID. The Notes Administrator selects FileIToolsIServer AdministrationICertifiersIRegister Flat... With this ID, certify the IBMLN01.ID file and your server's ID file. Send the nonhierarchically certified IBMNOTES ID file, along with a new safe copy of your server's ID file (created after it was certified with the newly created nonhierarchical certifier). This will be stamped and returned to be merged into your domain server's ID.)

Send the ID files to "IBMNOTES Administrator@IBMNOTES."

Finally, the Notes Administrator uses the nonhierarchical certifier to recertify the IDs of all users in the domain who wish to access IBM-NOTES.

NOTE: *If you are connected to the IBM_INTERNAL hub, mail routing will be automatic. If not, first contact your Notes administrator to have IBMNOTES defined as a Non-Adjacent Domain through IBM_INTERNAL.*

Additional Setup

There are two known ways to set up for Notes to access the IBMNOTES server through TCP/IP. First, try just using the server IP address "ibmln01.ims.advantis.com" while opening an IBMNOTES database. Should that fail, try either adding "ibmln01.whp.advantis.com" to the file on your system (see the HOSTS SAMPLE file included with this package) and/or create a Server document in the Name and Address Book as follows:

```
Server name:      IBMLN01
Server title:     IBMNOTES Repository
Domain name:      IBMNOTES
Network Configuration
Port      Notes Network    Net Address                  Enabled
TCPIP     TCPNET           ibmln01.ims.advantis.com     ENABLED
```

OPENING AN IBMNOTES DATABASE.   Select File|Database|Open. In the Server: field, enter the IBMNOTES server name (IBMLN01), and select Open. This should give you a list of the available subdirectories on the server. Feel free to "Add Icons" for any databases you see. Also feel free to copy any of the sample databases you find. Certainly, feel free to contribute any Notes applications you have to be shared by others in the company.

SERVER REPLICATING.   Feel free to replicate any databases of interest. If possible, post a message to the IBMNOTES Administration discussion to let us know, in general, how often and when you will be replicating.

Cascading Your Notes Domains

If you have several domains for your corporation, you can "cascade" the domains so that, for certain functions, they appear as one domain. Most Notes gateways (e.g., Lotus SMTP gateway and IBM Mail LAN Gateway) require that any Notes domain, with users that need to use the gateway, be part of the cascaded Name and Address Book on that gateway. Also, if you have cascaded Name and Address Books on your Notes mail servers, when users address mail, the addresses of users in any of the Name and Address Books will appear when you press the Address button when composing mail. The way to set up and use cascaded Name and Address Books is described in this section. Remember that the limit for the number of cascaded Name and Address Books in your NOTES.INI file is 10 (including the Name and Address Book for your own Domain, NAMES.NSF).

In order to cascade Name and Address Books in your Notes domain, it is first necessary to make replica copies of all desired Name and Address Books on your Notes server. Of course, because the original Name and Address Books that you want to replicate probably have a name of NAMES.NSF, it is necessary to change the name of those files when you replicate. Your domain can have only one NAMES.NSF file, and that belongs to your own domain's Name and Address Book. Let's say the filenames you choose for the replicated Name and Address Books are SITEA.NSF, SITEB.NSF, and SITEC.NSF. Then the next step is to edit the NOTES.INI file on your Notes server. Change the NAMES= parameter in the NOTES.INI file so that it has the form:

```
NAMES=NAMES, SITEA, SITEB, SITEC
```

Then bring up the Notes server again, and your users will be able to look at all four Name and Address Books when they use the "Address" selection in composing mail.

13

Network Tools and System Platforms for Lotus Notes

Lotus Notes is truly cross-platform.

Forrester, January of 1997.

There are many network tools, other than the tools that come with the Notes product, that can be used to help manage your Lotus Notes system. This chapter describes some of these tools. Also, discussed are aspects of the different system platforms that can be used with Lotus Notes. In addition, some of the tools that should be useful in developing customized applications for your Lotus Notes network are described.

Remote Control (PolyPM/2)

PolyPM/2 is a graphical remote-control program for OS/2 and its graphical interface Presentation Manager (PM). There is also a Windows version of PolyPM/2. This software allows an OS/2 workstation, called a *Teacher*, to access and control remote OS/2, DOS, and Windows workstations, called *Pupils.* The communications link between a Teacher and a Pupil can be:

■ Directly over serial lines through null-modem cables.

■ Over telephone networks through asynchronous or specialized ISDN modems.

■ Across the LAN through NetBIOS (IBM), NetAPI (MS), and IPX-SPX (Novell).

■ Across SNA backbones (SDLC, X.25, ISDN, and modem) using APPC or APPN.

■ Across native X.25 or ISDN networks.

■ Across TCP/IP LANs and WANs.

PolyPM/2 provides both complete text mode and graphical support such that the Teacher can control the Pupil's desktop as if it were locally available. IBM Global Network uses PolyPM/2 to remotely manage Lotus Notes servers running on the OS/2 platform. The TCP/IP protocol is used over the LAN and WAN for this management function. For example, if I am a Notes Administrator in White Plains and need to change the NOTES.INI file on a server in Schaumburg, I can use PolyPM to "login" to the server. On my workstation, it will appear as if my monitor is directly attached to the server. I can use OS/2 windows and the OS/2 desktop just as if I were in Schaumburg. I can even restart the Notes server to make the changes effective.

NetFinity Software

NetFinity is an IBM PC product that has various threshold and alert functions to provide constant bulletins on the status of your networked system. The product reports imminent hard disk failures, memory errors, and other mayhem. NetFinity reduces downtime by:

- Predictive failure analysis with alert management
- Security management
- Scheduled maintenance
- System monitor service

NetFinity is bundled with all IBM 320 and later servers as well as with 300- and 700-series workstations. Release 3.0 includes such enhancements as critical file monitoring, serial connection control for remote users, and an alphanumeric paging feature. It also includes software inventory capabilities. However, for accurate software scanning, administrators must manually input detailed software information. In addition, the hardware inventory tool lacks fundamental information, such as IRQ (interrupt request) statistics.

NetFinity works on a peer-to-peer basis and collects data from workstations and servers on demand. Because NetFinity Version 3.0 (and up) does not store data in a central repository, this can cut down on network traffic. To collect and track system data, both clients and servers must have NETFBASE.EXE installed. There is also an optional executable that administrators can install on client machines that lets users monitor hardware and software statistics for their own systems.

One of NetFinity's strongest features is its alert functions. Alerts can be set for a number of preset scenarios (such as low disk space) for both workstations and servers, or you can create your own scenarios. Administrators can store and execute Alert handlers in a variety of ways, such as in a log or a pop-up window. NetFinity provides the ability to generate and send messages to Lotus Notes, cc:Mail, or any Vendor Independent Messaging-compliant e-mail package. Also, administrators can generate a DMI (Desktop Management Interface) indication and send it to the DMI Service Layer. In addition, NetFinity includes an e-mail/pager gateway for generating and sending alert messages, including alphanumeric pagers.

Administrators can configure NetFinity to pass SNMP alerts onto such SNMP-based management consoles as IBM's NetView, Lotus NotesView, Hewlett-Packard Co.'s OpenView, and Computer Associates International Incorporated's UniCenter. With NetFinity's Process

Manager, you can see which modules were loaded on your server and can load and unload modules. You are able to use all NetFinity's management features using a dial-up link.

NetFinity allows you to create graphs for workstations and servers that include such information as the number of connected users, volume space used, and cache blocks in use. NetFinity supports Windows, OS/2, and NetWare servers. Management-station support includes Windows and OS/2. Network support includes NetBIOS, IPX TCP/IP, and serial connections for remote users.

Some important items to monitor with NetFinity are:

- Alert that the NOTES.INI file was modified.
- Alert that a Server Record in the Name and Address Book was modified.
- Alert that the size of the MAIL.BOX grew beyond a certain threshold.

Telnet

Telnet allows users of one TCP/IP host to log into a remote host and interact as a normal terminal user of that host. Telnet is user ID and password protected. Telnet is not considered very secure on the Internet; however, it is often used on corporate internal networks. Because telnet allows remote login, it can be used for most of the same tasks as mentioned for other remote access applications, such as PolyPM or NetFinity. PolyPM and NetFinity are not available for Unix systems, so telnet is especially useful for Unix systems. Like PolyPM and NetFinity, telnet brings you in at the operating system level and is generally used only in text-based mode. Telnet is part of most TCP/IP stacks, so it doesn't have to be ordered separately.

Telnet is quite useful as a management tool to check on Notes application availability. For example, the TCP/IP Port 1352 is used exclusively for Lotus Notes. A test for Lotus Notes server availability can be used with the following scenario (the format is for the OS/2 system, other operating systems have telnet tests that are very similar):

```
telnet -p 1352 hostname
```

where hostname is the host name of the Notes server. If this command fails, a connection "timed out" error will occur. Success will give a connected message. Note that this is not a true remote login, it's just to test availability.

File Transfer Protocol (FTP) Tips

FTP is part of most TCP/IP stacks. It is used for efficiently sending files over a network. First you must execute a FTP login with a user ID and password. Once connected, you can navigate through the directory on the remote server. The basic FTP commands are put and get. So you can easily get files from the remote host (server) and copy them to the local host. Or, you can put files from the local host to the remote host. For example, to copy Notes databases from one server to another, it is much (much!) quicker to use FTP than the Notes File¦New Copy procedure.

Like telnet, FTP is not considered very secure on the Internet. In fact, anonymous FTP, is used for public access on the Internet, so by design, anonymous FTP is wide open. The user ID and password are typically both guest for anonymous FTP.

FTP is not considered secure on the Internet; however, it is often used on internal corporate networks. As an example, FTP can be used with ADSM for Notes server backup and restore. Another example of the use of FTP is for the distribution of updated Notes software. From a central server, a new release of Notes software (or other system software) can be distributed to multiple remote servers via FTP.

FTP and Notes Databases

The quickest way to copy a Notes database from one Notes server to another is to use FTP. (FTP stands for "File Transfer Protocol" and is part of most TCP/IP protocol suites.) Using FTP is much quicker than using File¦New Replica in Notes. There is a lot processing overhead using Notes to make a new replica. FTP is part of nearly all TCP/IP stacks, and the command syntax is usually the same across the various operating systems.

FTP copies a file from a remote server to the local server. From a Notes viewpoint, this means the Replica ID is the same on the copied file as on the original. The copied file is indeed a new replica, and the two files can now replicate. Of course, if you intended to make a new copy instead of a new replica, you cannot use FTP directly. You can make a new copy in two steps by first FTPing a new replica to the local server, then using Notes to make a new copy locally. This will be quicker than using File¦New Copy with the remote Notes server across the network.

The next level of complexity occurs when FTPing between a Notes 3.x server and a Notes 4.x server. The Notes database on the Notes 3.x server has a .NSF file extension. FTPing a Notes 3.*x* database using the default FTP settings (which does not change the filename) will result in a file with a .NSF file extension on the Notes 4.*x* server. This will cause problems because, on the Notes 4.*x* server, a file with a .NSF file extension is treated as a Notes 4.*x* database. The file must be changed to have an .NS3 file extension to be handled as a Notes 3.*x* database. The file can be renamed during FTP. Alternatively, after FTP, rename the file using the operating system commands. Finally, the .NS3 database can be converted to a Notes 4.*x* database using the FileINew Replica command in Notes. This method is very useful for migrating from a Notes 3.*x* to a Notes 4.*x* infrastructure.

FTP can be used with a graphical user interface (GUI). For example, from an OS/2 prompt, the FTPpm command brings up a graphical interface for running FTP. The menus and prompts help guide you through the proper configuration. The files and directories for the local host and remote host are displayed together, which helps avoid confusion and common pitfalls when trying to keep track of which direction files are being transferred.

A graphical user interface (GUI) is more convenient to use than issuing a sequence of FTP commands at the command prompt. For remote monitoring, however, using the command-line FTP commands is usually preferred for the faster response time. The graphics contains so much data that it takes longer to send a screenful of data than to send a line of text in response to a command-line FTP command. The difference is noticeable when going over the wide area network, and especially (painfully!) evident when using dial access to the servers. It is also good to know the command-line FTP commands because, sooner or later, you will be confronted with a system that does not have a graphical user interface to FTP. See the CD-ROM for a sample FTP text-based session.

The FTP commands include standard UNIX commands (because FTP is part of the TCP/IP protocol suite, which was originally developed intimately with UNIX). These are not OS/2 or Windows commands but subcommands under FTP. To begin a FTP session, enter:

```
FTP <TCP/IP hostname>
```

or, enter:

```
FTP <TCP/IP numeric address>
```

You will be prompted for your user ID, which is case sensitive. You will then be prompted for your password, which is also case sensitive. If you make a mistake in entering either the user ID or password, you will receive an error message after you enter a password. There might be an error in either the user ID or password you entered, or an error in both. This is a feature to make it more difficult for somebody to guess valid user IDs and passwords. (Of course, it is more difficult for me to figure out if I typed the user ID wrong or the password wrong.)

Once you have established a FTP session, you use the FTP commands. The following sections cover the most commonly used commands.

pwd. This stands for "Print Working Directory." This means show the current directory name. (In the old days before video displays, you interacted with the computer via a teletype. Your input and the computer's output was printed out on a scroll of paper. You could not "display" the directory name, you could only "print" the name. The command name, pwd, has stuck around for historical reasons.) *Make sure* you issue the "pwd" command just before you issue the command to transfer files! It is too easy to be in the wrong directory. You can overwrite files unintentionally and have an awkward time recovering the original file. (Believe me. . . . I have done this before!)

cd. This stands for "Change Directory." You can change to the root directory of an OS/2 or Windows server with the command:

```
cd \
```

Of course, on a UNIX server the slashes are reversed from OS/2 and Windows. (It is rather unfortunate that the direction of the slashes were reversed when Microsoft developed the MS-DOS operating system, the ancestor of OS/2 and Windows. Those of us who manage a mix of UNIX servers and OS/2 and/or Windows servers are invariably using the wrong slash on the wrong server!)

You can change to the Notes data directory on the E: drive with the command

```
cd e:\notes\data
```

If you issue the pwd command now, you would see something like:

```
The current directory is e:\notes\data\
```

You can move up one level of the directory tree with the command:

```
cd ..
```

Issuing pwd gives:

```
The current directory is e:\notes
```

Caution: Notice that the cd command is like the standard DOS/Windows/OS/2 cd command for changing directories. However, the DOS/Windows/OS/2 cd command is also used to display the current directory. In UNIX and hence FTP, a separate command (pwd) is used to display the current, or working, directory. To make matters worse, cd without a parameter has special meaning in UNIX and FTP. It means to change to the "home" directory. This is usually the directory you started from after logging in using FTP. So if you meant to see what directory you are currently in and enter:

```
cd
```

you might get a message like this:

```
The current directory is c:\home\user\
```

Or worse yet, you might get no message at all, and you might think that you are in the directory e:\notes\data\. This brings me another important tip: Make it a good habit to issue the pwd command after *every* cd <directory> command to verify you are in the directory you think you are in! (Another unfortunate occurrence during the development of MS-DOS was to make the cd command take on the additional function of pwd! This also mixes up people who work with UNIX systems and DOS/Windows/OS/2 systems!)

On some operating systems, the command cd e:\notes\data might not work (I confess—I told a white lie). You might have to take two steps:

```
cd e:
cd notes\data
```

The first command changes to the E: drive and the second command changes to the desired directory. (Don't forget to do a pwd afterwards!)

dir.   Now, how do you list the files in the directory on the remote server? On servers running OS/2 or Windows, the usual dir command works. On UNIX servers, dir usually does not work. You must use the UNIX command ls to list the files (I have searched the UNIX lore, but have not found if ls is an abbreviation for anything other that "list." Let me know if you find out!).

ls. See "dir."

a. Before you transfer files you should specify if the file is ASCII or binary. The default might be ASCII, but you should be sure. It is a good idea to be explicit and issue the a command. On some systems, you might need to use asc instead of just a.

bin. See "a." b sets the transfer mode to binary. I often forget to change modes to binary. Of course, this always happens when I am dialed in on a slow line and transfer a large file. After an hour to transfer the file, I close the connection. When it comes time to use the file, I then discover the file is no good, and I realize I forgot to change to binary mode. You would think that, with all of the psychological reinforcement, I would remember. Well, with all my dwelling on this subject, please learn from my lessons and remember to change to the proper transfer mode!

get <*FILENAME*>>. This command (guess what?!) gets the file (with the name <filename>) from the remote server. The default is to keep the filename on the local server (or workstation) the same as the remote system. At times, you will transfer files from a UNIX server to an OS/2 or Windows 3.x system. The UNIX server and OS/2 server running HPFS support files with long filenames and long file extensions, whereas OS/2 running FAT file systems and Windows 3.x will only support filenames with 8 characters and file extensions with only 3 characters. In this case, you must rename the file with the following syntax:

```
get <remote_filename> <local_filename>
```

For example, to transfer the file How_to_FTP_files.readme.text from a remote UNIX system to a local Windows 3.x system with the name howFTP.txt, you issue the command:

```
get How_to_FTP_files.readme.text howFTP.txt
```

Usually you will get a warning error message if you forget to specify the local filename and the local filename is incompatible.

mget. This is a very useful command to get multiple files (in fact, this is how to remember the command: "multiple gets"). Say you want to get all five files in a directory. mget * gets all five files and saves you from typing get commands five times. You do not have to type the filenames either. (Saving on typing always appeals to people like me who get lazier with age!) Standard wildcard conventions apply. If you want to get just the files with the file extension ".tmp", you issue the command mget *.tmp. (get and mget change the file date and time to the current date and time. Keep this in the back of your mind when you are checking dates and times so that you do not get confused.)

prompt.   This command toggles the prompt off and on. When you use the mget command, the system will ask if you really want to get the next file (whose filename is presented in the prompt). This prompt is good. If you are dialed in to the network on a slow link, you might not want to get the large files. This prompt is bad. If you know what you want (and are really, really sure you know!), you can turn off the prompting by issuing the prompt command. This way you can get all the files in a directory and walk away. After your lunch hour, all of the files have been transferred over, and you did not have to answer "yes" a hundred times to transfer each file. Furthermore, you did not have to strain your eyes staring at the screen waiting for the next prompt. (I have strained my eyes many times before I learned about this prompt command!) The default for the prompt is to be turned on. Issue prompt to turn the prompt off.

I cautioned you to be really, really sure before you use the mget command and turn off the prompting with the prompt command. Remember that FTP overwrites files. The prompt will not warn that you are about to overwrite a file. The prompt only gives you a chance to think about whether you are about to overwrite a file. By turning off the prompt, you do not even have a chance to think about it. Be careful! (Again, I would not be warning you if I had not made this mistake myself many times. I confess!)

lcd   This command allows you to change directories on your local disk drive ("local change directory"). If you initiated your FTP session from the c: directory, then all of the files you get will be placed in this directory. If you want to use a different directory, such as the data, you enter:

```
lcd data
```

You can change to the E: drive with:

```
lcd e:
```

The lcd command is analogous to the cd command discussed previously. The hints, tips, and cautions for cd apply to lcd as well. For example, a plain lcd command might show you the current directory or it might bring you back to the local "home" directory. This latter case is frustrating because you have to do your lcd commands all over again.

The lcd command is not implemented on all systems. Instead, use !cd in the same manner as lcd. The exclamation point is the escape character that causes the command following it to be executed on the local system instead of on the remote system. For example, !pwd would show you the current local directory, and !ls or !dir would show you the local directory listing. The exclamation point is often called "bang" (from cartoon guns).

Just as it is a good idea to run a pwd command before transferring files (to confirm which remote directory you are getting your files from), it is a good idea to run an lcd or !pwd before you transfer files (to be sure you put files in the right local directory and do not overwrite files unintentionally).

?. It is always worth entering ? to get a list of valid commands. This is particularly helpful to determine if bye, exit, or quit is supported. I often forget which command gets me out of FTP. Applications like FTP, telnet, and nslookup all seem to use a different command.

Here is a sequence of commands that are typically used for a FTP session. Comments are in square brackets:

FTP apollo	[FTP to the server with the alias "apollo"] (Enter user ID and password)
pwd	[Show current directory on remote server]
cd j:	[Change to J: drive on remote server]
pwd	[Confirm the change of directories]
cd notes\data\mail	[Change to notes\data\mail directory on remote server]
pwd	[Confirm the change of directories]
dir	[List files in remote directory]
bin	[Set mode to transfer files as binary files]
prompt	[Turn off the prompt—don't ask if I want the file]
lcd	[Double-check the local directory is correct]
pwd	[Double-check the remote directory is correct]
mget *	[Get all the files in remote directory]
!dir	[List files in local directory to confirm successful transfer]
bye	[Quit out of FTP]

NotesView

Lotus NotesView is a graphical management product that lets administrators monitor and control enterprise-wide Lotus Notes environments, using industry-standard management tools.

The NotesView product is comprised of a Windows-based management station, which uses a color-coded, graphical display to enable administrators to gain real-time access to information about the status of their Notes environments. NotesView agents collect comprehensive system, database, replication, mail, and network information from servers running on the full complement of heterogeneous Notes platforms, and automatically pass it to the management station for display and analysis.

Features

With its graphical user interface (GUI), NotesView provides the Notes administrator with a logical layout of the Notes environment (including mail routing, replication, database, and servers) so that they can detect potential operational fluctuations before they seriously impact the Notes network. Leveraging the capabilities of NotesView allows a Notes administrator to become proactive instead of reactive to Notes network issues.

NotesView can be used to monitor and control an entire Notes backbone from a single location. In larger networks, multiple management stations can be established and management tasks delegated.

Using NotesView, normal operating levels for system statistics and events can be established, and then the performance can continually be monitored relative to that criteria. In the event that a threshold is exceeded, indicating a situation that requires administrator intervention, NotesView will trigger an event or alarm notifying the administrator so that he/she can take corrective action.

Furthermore, the statistics and events gathered from Notes servers are stored on the management station in an ODBC-compliant database. Customized reports can be created for analyzing specific trends.

Using the statistics and events that are instrumented in Notes, NotesView provides:

- Control over Notes servers, replication, mail routing and Notes databases.
- Graphical Notes network maps:
  - Network topology map shows the network each server is connected to as well as the status of the connection.

- Four distinct Replication maps including the replication topology of all the servers and a map that shows where each database is replicated
- Three Mail routing maps
- Isolation maps that allow you to customize a map-view to suit your own criteria.

■ The NotesView Poller collects information from the Notes servers, such as number of users, replication traffic, document changes during replication, available disk space, and mailbox size. The Poller can be configured to run a series of diagnostic tests on the servers to check connections, ports, and more.

■ Real-time monitoring and statistics using InfoBox technology, users get tabbed information about:
- Server statistics
- Mail monitoring
- Mail probe
- Replication
- Notes databases
- Network

■ Using data collected from the Poller, network-wide graphs can be created that show trends for the Notes server, mail, replication, and network statistics. This information is critical to better understand and optimize capacity planning.

■ Lotus NotesView is designed to monitor and control all Notes R3 (or greater) environments including OS/2, Windows NT, Novell NetWare, Solaris, HP-UX, SCO, and AIX.

■ NotesView communicates with Notes servers either using the Simple Network Management Protocol (SNMP) or Notes protocols.

System Requirements

NotesView Management station:

■ 486 PC with 16MB RAM, or better.

■ Minimum 300MB hard disk space.

■ Optimized for 1024×768 display.

■ HP OpenView for Windows R7.2 or HP OpenView for Windows WorkGroup.

■ Node Manager R1.0.

■ Notes client Release 3.31 or greater.

- Microsoft Windows 95, Windows NT, Windows Version 3.1, or Microsoft Windows for Workgroups 3.11.
- Winsock-compliant TCP/IP is optional to utilize SNMP management capabilities.

Notes servers:

- Notes server software.
- NotesView SNMP Agent (optional) available for OS/2, NT, and NLM.
- NotesView Mail Reflector Agent (optional) available for OS/2, NT, and NLM.

Network/System Platforms (with Hints and Tips)

This section discusses the different network and operating systems used for Lotus Notes and some of the performance considerations for each operating system.

OS/2 System

OS/2 was the original operating system for Lotus Notes, and many Notes add-on programs only run on this system, although Windows NT is quickly becoming another favored system for Notes.

OS/2 NOTES.INI TUNING. Here are recommended parameters to be used in the NOTES.INI on your OS/2 Notes Server. (The parameters run to the end of this section. The text between the parameters are comments.)

```
NIF_Pool_Size=524288
NSF_Buffer_Pool_Size=8388608
Default_Index_Lifetime_Days=15
MailMaxThreads=2
No_Mail_Broadcast=1
MinNewMailPoll=15
Server_Session_Timeout=45
```

If there is 32MB of memory or more, then:

```
NIF_Pool_Size=524288
NSF_Buffer_Pool_Size=8388608
Default_Index_Lifetime_Days=15
```

(The default is 45.) This is overridden by individual view settings.

```
Update_No_Fulltext=1
```

Will only stop updates, won't delete the indices.

```
MailMaxThreads=2
```

If you are using multiple Router tasks, you should monitor the performance and use of each of the tasks to be sure that none of them remains idle the majority of the time.

```
No_Mail_Broadcast=1
```

and remove the STATS (and LOGIN) tasks from ServerTasks=.

Disable the Statlog server task (installed by default to run daily at 5 A.M.). This task updates database statistics in a server's log file (LOG.NSF) such as those found in the "Database Sizes" view of the log file. If you are not concerned with collecting these statistics, you can disable the Statlog task.

Disable the Design and Catalog tasks (installed by default to run daily at 1 A.M.) if they are not being used.

```
MinNewMailPoll=15
```

(The user default is 15, but they can lower it. The server set default is 1 or 2.)

```
DailyMacrosHour=6
WeeklyMacrosDay=1
WeeklyMacrosHour=7
Server_Session_Timeout=45
```

(The default is 240, Lotus says that 45 is about the optimum, as it takes resource to re-open them again.)

```
Log=log.nsf,1,0,3,4000
```

Parameters for Notes Log:

- Variable 1 = Log's filename

- Variable 2 = Logging options to be added together

- 1 = Log to console

- 2 = Force database fix-up when opening log

- 4 = Full document scan (as opposed to quick scan or open)
- Variable 3 = Not currently used
- Variable 4 = Number of days at which to delete log documents
- Variable 5 = Size of log text in event documents

```
Log_Replication=0
Log_Sessions=0
PhoneLog=0
```

WINDOWS CLIENT IN WIN-OS/2.   The Windows version of Notes can be run in a WINOS2 session using the virtual NetBIOS device driver support of OS/2 2.x supplied with LAN Requester 2.0 or 3.0 or NTS/2. All you need to do is enable netbios VDD support.

If you are running LAN Requester 2.0, then from an OS/2 prompt:

```
x:\IBMCOM\SETUPVDD
```

where x: is the boot drive. This adds two device driver statements to CONFIG.SYS:

```
DEVICE=C:\IBMCOM\PROTOCOL\LANPDD.OS2
DEVICE=C:\IBMCOM\PROTOCOL\LANVDD.OS2
```

If running LAN Requester 3.0, you might have to add these manually, due to a bug in the install. (It depends on how you installed.)

Reboot, then configure the VDM for Name Number 1 support (plus a few sessions commands and names) from the DOS session prompt, AUTOEXEC.BAT, or batch file:

```
x:\IBMCOM\LTSVCFG N1=1 S=10 C=14 N=4
```

where x: is the boot drive.

NOTE: *This allocates more resources than are probably needed.*

Virtual NetBIOS support is documented in the "IBM OS/2 LAN Server Network Administrator's Reference Supplement for OS/2" (S04G-1080).

WARNINGS.   Only one program can use NameA1 support (the N1=1 parameter) on the adapter at a time. The first DOS program to request it gets it. In OS/2 2.1, you might want to use the DOS setting that lets you specify a different AUTOEXEC.BAT and put the command in there. That way, only the Notes VDM gets NAME NUMBER 1.

The OS/2 and Windows versions can run at the same time if the OS/2 version is using the NETBIOS driver (LAN requester started) as opposed to the IBMEENB driver, which I think requires the Name #1 support. However, *never* let them access the same databases, DESTOP.DSK file, or ID file *at the same time.* It'll corrupt them. Better if they have their own data directories.

UNIX System

Unix systems come in several different flavors. There is the Sun Solaris system, the Hewlett Packard HP-UX system, the IBM AIX system, etc. There are some applications that work for any of these UNIX systems, and some that work only for specific systems.

OVERVIEW. Lotus Notes supports HP-UX, IBM AIX, SCO-ODT, and Sun SPARC platforms, using Motif as the user interface. Notes for UNIX is designed to exploit the multitasking and multiprocessing capabilities of these platforms.

You can use Notes for UNIX to track product and development issues, review documentation, and create libraries of information. You also can use Notes to quickly update a customer service or account management database and access the specific customer information you need.

To make it easier for workgroups to exchange information across systems and platforms, LEL (Link, Embed, Launch-to-edit), a new technology that provides OLE (Object Linking and Embedding) functionality to UNIX applications, was added. With LEL, you can view and edit linked and embedded documents on UNIX systems.

In addition, LEL works across the network: You can edit an object in a Notes document even if the object's source application is running on a different UNIX machine on the network. For example, when using a workstation that does not have Ami Pro to view a Notes document containing an Ami Pro object, the user can invoke Ami Pro running on a different workstation on the network.

HOW DOES NOTES FOR UNIX WORK? Lotus Notes is a true client/server environment in which users (clients) communicate over a local area network (LAN) or a telecommunications link with a document database that resides on one or more shared Notes servers. Notes for UNIX makes intelligent use of multiprocessing to maximize the performance of many operations and to handle larger numbers of users.

For Notes for UNIX servers to connect to other Notes servers on any platform, they must be running the TCP/IP protocol or SPX/IPX or they can connect using Notes' dial-up capabilities. All clients—whether they are on the UNIX, Windows, OS/2, or Macintosh platforms—can access information from a UNIX server directly from a TCP/IP connection over the LAN, via Notes' dial-up capabilities, or by connecting to a server on a different platform that replicates with the UNIX server. Notes for UNIX uses the same pricing and licensing as other Notes platforms.

KEY NOTES FOR UNIX FEATURES.   To help you quickly set up your applications, more than 25 templates and example files are included with Lotus Notes for UNIX. Other features are:

- Online help and free client support for a period of 30 days.
- Installation using the UNIX graphical user interface (GUI) installation program.
- Integration within the desktop environment of the various UNIX platforms; the user can select Lotus Notes from the Root menu or launch Notes from the File Manager.
- SmartIcons, one-click shortcuts to frequently used commands.
- Full text search on server databases and across multiple databases.
- LEL (Link, Embed, Launch-to-edit), a new technology that provides Object Linking and Embedding (OLE) functionality.

APPLICATION DEVELOPMENT TOOLS.   There is a unique application framework for rapidly customizing both the look and the behavior of business process applications. Notes Forms serve as compound document containers, and Views provide a framework for browsing ordered collections. Server-based background macros can contain the process logic for workflow. There is multi-level security at server, database, form, document, section, and field levels. API (Application Programming Interface), a library of C language subroutines and data structures, allows comprehensive programmatic access to Notes services.

NETWORKING CAPABILITIES.
- Database replication across servers on different platforms or remotely via modem.
- Network support: When connected to a Notes UNIX, NLM, Windows NT, or OS/2 server that is running the TCP/IP protocol (or via modem), UNIX, Windows, OS/2, or Macintosh clients can all share information with other clients
- Remote use features for dial-up capability.
- Notes for UNIX servers can take advantage of the scalability and performance characteristics of UNIX systems.

- (Optional) The Lotus Notes Mail Gateway for SMTP allows Notes mail users to communicate with users of a variety of mail systems. MIME support allows file attachments to be handled between different mail system types.

- Use of the VIM developer's toolkit to mail-enable other applications to take advantage of Notes mail: The user can send documents created in a VIM-enabled application, using Notes mail messaging, without leaving the application.

- OCR, Incoming Fax, and Outgoing Fax are available for UNIX clients.

- Remote administration.

SYSTEM REQUIREMENTS FOR NOTES FOR UNIX.

- Intel 486/33 systems or above that are supported by SCO-ODT; HP 9000 Series 700 and 800 (PA RISC 1.1 chip sets); all IBM RS/6000s; or all Sun SPARC systems, IPC or above.

- RAM: At least 16MB for workstations; 32MB is recommended. At least 32MB for servers; 64MB is recommended.

- Swap space: Twice the amount of RAM, with at least 32MB for clients and at least 60MB for servers.

- Disk space: 100MB for the installation directory or distribution directory; 20MB for the server directory; 20MB for each workstation directory. In addition, the server directory needs enough space for databases and mail files (300MB or more is recommended).

- Removable media: Available on a CD-ROM for Sun SPARC and HP-UX, QIC 150 for SCO-ODT, 8mm DAT for IBM AIX; also media exchange to 4mm for HP-UX and IBM AIX, QIC 150 for Sun SPARC, and CD-ROM for SCO-ODT.

- Lotus Notes 3.0b or higher for each client; Lotus Notes 3.2 for each server with TCP/IP support.

- Operating system: Sun Solaris 1.1, Solaris 2.3, HP-UX 9.03, SCO-ODT 3, and IBM AIX 3.2.5.

- Motif 1.1 for X/11R4 or 1.2 X/11R5 interface. Notes for Sun SPARC will also run using Motif (under Open Windows 3.0).

- TCP/IP protocol software.

- Support for the following X-terminals: HP 700/RX and HDS FX, IBM XStation, NCD, Tektronix, Visual Network interface card, and appropriate driver software compatible with supported local area networks.

■ If connecting the Notes client to the Notes server via remote dial-up, 9600 to 33.6K bits per second recommended for daily use to access large databases. Over 60 popular modems supported; instructions are provided for additional modems.

UNIX (AIX) HINTS AND TIPS. Here is a list of commonly used UNIX commands to get you started. UNIX commands vary somewhat by the version of UNIX. These commands are based on IBM's AIX version of UNIX.

UNIX COMMANDS: QUICK REFERENCE

pwd	Lists the current subdirectory ("print working directory").
ls	Lists the files in the current subdirectory.
ls -l	Lists the files in the current subdirectory with file size and modification date ("long" format).
cd	Change directory.
df	Shows how much disk space is available, by file system ("disk free").
lmore	Displays one screen at a time; hit the spacebar to see the next screen.
lpg	Like lmore.
/	AIX uses only forward slashes, not backslashes.
kill -9 -1	While logged in as "notes," it kills all processes under Notes.
ps -ef l grep notes	To view all notes tasks running on AIX (not Notes).

AIX is case sensitive. Commands and filenames must have the matching capitalization. All databases defined within commands must match with the filename case in order to work.

UNIX NOTES HELP All Notes databases are located in the /home/notes/notesr4 subdirectory/.

Here is how to start Notes (using the default installation parameters):

1. Login.

2. Type xinit to start X-Windows.

3. At the $ prompt, type pwd to see what subdirectory you are currently in. You will be in the home directory for your user ID /home/notes. All Notes databases are located in the /home/notes /notesr4 subdirectory. All Notes executables are located in the /opt/lotus/bin directory.

4. Go to the /opt/lotus/bin subdirectory.

5. Type server to start the Notes server.

6. Use Ctrl—Alt—Backspace to end X-Windows.

SHUTTING DOWN NOTES There are several command options for stopping a Notes UNIX server, depending upon the situation:

■ In the Notes server console window, type q to quit.

■ Use the Notes remote server console, and issue the command q (for quit).

■ With a telnet session into the server, type server -q. *Note:* You must have Notes in this user's path for this to work.

■ If server -q does not work, another option in a telnet session is to try the following:

1. Issue the command ps -elf Igrep notes. This command generates a list of processes pertaining to Notes that are currently running, and each will have a process ID number associated with it.

2. At this point, you would want to kill these processes. Issue the command kill -9 xxx. In this command, xxx represents the process number associated with Notes. One can string these process numbers in the same command by separating them with a space.

3. Then, stop the semaphores and shared memory handles in order to be able to get the scrver back up. Issue the command ipcs. This command returns an output of memory and semaphore processes. These processes will also have an associated process ID assigned to them.

4. Look for the processes associated with Notes. Then issue the command ipcrm -m xxx. In this command, xxx is the process ID. These can also be stringed by separating them with a space, and you also need to specify the -m with each process being specified.

5. Issue the command ipcrm -s xxx. Again, xxx is a process ID number associated to each semaphore processes that is running.

6. After all processes are cleaned up, restart the server by typing server in any shellwindow.

DISABLE ROOT LOGIN: HARDER TO HACK Root is the UNIX ID used for system administration. The root ID can be configured to disable login from the server console and telnet sessions. To get root authority, you must log in with your own user ID and password and then "become" root by entering the su (become the "superuser") command and the root password. In this manner, there is no direct access to the administration functions of root from "outside" of the server. A hacker must break in to a regular user's account and then try to hack the root password from within the UNIX server. The hacker's programs must run on the UNIX server instead of on the hacker's own workstation. Hacking from within the server is more difficult for the hacker than from outside of the server. Hacking from within a UNIX server is also riskier for the hacker because it is easier to detect and track.

DISABLE ROOT LOGIN: IDENTIFY WHICH ADMINISTRATOR IS LOGGED IN I have been logged in as root and changed the "message of the day" (motd) that displays when any user logs in. When I tested it, I noticed that somebody else had changed the message right after me. Entering the who command (to show who is logged in), I saw that other administrators were also logged in as root from remote mobile workstations. If I did not disable root login, who would only show multiple entries for root logged in. I would not be able to know who else was logged in as root and making changes that conflicted with my changes. By disabling root login, who shows the user IDs logged in, and I could call or page my administrator colleague. This example demonstrates that disabling root login is good not just from a security viewpoint, but also from a collaborative system administration viewpoint.

I sent a broadcast message to all users using the wall command (which stands for "Write to ALL users' displays") asking who else was logged in as root. However, the other person had stepped away or just logged off and did not see my message.

DISABLE ROOT LOGIN: TRACK ADMINISTRATOR ACTIVITY Sometimes you want to ask an administrator colleague why a change was made. To find out who made changes as root, you must have root login disabled so that each administrator must log in to their own ID first. A log file, /usr/adm/messages (this stands for the "messages" file in the "administration" subdirectory of the "user" directory), tracks who became root at what time. By comparing the time a file was changed (using ls -l <file-name>) with the time people became root, you can determine who made the change and who you need to contact. Tracking administrator activity is one more reason to disable root login and require each administrator to log in with his or her own ID.

STARTING THE NOTES SERVER REMOTELY To start the Notes server locally, you use the server command described earlier. To administer the Notes server remotely, you can telnet to the UNIX server, start up X-

Windows (exporting the display to your host) and start the Notes server in a window with the server command. The problem with this method is that the Notes server will go down if you close the telnet session. It is not practical to leave your workstation on at all times (it could be a mobile laptop) to keep the Notes server running. Any network interruption between your workstation and the Notes server would also cause the Notes server to stop.

An alternate method for starting the Notes server remotely is to telnet to the UNIX server and issue the server & command to run in the background. You can then close the telnet session, and the Notes server will continue to run.

STOPPING THE NOTES SERVER REMOTELY   Stopping the Notes server remotely is similar to stopping the Notes server locally. From a telnet session, enter server -q. If you have a Notes client and your Notes ID is authorized, use the Notes remote server console to issue the command q (for quit). If these methods do not work, use the kill -9 xxx method described previously together with the ipcrm -m xxx and ipcrm -s xxx commands.

REMOTE NOTES SETUP THROUGH A FIREWALL   After the Notes server code is installed onto the server, the Notes server must still be configured. If the Notes server is the first Notes server in its Notes domain, the setup can be completed locally. If the Notes server is not the first Notes server in its Notes domain, the Notes server must connect to the first Notes server in its Notes domain to complete its setup. To setup a new Notes server outside a firewall, this method will not work because the external Notes server will not be allowed to initiate the connection back to the internal Notes server. The workaround is to FTP the SERVER.ID and NAMES.NSF files from the internal to the external Notes server through the firewall.

You do not have to go through a firewall to use this FTP method. This FTP method works for setting up any Notes server (or client) on the same local area network or wide area network. I prefer using FTP instead of using the standard Notes setup because FTP is much faster.

REMOTE NOTES ADMINISTRATION THROUGH A FIREWALL   You can telnet through the firewall to administer the UNIX system and the Notes server. Another useful method is to use the Notes Release 4.5 (or greater) client or Domino 4.5 server to access the external Notes server through a Socks firewall. Starting with Notes/Domino 4.5, the client and server are "socksified" applications and can work with a Socks firewall. You can use the Notes remote server console to administer the remote external Notes server from an internal Notes client.

REPLICATING FROM AN INTRANET DOMINO WEB SERVER TO AN INTERNET DOMINO SERVER   Domino 4.5 is socksified, so using a Socks firewall, it is possible to replicate between your Domino intranet Web site and your Domino Internet Web site. Because your intranet Web site contains public information (like press releases and product announcements) and private information (like new products schedules and downloads for internal programming tools), you will definitely want to use selective replication. You will need fewer people to manage both the intranet and Internet Web sites because they will have much in common and replication handles the updates. Remember that all replications must be initiated from the internal Notes server. Schedule the frequency of your replications appropriately. If your external Web site allows people to fill out forms, replicate often enough to process the forms with the desired frequency.

WHY THE WEB SITE IS NOT AUTOMATICALLY AVAILABLE WHEN THE DOMINO SERVER STARTS   Many people who start the Domino server for the first time are often puzzled as to why the Web site is not automatically available. You must edit the NOTES.INI configuration file to start the Web server automatically. Add http to the end of the ServerTasks=... line of the NOTES.INI file to start the Web server every time you start the Domino server. If you need to stop the Web server, enter the command tell http quit at the Notes server console. To start the Web server manually, enter the command load http at the Notes server console.

LOAD THE UNIX ONLINE HELP ("MAN" PAGES) AND USE IT OFTEN   UNIX comes with a good amount of online help kept in the "man" pages (short for manual pages). If you need help on using the command grep, enter man grep, and you will get the online help for this command. The syntax, options, and examples are usually available for each command. Enter man man to find out more about the manual.

MORE HELP AVAILABLE FOR AIX AND NOTES   IBM offers a Redbook with more help for AIX and Notes. Although the IBM Redbook is specific for IBM's version of UNIX, AIX, much of the information is helpful for all UNIX platforms. See the Bibliography for how to obtain a copy of the IBM Redbook.

LOTUS SOMETIMES USES PROXY WHEN IT MEANS SOCKS   Lotus sometimes uses the term *proxy server* when it means *socks server*. This is confusing. See chapter 7 for the distinctions between a proxy server and a socks server.

Windows NT System

Some of the main network tools for the Windows NT system are described in the following sections.

OVERVIEW. Lotus Notes for Windows NT provides a client/server platform for developing and deploying applications that facilitate the sharing and exchange of information and the monitoring and managing of work processes. Notes creates a true client/server environment: Users (clients) communicate over local or wide area networks with a document database that resides on one or more shared Notes Servers.

Lotus Notes Server for Windows NT is a version of Lotus Notes designed for the Windows NT operating environment. The product consists of the Notes Server and the Notes Administration Client.

With the Lotus Notes Server for Windows NT, Lotus has extended its industry-leading groupware solution to yet another platform, part of its ongoing effort to expand the products support for multiple platforms. A key to the power of Lotus Notes in today's heterogeneous corporate systems environment is its extensive multiplatform support. For Notes, the supported platforms are: IBM OS/2, Microsoft Windows, Macintosh, Novell NetWare NLM, Hewlett-Packard HP-UX, Sun Solaris-2, SCO UNIX, and IBM AIX/6000. All clients, regardless of the platform, can access all servers, exchange information, send and receive electronic mail, and participate in the full range of Notes services.

The Notes Server provides database-sharing and mail-routing services. The Notes Client, which is installed on client workstations, provides each user with a graphical interface to Notes. Shared databases reside on Notes servers, and users open those databases by moving icons to their Notes workspaces.

KEY FEATURES OF LOTUS NOTES SERVER FOR WINDOW NT. Lotus Notes Server for Windows NT is a full-featured Notes server, providing all of the functionality of the product as it is found on other platforms. Lotus Notes databases store text, rich text, images, motion picture images, graphics, sound, and numerical and structured data. Within the users workspace, Notes provides ready-to-run application templates for customer service, meeting tracking, and other common groupware applications. The Notes graphical user interface features scroll bars, pull-down menus, and context-sensitive help screens. Figure 13.1 shows the server and client platforms available for Lotus Notes.

Notes applications are mail-enabled automatically through NotesMail. Notes macros (agents for R4 and R5) can carry out commands, monitor

Figure 13.1

Lotus Notes server
and client platforms.

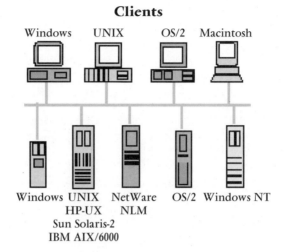

Clients

processes, and generate alerts while operating in the background, without user intervention.

The Notes Server for Windows NT will run on symmetrical multiprocessing (SMP) computers and can take advantage of up to two processors.

ADMINISTRATION. Lotus Notes can be run as a Windows NT service. This lets users start the Notes server either manually or automatically. With automatic service, the Notes Server Program starts when the server machine starts, without a logon procedure, so users do not have to be physically present to restart the server.

The Notes Administration Client can be run remotely from a separate Windows workstation while the server is running. The Notes Administration Client can also be run on the Windows NT Server, but not while the server is running.

SECURITY. Lotus Notes employs a Public-Key/Private-Key encryption scheme to provide the highest level of security available for client/server applications. Notes security performs the functions of authentication, access control, maintenance of confidentiality, as well as source verification and integrity. Electronic signatures can be used to verify the authenticity of objects, ensuring that the objects have come from a valid source and are unmodified. All this is provided in a manner that is extremely flexible.

With Lotus Notes, security can be added at several levels: server, database, view, form, document, section, and field. Access authority is defined through an access control list, with users identified as Managers, Designers, Editors, Authors, Readers, or Depositors.

To further secure the Notes Server Program in the Windows NT environment, users can invoke the SET SECURE command from the server console.

SYSTEM REQUIREMENTS
WINDOWS NT SERVER
- Hardware
  - Processor: Intel Pentium, 80486, or 80386 PC (80486 or higher recommended)
  - Memory: 16MB of RAM minimum (32MB recommended)
  - Hard disk: 120MB minimum (340MB or more recommended)
  - Graphics adapter and monitor: Windows-supported display (includes EGA, VGA, mono VGA, SVGA, IBM 8514A, CGA, and Hercules)
  - Mouse: (Optional) Highly recommended
  - Modem: (Optional) Needed to access Notes server from nonlocal area network machine; 28.8K bps recommended
  - Printer: (Optional) Must support Windows

- Server software
  - Windows NT Advanced Server Version 3.1, or higher, installed

NOTE: *If you are using the Standard Edition or a later version of Windows NT, check the side panel on the Lotus Notes Server for Windows box for a list of supported software*

WINDOWS WORKSTATIONS
- Hardware
  - Processor: Intel Pentium, 80486, 80386, or 80286 PC (80386 or higher recommended)
  - Memory: 4MB RAM in enhanced mode, 2MB in standard mode (6MB or more recommended)
  - Hard disk: 40MB (60MB or more recommended)
  - Graphics adapter and monitor: Windows-supported display (includes EGA, VGA, mono VGA, SVGA, IBM 8514A, CGA, and Hercules)
  - Mouse: (Optional) Highly recommended
  - Modem: (Optional) Needed to operate away from network; 28.8K bps recommended
  - Printer: (Optional) Must support Windows

- Software
  - MS-DOS or PC-DOS Version 3.0 or later
  - Microsoft Windows Version 3.0 or later (3.1 recommended)

NETWORKS Systems running Notes server or workstation software must connect to the LAN through a network interface card (NIC) that supports the network operating system. Each NIC must use cables and connectors compatible with the LAN cabling system.

NETWORK PROTOCOLS Notes Server for Windows NT runs with the following:

- NetBIOS (NetBEUI)
- TCP/IP
- XPC (for dial-up connections)

WINDOWS NT HINTS AND TIPS. Windows NT has a graphical user interface similar to Window 3.*x*, but otherwise, there are a lot of differences under the cover. Windows NT is a full 32-bit operating system, whereas Windows 3.*x* is a 16-bit operating system. Windows NT has many features in common with UNIX, such as multiple users with user profiles, groups and group profiles, file and directory security settings, long filenames, and a system administrator user ID.

LOGIN Windows NT allows multiple users, and each user has his/her own user ID and password. There is a special user, the system administrator, with the default user ID "Administrator." After the operating systems completes the boot-up process, the Administrator logs in by first pressing Ctrl—Alt—Del. (I tell you this scared the heck out of me the first time I logged on to a server that was already up and running. I thought I was going to reboot the system. For DOS, Windows 3.*x*, and OS/2, Ctrl—Alt—Del reboots the system, but for Windows NT, this begins the login process. It took me a while to get used to it.) A login window pops up onto the screen prompting for the user ID and password.

RENAME "ADMINISTRATOR" ID As a default, all Windows NT servers have a user ID called "Administrator," which is the all-powerful user ID used to administer the Windows NT servers. This Administrator ID should be renamed to something else to make it more difficult for hackers to break into the server. The Windows NT Administrator ID is similar (but with significant differences) to the "root" ID on UNIX machines.

CREATE ANOTHER ADMINISTRATOR ID Create another ID for performing server administration. You should have at least two administrator accounts in case one breaks for whatever reason. For example, you might have set the Administrator ID to be locked out after three bad login attempts. If you set the "Lockout Duration" to "Forever (until admin unlocks)," you will be locked out and cannot log in again! In this situation, it is helpful to have another administrator account to log in with.

HOW A USER ACQUIRES ADMINISTRATOR POWER Any user assigned to the "Administrator" group has the power of the Administrator. The administrator(s) do not have to be named "Administrator." In this manner, the user Zeus who is in the "Administrator" group logs in to his own account with the name Zeus and his password. Likewise, Hera, who is also in the "Administrator" group, logs in to her own account with the name Hera and her password. Zeus and Hera have their own accounts, and each has administrator power, but neither account is named "Administrator." Either Zeus or Hera could create a new account for Odysseus in the Hero group, but it is necessary to know who did what when. The logs show what actions each person took, and this provides an audit trail. (You do not have to be of divine origins to be a Windows NT administrator, but sometimes you might be regarded as godlike by your users!)

HOW THE WINDOWS NT ADMINISTRATOR ACCOUNT DIFFERS FROM THE UNIX ROOT ACCOUNT UNIX differs from Windows NT in handling administration accounts. Whereas Windows NT has an "Administrator" group, UNIX has a single "root" account. In UNIX, Zeus and Hera would share the single account and password. If login to root is allowed from the server console or from a telnet session (Do not do this!), then you could not tell if Zeus or Hera made the change to add Cupid's user account with permission to execute certain lovely programs. By disabling root login, Zeus and Hera must first log in to their own account and then, essentially, log in to the root account (using the su command and the root password). The logs then show who is acting as root. (Only a few users can log in to the root account, so hacking Cupid's account name and password will not help get to the root account.)

On UNIX systems with root login disabled, there is no direct access to the administration functions of root from "outside" of the server. A hacker must guess the user name of a regular user's account, guess the password, then guess the root password from within the UNIX server. The hacker's programs must run on the UNIX server instead of on the hacker's own workstation. Hacking from within the server is more difficult for the hacker than from outside of the server. Hacking from within a UNIX server is also riskier for the hacker because it is easier to detect and track.

On a Windows NT system, hacking can be attempted from outside the server using programs running on the hacker's own workstation. Only one user account name and one password needs to be hacked to get administrator access. The hacker needs to guess or know which user name is the administrator, unless you did not rename the default Administrator account as I told you to do! In this latter case, the hacker only needs to guess your password—you made it much easier for the hacker. (This is the same reason for disabling root login for UNIX systems. Root is the first user name the hacker will try. Disabling root login makes the hacker guess account names and guess which one can log in to root.)

HIDE THE USER NAME FROM THE LOGIN WINDOW   The Windows NT login window shows the user name of the last person to log in. This is the default setting. Change this! Do not show the user name. Make the hacker guess the user name.

BACKUP THE REGISTRY!   Backup the Registry file on a regular basis. This is a vital file for Windows NT. It is like a combined AUTOEXEC.BAT and CONFIG.SYS for Windows or OS/2 systems. If the Registry file becomes corrupted, applications, and Windows NT itself, might no longer function. Be sure to have a backup copy (not on the same disk as the Windows NT server!).

BEGIN A COORDINATED WINDOWS NT DOMAINS DESIGN NOW Windows NT has the concept of Windows NT domains. These are analogous to flat Notes domains. There is no hierarchy for Windows NT domains. This is deja vu all over again. Remember how flat Notes domains began to spring up all over the organization? (Well, you will remember if you were working with Notes Release 2.x or the early days of Release 3.x.) Then, we wanted Notes domains to communicate with each other. What a problem there was. People's names were not unique, so mail was not delivered properly. Worse yet, security based on Notes Access Control Lists (ACLs) could not be properly enforced because people's names were not unique. Well, we will relive these problems with Windows NT. Windows NT domains are springing up all over the organization. People in different Windows NT domains are beginning to want to communicate with each other, but names are not unique. Hierarchical Windows NT domains will be available in future releases, but we have to live with what we have for now.

The solution for handling Windows NT domains will be similar to the solution for handling Notes domains: Start early to coordinate all domains. The actions include getting all of the administrators together, deciding on a naming convention, and setting up hub and spoke domains. Sound familiar?

WHY THE WEB SITE IS NOT AUTOMATICALLY AVAILABLE WHEN THE DOMINO SERVER STARTS Many people who start the Domino server for the first time are often puzzled as to why the Web site is not automatically available. You must edit the NOTES.INI configuration file to start the Web server automatically. Add http to the end of the ServerTasks=... line of the NOTES.INI file to start the Web server every time you start the Domino server. If you need to stop the Web server, enter the command tell http quit at the Notes server console. To start the Web server manually, enter the command load http at the Notes server console.

DIRECTORY AND FILE PROTECTION Directory and file access is controlled by the system administrator. Directories and files can be given read/write access, read-only access, or no access (except by the system administrator). The access can be controlled at the individual user level, but (as in Notes) it is usually more convenient to control access at the group level. The access is controlled using the graphical user interface (which is a bit easier to use than what's available on most UNIX systems). You do the following:

1. From the Program Manager window, double-click on the Main icon.
2. From the Main window, double-click on the File Manager icon.
3. From the File Manager window, click on the drive your file is on.
4. Click on the file.
5. From the menu bar, click on Security.
6. From the pull-down menu, click on Permissions.
7. Click on the Name of the user (e.g., Everyone).
8. From the pull-down menu for Type of Access, select Change.
9. Click on the OK, button and you are done. "Change" allows the user to read/write/execute/delete.

While in Notes, I had the following error pop up: "Access to data denied." I changed the Access Control List (ACL) on the Notes database so that I had manager access, but I still could not edit the database. After being puzzled for a while, I realized that the file was protected at the operating level of Windows NT. You need to keep this in mind when you are working with Windows NT and UNIX if you have been used to working with OS/2 and Windows 3.x operating systems.

EASE OF USE VS. SECURITY Features that are easy to use tend to be not so secure. Windows NT was designed more from an ease-of-use approach. For example, having the login window remember and display the user ID name is a nice ease-of-use feature for end users, but is not so nice from a security viewpoint. Check all ease-of-use features of Windows NT for possible security exposures. The developers had end users in mind, not system administrators, when they developed Windows NT.

NetWare System

OVERVIEW. Lotus Notes Server for NetWare is a full-featured Notes server that runs on a Novell NetWare (3.11, 3.12, 4.01, 4.02, or higher) file server as a NetWare Loadable Module (NLM). NLMs are integrated into the NetWare operating system and share operating resources. Notes is easy to deploy and manage because it operates as an integral part of the familiar NetWare environment.

Lotus Notes Server for NetWare takes advantage of NetWare features that companies rely on to build high-performance, secure, reliable networks. Additionally Notes Server for NetWare utilizes NetWare protocols, administration facilities, and benefits from Novell's high file-access performance.

WHAT IS NEW WITH THE NOTES SERVER FOR NETWARE?

- Customer service applications that are ready to use out of the box.
- New administration features to easily deploy and monitor Notes clients in a NetWare environment.
- Support for NetWare 4.01, 4.02, and higher.
- Performance enhancements: full-text search, multi-instance compactor, and multithreaded routing.

HOW DOES THE NOTES SERVER FOR NETWARE WORK?

Lotus Notes is a true client/server environment in which users (clients) communicate over a local area network (LAN) or a telecommunications link with a document database that resides on one or more shared Notes servers. With Notes, the information stored in the database is available to all Notes clients, including the new Notes client offering Lotus Notes Express, and servers, whether they are using Windows, OS/2, Macintosh, UNIX, or Novell platforms.

The Notes Server for NetWare can directly connect over a LAN with Windows, Macintosh, UNIX, and OS/2 clients and servers, using SPX/IPX, TCP/IP, or AppleTalk. All clients and servers can connect remotely with the Notes Server for NetWare via modem using the XPC protocol.

Novell's NLM interface provides a means of installing server applications directly on the NetWare operating system. The Notes Server for NetWare is loaded on the NetWare file server as a NLM and can run on the same hardware as NetWare file and print services. For optimal performance, it is recommended that the Notes NLM server be installed on a dedicated server.

Customers have the freedom of platform choice for their Lotus Notes Servers. Having the Lotus Notes Server available as a NetWare Loadable Module (NLM) provides an easy way for NetWare LANs to become the centerpiece of a whole new class of powerful, solutions-based business applications. Lotus Notes increases the value of the entire NetWare network by providing applications that radically improve the quality of strategic business processes, like office automation, customer service, account management, and product development. Companies can have a bigger return on their LAN investments by providing those networks with solutions-based Notes applications.

KEY FEATURES OF LOTUS NOTES SERVER FOR NETWARE
GETTING STARTED AND EASE OF USE

- Adds value to the entire NetWare network. New ready-to-use Customer Service Application+ out of the box.

- Easily add Lotus Notes Express Clients to your growing Notes network.

- Allows LAN administrators a platform choice for their Notes server.

- Easy to deploy and manage in the NetWare environment.

- Provides an easy migration path, and reduces training cost.

- Creates a natural extension into the familiar NetWare environment.

- Takes advantage of the Notes graphical user interface for installation.

- Powerful application development tools, over 25 template and example files.

- Online help is provided

NOTES SERVER FUNCTIONALITY

- New administration features to easily deploy Notes Express and Notes clients in a NetWare environment.

- Provides NetWare networks with an object store that enables workgroups to assemble, manage, and share network-wide information in the form of compound documents via Notes databases.

- Database Replication, mail routing, and workflow across Notes servers.
- Multithreaded routing.
- Notes API programmability.
- Advanced security in the form of access control lists at the server level, as well as security at the database, form, document, and section levels.
- Remote use features for dial-up using Notes asynchronous capabilities.
- Support for larger databases and directories.
- Support for the X.500 hierarchical naming standard.
- Manual database compression.

SYSTEMS MANAGEMENT FEATURES

- Administration features to monitor and upgrade Notes Express users to full Notes clients.
- Notes has a remote administration facility that, combined with the NetWare Remote Console, allows complete control over the Notes Server for NetWare from a remote system.
- Administrators can use the remote NetWare console to shut down and restart the Notes server.

APPLICATION AND DOCUMENT MANAGEMENT SERVICES

- Tight integration with popular desktop applications like WordPerfect, Lotus 1-2-3™, and Microsoft Excel using OLE, Publish & Subscribe, and LEL for UNIX.
- Full-text search on server databases and across multiple databases.
- Automatic document versioning.
- Background server agents for monitoring workflow processes or other unattended user tasks.
- Multibyte character support for internationalization.

NETWORKING CAPABILITIES

- To communicate with Windows or OS/2 clients, the Notes Server for NetWare uses the SPX/IPX or TCP/IP protocol.
- To communicate with Macintosh clients, the Notes Server for NetWare uses the TCP/IP protocol, and Lotus will soon support the AppleTalk protocol.

■ To communicate with OS/2, the Notes Server for NetWare uses the SPX/IPX or TCP/IP protocol.

■ To communicate with UNIX servers, the Notes Server for NetWare uses the TCP/IP protocol.

■ All clients and servers can connect remotely with the Notes Server for NetWare via modem using the XPC protocol.

■ Gateways and companion products can connect to the Notes Server for NetWare while running on an OS/2 machine.

SYSTEM REQUIREMENTS FOR NOTES SERVER FOR NETWARE
HARDWARE REQUIREMENTS

■ Dedicated server is recommended

■ 80386, 80486, or Pentium IBM-compatible PC; 80486 and above recommended

■ 200MB high-capacity hard disk (minimum); 300MB or larger recommended

■ 24MB of RAM (32MB recommended) above the file server requirements

■ LAN adapter fully compatible with supported networks

■ Any printer supported by NetWare 3.12 or above.

■ For optional remote connections, one or more of a variety of commercially available modems

SOFTWARE REQUIREMENTS

■ This server supports all Notes clients

■ Windows 3.1 or Windows 95

■ Novell NetWare 3.11, 3.12, 4.01, 4.02, or higher

■ Lotus Notes

NETWORK PROTOCOLS

■ For communication with Windows, OS/2, or UNIX clients: TCP/IP or SPX/IPX protocols.

■ For communication with Macintosh clients: AppleTalk or TCP/IP protocols.

■ For communication with OS/2 servers: TCP/IP or SPX/IPX protocols.

■ For communication with UNIX servers: TCP/IP.

Hardware Platform Considerations

Because Notes and Domino run on many hardware platforms, you have a lot of choices. Some of the considerations should be:

- RISC vs. CISC
- Nontypical systems such as the IBM AS/400 or IBM S/390
- SMP vs. Uniprocessor. With an IBM 720 A SMP processor, you could have 1500 concurrent Notes users, as opposed to 1000 concurrent users using a similar uniprocessor system. Performance estimates like these are usually made using the Lotus NotesBench.
- Massively Parallel Processing (MPP) systems. The IBM SP2 used for IBM's internal Notes roll-out falls into this category.
- RAID storage. Here the considerations are:
  - Hot swappable disks
  - Rebuild of problem disks from other disks

Application Development Tools

This section lists some Lotus Notes tools that should be useful in developing customized applications for your Lotus Notes network. Note that the tools described are not a complete list of all the Notes development tools available. There are many tools available, and the number is growing as the popularity of Lotus Notes increases.

Lotus Notes ViP

ViP is Lotus Development Corporation's Windows-based visual programming environment for Notes. ViP is used by developers who need to build custom applications that use information in Notes databases. (*Note:* ViP is currently being developed by Revelations Technology, Inc., under an exclusive license from Lotus.)

Lotus Notes ViP is a visual programming environment for developing client/server applications. It provides the following:

- A rapid application development technology called Visual Linking. Visual Linking is a unique capability of Notes ViP that allows a developer to create applications by sketching or drawing the relationships between parts of the application. Each Visual Link encapsulates behaviors and events that speed the development of client/server applications. There are over 70 predefined behaviors in Notes ViP that, for example, specify what is to happen to a list box when a button is clicked on with a mouse. Visual Linking increases the speed with which client/server applications can be prototyped, developed, and delivered to end users. Visual Linking also makes the Notes ViP visual programming environment easier to learn than traditional visual development tools. Figure 13.2 shows ViP in design mode with links drawn.

- A complete set of tools within the programming environment, including a debugger, script editor, and toolbox containing development objects.

- LotusScript, a BASIC-compatible language with object-oriented extensions and the ability to create user objects.

- The ability to visually prototype the GUI. Developers can use Notes ViP objects to create a user interface and optionally have complete access to the Windows controls.

- Comprehensive support of 14 different SQL and relational databases.

Lotus Notes ViP also provides additional facilities for the development of groupware applications:

- Comprehensive access to the Notes environment, including the initiation of replication of Notes databases, access to Notes database

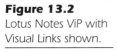

Figure 13.2
Lotus Notes ViP with Visual Links shown.

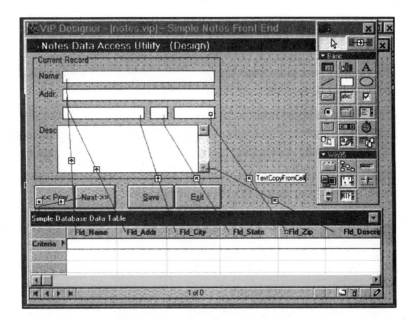

document hierarchies, access to Notes full-text search engine, use of Notes-specific databases fields, and optimized queries for Notes databases.

■ Use of the replication facilities in Notes as a container for Notes ViP applications and as a distribution method for Notes ViP applications.

■ Use of Designer Tool Extensions (DTXs). These DTXs allow the Notes ViP environment to be customized to the specific needs of an individual developer or to the standards of an organization. Tools that extend Notes ViP can access attributes of the application under design and therefore can contribute to the speed with which applications are developed.

InterfloX

InterfloX is a programmer's toolkit that provides a REXX application program interface (API) to Lotus Notes. It enables a REXX program to access, modify, create, and delete Lotus Notes databases and documents. This toolkit allows the creation of complex applications and the automation of routine tasks. It also facilitates the connection of Lotus Notes to host systems and other databases such as DB2/2. The toolkit is easy to use and very robust.

The simplicity of programming when using this toolkit is found in the fact that it uses about 60 high-level functions instead of the about 300 low-level C functions. Most programming needs are found in 7 of the 60 functions. Enhanced error checking and data validation provide improvement over errors common with C-API programming. A REXX background and familiarity with Notes are all that is required to begin using InterfloX. No C coding skill is needed.

Some functions of the InterfloX REXX API are:

■ Read and write documents and response documents.

■ Mail documents.

■ Perform searches for documents.

■ Create doclinks, rich text, and buttons.

■ Import and export data.

■ Attach, detach, and extract files.

■ Get and set Notes environment variables.

■ Create, copy, and delete databases.

TYPES OF INTERFLOX PROGRAMS

- *Configuration programs*—One time use. Examples are: automated database creation, preloading of databases, and database conversion.
- *Scheduled batch programs*—Examples are: report generation, administrative tasks, and mail generation.
- *Polling applications*—Examples are: executive information systems, incremental numbering, and automated continuing inheritance.

INTERFLOX VS. THE NOTES C API. InterfloX provides a programmer-friendly development environment such as:

- A few high-level REXX functions rather than many low-level C functions (with minimal loss of function).
- 90% of your programs with just 7 functions.
- Data validation that prevents the PANIC messages and traps encountered in the C API.

BENEFITS OF TH\E REXX LANGUAGE. The benefits of using the REXX language are quick prototyping, easier maintenance, and lower skill level required. There are performance considerations, however. C code runs up to four times faster than REXX code. Also most API programs are batch processes or poll for changes.

IBM PRODUCTS USING INTERFLOX

DATAGUIDE FOR LOTUS NOTES This product uses InterfloX to import DataGuide tag files into a Lotus Notes database. This provides the information catalog of DataGuide to Lotus Notes users to share and reuse business information in an organization.

LOTUS NOTES EXTRACTOR This program is shipped with DataGuide. The Lotus Notes extractor uses InterfloX to read the Lotus Notes Database Catalog. The extractor creates a tag language file with information about the databases on a server and the policy document for each database that can be imported into the DataGuide Information Catalog.

TIME AND PLACE/2 EVENT GATEWAY This product uses InterfloX for a gateway, which enables Notes users to add meeting notices they've received in their mail to their TaP/2 Calendar. The Notes Administrator installs the gateway on one machine, then updates the server's Mail template so that users will have an additional Add to TaP/2 Calendar push button in their Memo form.

LABEL MAKER   This is used by the InterfloX team to create printed labels for diskettes. A Lotus Notes database allows users to create new labels or select from a list of saved labels. InterfloX scans the database for new or updated labels and sends the labels to the printer.

INTERFLOX FUTURE ENHANCEMENTS.   The following further enhancements are being considered for InterfloX:

- Decrypt documents (if you have the correct key)
- Macro support (create/modify/execute)
- Add new "Notes 4" APIs
- Have InterfloX return all field names in uppercase
- Porting to other operating systems

Notes C API

The Notes Application Programming Interface (API) Toolkit is a C language subroutine library that provides programmatic access to Lotus Notes services. With the Notes API, developers can create programs to:

- Take advantage of Notes' document management services.
- Create new Notes applications or modify design elements of existing applications.
- Extend Notes' functionality.
- Apply Notes' system administration and management services.

All applications that are built using the Notes API benefit from the security, replication, and other multiplatform features of Lotus Notes. The 3.1 API Toolkit contains:

- Header (source code) and library files.
- Documentation: The Notes API User Guide and the Notes API Reference, providing both an overview of the product and detailed documentation, along with a Microsoft QuickHelp file.
- Example programs: More than 90 programs, representing tens of thousands of lines of code and many different types of example applications (from import/export filters to mail gateways), including examples for building a Microsoft Visual Basic "wrapper" on top of the Notes API are also provided.

The API Toolkit does not contain any executable files; you must have Notes installed to use programs created with the Notes API.

LotusScript

LotusScript is a Lotus Product used to program sophisticated Notes applications. LotusScript is a superset of the familiar BASIC language. Key enhancements over standard BASIC include object-orientation and extension for event-driven environments. LotusScript is also multiplatform. Notes applications that contain LotusScript run on all Notes clients and servers, no matter which operating system the application was originally developed on. The LotusScript programming environment has a script editor, debugger, and variable/property inspector.

Network Links to Host and DBMS for Notes

Lotus has many connectivity options to host and database management systems (DBMS). These products are used to improve connectivity between Notes and legacy back-end systems. Several of these products are described in this section.

MQSeries Link for Lotus Notes

IBM's MQSeries Link for Lotus Notes allows front-end Notes databases to access back-end mainframe and minicomputer systems, including DB2 databases, Customer Information Control System (CICS) transaction-processing applications, and VAX/VMS platforms. The Link runs on OS/2, AIX, HP/UX, and Windows NT.

MQSeries is a very powerful message-oriented middleware that has the potential of integrating more advanced IBM technologies like SOM (System Object Model) and DSOM (Distributed SOM) and workflow technologies with Notes. The goal is to leverage these technologies off one another, rather than just have gateways.

MQSeries Link ties structured transactional data stored in tables in host systems with unstructured notes data. This allows a Notes user, for example, to bring data from a DB2 human-resource database into a Notes document that contains a scanned-in resume and photograph of an employee.

The IBM software, which ties the Notes API with the MQSeries API, allows application developers to embed MQSeries procedures for accessing host data within Notes applications. In addition, the software switches a Notes request into MQSeries format and routes it to the host system. The requested information is then placed in a MQSeries message, where it is posted back to a Notes database or application, and the user is notified.

The Link also allows mobile users to initiate requests remotely. In addition, users can make "live" requests of the transaction-processing systems from Notes over a telephone line. The Link allows Notes administrators to set up security, transaction IDs, and mapping for notes users to legacy systems. This information is managed by the MQSeries Link.

Also, the MQSeries Link software handles field requests, security, error handling, updating of Notes databases, and communications with host systems. The MQSeries Link is offered as a separate product. However, IBM plans to build it into the MQSeries middleware.

CURRENT METHODS.　Current methods of accessing transaction system data by Notes usually fall into one of three categories. These are terminal emulation, staged data, and direct access. These methods usually require manual intervention for transfer or a significant investment in middleware and ignore transaction integrity, security, recovery, and logging that are the hallmark of transaction systems.

THE IDEAL SOLUTION.　The ideal solution would be one that leverages the transaction processing systems, is transparent to the end user, supports all operating systems, provides mobile support, is cost effective, and is coded to industry standards for future compatibility.

IBM Transaction Systems' MQSeries and CICS links to Lotus Notes SupportPacs provide this solution and enable the connection of Lotus Notes to all major IBM transaction systems platforms (i.e., OS/2, MVS, AIX, OS/400, CICS, and IMS). These IBM links significantly simplify data access and management requirements by providing the Notes user with access to a single managed copy of data rather than the multiple copies required via currently used methods available to Notes users, such as Extract, Copy, Format, and Resynchronize.

The IBM MQSeries and CICS links for Lotus Notes work with and through existing transaction systems. The technology that IBM has developed is a Notes server task written to the standard Notes APIs. This task receives data requests from Notes, switches them to a transaction format, posts the request to the transaction system, and handles the reply. The host system sees the request as an ordinary transaction normally generated by the system. No changes are required on the transaction system. All current facilities—such as logging, security, backup, and recovery—remain intact and unchanged.

Because this IBM technology is written at the Notes server level, any Notes client can have access to the transaction data. The mobile Notes user was considered when designing these products, and they can be used with a Notes client in connected or disconnected mode.

A WIDE RANGE OF CONNECTIONS. The MQSeries link utilizes IBM's MQSeries messaging product family to connect Notes users to the subsystems and application bases available on a wide variety of IBM and non-IBM platforms, including HP-UX, Sun Solaris, and VMS. The CICS link connects Notes servers directly to any of six CICS transaction system platforms. Thus these links connect the groupware world with that of traditional transaction processing systems. Its CICS and IMS connections provide Notes access to the most widespread mission-critical OLTP systems in the industry.

The CICS connection alone provides business partner opportunities in approximately 30,000 customer sites for integration of Notes and CICS applications and data.

Messaging and queuing services are interconnected across various communication networks such as TCP/IP and SNA, providing application access to other applications anywhere in the network.

Overall, IBM's links to Lotus Notes significantly extend the reach, value, and return of your CICS, MQSeries, and/or Lotus Notes investments and represent a substantial new business opportunity for our business partners.

LOTUSSCRIPT EXTENSION (LSX). With the LotusScript Extension (LSX) to MQSeries Link for Notes, corporate developers can connect Notes to 18 transaction-processing platforms supported by IBM's MessageQueue Series middleware. Available as a free upgrade, MQSeries link can be downloaded at www.hursley.ibm.com.

NotesSQL Release 2.0

The new version of the Notes Open Database Connectivity driver includes improved 16-bit performance and support for 32-bit Microsoft Windows clients and IBM's OS/2. This enhancement can be downloaded at www.lotus.com/devtools.

Lotus NotesPump Release 2

This enhancement to Lotus NotesPump provides scheduled, high-volume replication sessions between Notes servers and DBMSs. The enhancement shipped in 1996.

Oracle LSX

This product lets Lotus Notes clients read and write to Oracle databases using SQL*Net. Oracle LSX is available free from Lotus and Oracle.

LSX Toolkit

Like Oracle LSX, developers can let any data object be accessed natively from Notes clients with applications developed in LotusScript. LSX Toolkit is available free to Notes users.

Other Notes Services Over the Network

For the IBM user, the Lotus Notes desktop is the gateway to all required services from printer to pager to fax.
—Albert Schneider, Lotusphere '97, January 28, 1997.

There are many other services related to Lotus Notes that a company can provide over the network. Besides dial access to Notes via the company's internal network or the Internet, and gateway services, which were mentioned earlier in this book, there are also fax services, pager services, company-wide name and address book (directory) services, and Notes database repository services. These services are offered within the IBM Notes environment and are discussed in this chapter.

Notes Fax and Pager Services

Lotus supplies fax and pager add-ons to Lotus Notes that allow Notes users to send fax and pager messages directly from Notes. The addressing is quite simple. Notes users in the ADVANTIS domain address data to be sent to a FAX machine with the address:

```
User Name @ Fax number @ FAX
```

For example:

```
Taylor McKenna @ 9-684-4542 @ FAX
```

The User Name is just for the fax cover sheet. The "9" in front of the number is to get out of the building where the Fax Gateway is located. Any document that can be mailed from within Notes can be faxed to any fax machine in the world.

It should be noted that sending an "outbound" fax is easy, but receiving an "inbound" fax is difficult because of routing problems to the correct end-user. Unless you use DID (Direct Inward Dial) numbers, the routing for inbound fax must be done manually. In order to use DIDs, you must have a PBX, or equivalent, with which you can assign each end-user a unique number.

Pager addressing is similar to fax addressing. A typical pager address is:

```
Pager Number @ SKYTEL
```

Lotus Fax Server

The Lotus Fax Server (LFS) is a fully-integrated fax solution for Lotus Notes and Lotus cc:Mail that allows users to send, receive, view, and manipulate rich-content faxes without leaving their familiar Notes and cc:Mail environments. The server makes faxing as simple as sending electronic mail.

The LFS combines both incoming and outgoing fax capabilities in one common product that is accessible through all Notes and cc:Mail client platforms. With the LFS, Lotus offers an alternative to the traditional networked fax pricing model where each user pays for access. The LFS user only needs electronic mail; additional software isn't required. This allows fax hardware and phone lines to be shared by all users.

The LFS allows users to view and manipulate their fax documents using a new Lotus fax viewer that can be distributed throughout the enterprise. The LFS supports Direct Inward Dial (DID) and Dial Tone Modulation Frequency (DTMF) for easy, confidential routing of faxes right to the desktop with all other electronic mail. The LFS also supports a print-to-fax driver that enables faxes to be sent directly from any mail-enabled Windows application. For example, faxing a spreadsheet can now happen directly from 1-2-3, or memos can be sent right from Ami Pro.

SETUP AND ADMINISTRATION. For administrators, the LFS increases the ease of managing and administering the flow of information on the network. The LFS is available in two editions built from a common code base. The Lotus Notes Edition and the cc:Mail Edition offer the same feature set, thus simplifying administration for both products. In addition, the LFS comes with easy-to-use management tools so that setup, administration, and control are simplified. As an example, a fax phone number can be incorporated into the common e-mail directory, eliminating the need for a separate fax phone book. For less frequently needed or *ad hoc* requirements, the fax phone number can be entered directly in the address field. Finally, in addition to Lotus' strong relationship with GammaLink, Lotus will add support for Optus Software, Inc. technology, expanding support to more than 50 additional fax cards and modems.

The LFS combines the rich functionality of both the Notes Outgoing and Incoming Fax Gateways into one common product, as well as replaces cc:Fax 1.2.

SYSTEM REQUIREMENTS.   The LFS requires a PC/AT or compatible (a 386 or 486 running at 33-MHz or faster is highly recommended) running Microsoft Windows 3.1. It requires at least 16MB of memory and at least 3MB of free hard disk space for the Lotus Fax Server program files. A minimum of 3MB of available hard disk space is recommended on Notes servers for the LFS mail database. 5MB of available hard disk space is also recommended on the LFS for the other LFS databases. For the Notes Edition, a Notes client license will be provided. For the cc:Mail Edition, Notes client software and a Notes client license are required and will be provided. One or more of the fax cards or modems supported also are required.

Pager Gateway

Lotus Development Corp. and SkyTel have an agreement to jointly develop technologies and applications to meet the demands of a growing wireless messaging market. The first product delivered as a result of this agreement was a wireless messaging gateway for Lotus Notes. This new gateway extends Notes beyond the desktop and into the realm of wireless communications, representing the first national wireless gateway for workgroup computing software.

Combining Lotus Notes and a SkyWord alphanumeric receiver, the gateway creates a Notes "thinking mailbox" that can forward changes in mission-critical information from a Notes server to SkyWord receivers. Notes can filter and forward messages by category or user and can issue alerts via the gateway whenever key documents are received. Additionally, cc:Mail customers, using the high fidelity Message Exchange Facility between cc:Mail and Notes can take full advantage of the wireless gateway. Lotus and SkyTel jointly promote and market the gateway through direct channels and Lotus Business Partners.

As part of the joint development efforts, Lotus and SkyTel are working with several large-scale Lotus customers, including Andersen Consulting and Mobil Natural Gas, to develop workgroup applications that exploit the functionality of the wireless messaging gateway for Notes and address the diverse and changing needs of Notes customers.

One of the applications developed allows users to selectively manage incoming and outbound alphanumeric messages over SkyWord receivers or Hewlett-Packard 95/100LX systems equipped with SkyStream data receivers. The application leverages the power of Notes to filter and screen messages continually according to priority lists of key Notes users. The lists are designated by the user for filtering and transmission of important messages directly from Notes to SkyWord receivers.

Additional applications under development include an application that can issue automatic pricing updates to companies' field sales staff that will enable staff to monitor changes. Other potential applications that could be based on this wireless pager gateway include a SkyTel feature known as Direct Data Deposit, offering users the ability to receive updates to built-in schedule, calendar, and spreadsheet functions when used with a SkyStream receiver and Hewlett-Packard 95LX palmtop computer.

Company-Wide Lotus Notes Address Book

Internally, IBM provides a company-wide Lotus Notes directory service (i.e., Name and Address Book). Copies of the Lotus Notes Name and Address Books of all domains connecting to the IBM_INTERNAL domain are kept on the hub servers in IBM_INTERNAL. Also a copy of the "IBM Name and Address Book," which is the consolidation of all of the individual Name and Address Books, is kept on each hub server in the IBM_INTERNAL domain. Normally, a site would replicate the IBM Name and Address Book onto the site hub server. This is an example of what IBM is doing internally that might be of use in designing your own Notes network.

Figure 14.1 shows the customized structure in the ADVANTIS domain Name and Address Book. To save space in the consolidated IBM Name and Address Book, much of the individual information contained in the ADVANTIS Name and Address Book is discarded. Only the user name and Notes shortname are included.

IBMNOTES Database Repository

The IBMNOTES Database Repository is an example of an enterprise Notes repository. IBMNOTES is accessed by IBM Notes users from all over the world, including users in Canada, Japan, Australia, Sweden, Spain, and the UK. The repository is on a server called IBMLN01 and contains information from several sources:

■ [LOTUS]folder contains databases for the Lotus Corporation's LOTUS Domain. These include the "Notes Knowledge Base" and "Lotus Partner Forums."

Figure 14.1

A "Customized Name and Address Book."

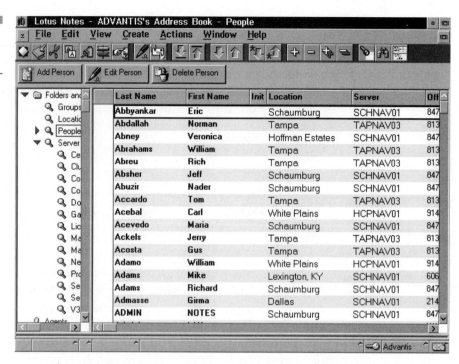

- [PACKAGES] folder contains Notes applications written by IBMers and provided as shareware. Open the "Lotus Notes Packages" database to see these. They are mostly stored as attached executables.

- [EXAMPLES] folder contains database templates and examples form the Lotus Corporation. The examples are much more extensive than those provided with the Notes product.

- [DISCUSS] folder contains the IBMNOTES Discussion database among others.

- [MSCTOOL] folder includes a "recipe" database and a "homebrew" database.

- [EXTERNAL] folder contains the IBM Announcements and IBM Press Release databases. These are downloaded from the Internet by IBM Atlanta.

- [VMFORUMS] and [CFORUMS] folders contain hundreds of VM Forums and Customer Forums downloaded from VM.

A separate server, AUSNFORM, in Austin, Texas, spends all its time downloading the Forums from VM machines at IBM Research and converting them to Notes databases. The access to the VM machines is via FTP, and the conversion program uses InterfloX and REXX APIs. In the [VMFORUMS] folder, users can find several forums on Lotus Notes; for example, the Notes Application Development and Notes Server databases. In the [CFORUMS] folder, two forums of interest to IBM users are the OS/2 WARP and OS/2 WARP BONUS PACK databases. A description of the conversion process appears later in this chapter, in the section "Enterprise Conferencing with VM and Lotus Notes."

A good way to look at all the databases is through the "IBMNOTES Catalog." Catalogs are provided by the Lotus Notes product and allow for an organized way to browse or add icons to a user's Lotus Notes desktop.

Calendar and Scheduling Options

There are several calendar options that can be used with Lotus Notes. Lotus, themselves, produces Lotus Organizer. Organizer was originally designed for personal organizing but was extended for workgroups.

Lotus Organizer with R4.5

Lotus Organizer 2.0 is a personal and group scheduler that allows easy integration of LAN-based group scheduling and daily task management features that include calendaring, to-dos tracking, address keeping, an additional call logging section, a notepad for storing text and graphics, a yearly planner, and anniversary reminder.

Organizer 2.0 offers major enhancements, reflecting suggestions made by Organizer PIM and scheduling customers. Many of these new features make Organizer 2.0 a more customizable and flexible PIM, while maintaining the product's familiar notebook-style user interface. Some of the most requested customer suggestions in Release 2.0 include:

- Group scheduling for Lotus Notes and cc:Mail users
- 30-day calendar view
- Ability to book conflicting and overlapping appointments
- Ability to attach files when scheduling a meeting with others
- "Calls" section for management of incoming and outgoing phone calls
- On-the-fly views across all Organizer sections
- Ability to categorize all entries
- Over two dozen new print layouts

- New Windows-based Organizer Administration Program

  Highlights of product features listed by Organizer sections include:

- *Calendar*—Shared calendar for multiple users; simultaneous access to the same file.

- *To Do*—Repeating tasks; specify completion dates on screen; display tasks by priority, status, date or category.

- *Address*—Home and business address records for each contact; customizable field labels.

- *Calls*—Display call entries by name, date, company, category, or status; linked to Address.

- *Planner*—Yearly and quarterly views; overlapping events.

- *Notepad*—View pages by page number, creation date, alphabetically by title, or by category.

- *Anniversary*—Display by year, month, category, or zodiac sign; set alarms, create links.

MAIL-BASED GROUP SCHEDULING.   When scheduling a meeting over the LAN, Organizer retrieves individual names from the cc:Mail directory or Lotus Notes' Name and Address book, eliminating the need for MIS to manage a separate directory for group scheduling. A meeting invitee has the option to accept, decline, or delegate the meeting request. All invitees can send a reply to the meeting requester with details regarding their response. The advanced Organizer 2.0 scheduling engine also permits repeating group meetings to be scheduled, not only on a regular daily, weekly, or monthly schedule, but also according to a user-defined, customized repeating schedule (e.g., the third Thursday of every month).

If an Organizer user schedules a meeting with a non-Organizer user, the meeting request will still be received by the recipient as a standard e-mail message, regardless of platform, thus eliminating the need to send a separate e-mail message to non-Organizer users.

SHARED CALENDARING.   The "User Access List" allows a user to grant different levels of access to the same Organizer file. The new multi-user, multi-access Organizer 2.0 database lets an assistant schedule personal appointments as well as group meetings for an executive while the executive is editing the same file.

ORGANIZER 2.0 EXTENDS SMARTSUITE'S WORKGROUP PRODUCTIVITY.   SmartSuite for Windows is the complete suite of desktop applications for users who work alone or collaborate in teams. By offering the industry's most popular networked personal information manager in its suite, Lotus is addressing the most common needs of business managers, such as maintaining contact and schedule information, as well as task management. Organizer 2.0 further extends SmartSuite's personal and workgroup productivity to offer unprecedented team computing functionality, such as group scheduling and shared calendaring, for Lotus Notes and Lotus cc:Mail users.

Designed to work the way users work, Organizer can be accessed via SmartSuite's SmartCenter, an integration and support feature that enable users to put the whole suite to work for them. With SmartCenter, users can manage their application integration, switching, and support needs from one central location. The SmartCenter icon palette supports drag and drop between OLE 2.0 applications and provides access to SuiteAnswers, a three-part support system including the SmartSuite Guided Tour, cross-application Help Cards, and ScreenCam movies. Organizer 2.0 is available to previous purchasers of SmartSuite 3.0 for a nominal shipping and handling fee.

SUPPORT FOR IBM PROFS AND OFFICEVISION/VM THROUGH ATTACHMATE. With Attachmate's ZIP! OfficeServer, Organizer users can schedule meetings and exchange information with PROFS and OfficeVision users.

SYSTEM REQUIREMENTS.

STANDALONE INSTALLATION An IBM PC or compatible with an 80386 processor or higher, 4MB of RAM (6MB recommended), Microsoft Windows 3.1 or higher, 8MB hard disk storage, VGA monitor, and mouse or pointing device (optional).

NETWORK SERVER INSTALLATION Same resource requirements as the Standalone Installation. Supports most popular networks including Novell NetWare 386, LAN Manager 2.0 or higher, and other 100% MS-NET compatible networks.

MAIL-BASED GROUP SCHEDULING Lotus cc:Mail 2.0 or higher or Lotus Notes 3.1 or higher, plus 6MB RAM and 3MB (minimum node installation) or 8MB (local installation) hard disk storage per workstation.

ORGANIZER SCHEDULING AGENT An IBM PC or compatible with an 80386 processor or higher and Microsoft Windows 3.1 or OS/2 2.1.

WINDOWS SCHEDULING AGENT   4MB RAM (6MB recommended) and OS/2.

SCHEDULING AGENT   8MB RAM (12MB recommended) and 2MB hard disk storage.

14.4.2 Time and Place/2

The IBM Time and Place/2 (TAP/2) program is a client/server network-based calendar application that includes individual and group scheduling, "to do" lists, calendar printing, and delegation of control for update purposes. Vendor Independent Messaging is supported with communications with other applications, such as Lotus Notes. The best features of TAP/2 will be combined with Lotus Organizer now that Lotus is a subsidiary of IBM.

PhoneNotes

PhoneNotes was designed to make the telephone a limited remote client for Notes. It allows users to dial into a Notes database and perform a variety of functions, such as creating documents and databases and playing back data via text-to-voice conversion technology. It also enables users to embed voice messages in Notes databases and gives them the ability to access Notes, with limited function, from a telephone's touch-tone keypad.

Overview

Lotus Phone Notes Application Kit, or Phone Notes, is a powerful development environment for Notes business partners and corporate developers. Phone Notes applications enable callers to access Lotus Notes data from any touch-tone telephone, expanding access to Notes workgroup applications to colleagues, business partners, and customers.

Typical Phone Notes applications include customer support help desks, information fax-back systems, human resource benefit enrollment or information hotlines, healthcare patient information systems, scheduling systems, and project status reporting applications.

When used in conjunction with the Lotus Notes Fax Gateway, Phone Notes applications can be used to fax information to the caller's local fax machine. Information can also be redirected to an alphanumeric pager using the Lotus Notes Pager Gateway.

Extend the Reach of Notes Applications

Nearly everyone has used Interactive Voice Response (IVR) systems. These systems walk callers through menus with voice prompts allowing them to get information (a stock quote), record a message (make a reservation), or initiate an automated process (request a service call). Phone Notes lets you use Notes as an IVR platform. With Phone Notes, a developer creates a simple touch-tone menu as a "front-end" to any Notes database using special Phone Notes forms.

For example, you can turn your Notes-based customer service application into an IVR system. Customers make a call, identify themselves by name or number, record a voice message describing their problem, and assign a priority level to determine what kind of response is needed. Your sales representatives can call in to the same application but has access to reporting data, such as the status of recent calls made by her top accounts. It's even possible to create a Phone Notes application that makes outbound calls for applications like seminar evaluations, status reporting, customer surveys, or data collection.

Turn the Telephone into a Notes Client

Phone Notes removes the barriers to working whenever and wherever you want and provides instant access to the timely, shared information you need.

With Phone Notes, Notes users, who are accustomed to accessing Notes applications on a computer, will find that it's easy to use many of these applications from a touch-tone telephone. Check on the status of a lead in your Notes-based sales force automation system, add your input to a discussion on your company's latest product, or publish information to your customers. Now, any information in a Notes database can be accessed via the telephone. All the Notes data that you rely on is literally at your fingertips and just a phone call away.

Sample Applications

The Phone Notes Application Kit comes with several sample applications that make it easy for developers to get started. Sample applications include:

- *Customer Support Help Desk*—Lets a customer access information and deposit problems into a Notes Help Desk through the telephone.

■ *Notes Document FaxBack*—Lets callers request information from a library of product literature. The application uses Notes Fax Server to fax requested information to the caller's local fax machine.

■ *Human Resources Benefits Selection*—Used by employees to understand, select, and change their healthcare benefits.

■ *User Validation*—Adds an additional level of password security to any Phone Notes application.

■ *Auto Attendant*—Takes incoming calls into a workgroup or small business and transfers the caller to a live representative, or records a message.

How Phone Notes Applications Work

Developers create Phone Notes applications that answer the phone and present the caller with a voice menu. The application then branches to one of several functions depending upon the touch-tone input.

Phone Notes scripts are then read by a Phone Notes-compatible voice-processing engine and translated into hardware-specific telephony commands. The voice-processing engine acts as an interpreter, stepping through forms, following branch instructions, and playing voice prompts. The application can direct the engine to access Notes databases to read, write, or delete field entries or to create, delete, or forward documents.

Phone Notes provides a forms-based development environment that makes it easy to create telephone-based applications for Lotus Notes. Each command is represented by a form containing fill-in fields that define its operation. A Phone Notes application can be written by simply filling in the right forms.

The Phone Notes scripting language contains the following commands: Assign, Call Subroutine, Case, Convert DTMF to String, Copy Document, Create Document, Decision, Delete Document, Dial, Execute Notes Macro, Execute OS Task, Forward Document, Get Next Document, Hang Up, Play, Record Voice, Restart Application, Select Documents, Touch-Tone Menu, Touch-Tone String Input, User Info, User Lookup, Wait for Call, **and** Wait Time.

Security

To extend Notes security to telephone clients, user proxy support ensures that access to data is secure by providing password authentication for telephone access to data. Access can be directly set in an Access

Control List, via Phone Roles. Phone Notes also provides two command forms (User Info and User Lookup) that can act as security filters to prevent unauthorized telephone access to Notes databases. These are sold separately. Because the information created by the Phone Notes applications is stored in a Notes database, Phone Notes applications inherit basic Notes features, such as database access, workflow routing, security, and replication.

Multimedia Compatible

Phone Notes leveraging technology from telephony engine providers allows audio data to be recorded and played back interchangeably between phones and multimedia-capable personal computers. A multimedia PC, which is capable of high fidelity audio (up to 44.1kHz), and a telephone with toll-quality voice (8kHz or 11kHz) can serve as sources of audio for Phone Notes applications. Translation between higher and lower resolution audio is handled automatically.

An Open Strategy

Phone Notes brings the ease of Notes application development to the voice-processing market. This simplified development environment makes it possible for Notes application developers to create or modify their own Notes-based voice processing applications, broadening the range of applications for Notes, and creating a whole new breed of voice-processing systems. This, in turn, will encourage voice-processing engine providers to support Notes. Lotus maintains an open-door policy toward telephony engine providers and voice board suppliers, ensuring that Notes telephony solutions will be available on many hardware platforms.

System Requirements

- LANs: All Lotus Notes supported LANs, including Novell, IBM, Microsoft, and Banyan
- Operating systems: Microsoft Windows 3.1 and IBM OS/2 2.1 or higher
- Software: Lotus Notes client 3.0 or higher
- Phone Notes certified telephony servers: Big Sky Technologies Remark!™ Phone

- Client for OS/2 version 2.1 (available from Big Sky Technologies, 1-800-488-4188)
- Big Sky Technologies STS3000 Voice Server for OS/2 version 2.1 or higher (available from Big Sky Technologies, 1-800-488-4188)

VideoNotes

VideoNotes integrates full-motion video into Notes. Full-motion video is a very demanding data type. Sound must be synchronized with images. Digital video files can be huge. VideoNotes solves the problem of large-object storage and provides enterprise-wide distribution and management of video. It will support the full Notes model, especially the routing and replication features that give Notes its dominant position in workgroup computing. With VideoNotes, Lotus offers developers and end users effective tools for leveraging this data type. For information providers and others looking to Notes as a publishing platform, VideoNotes supplies both a conduit for the delivery of multimedia content to customers and ready-made distribution system at the customer's site.

This new multimedia extension to Notes allows workgroups of all sizes to capture, record, play, edit, and distribute full-motion video in Notes applications, using intuitive buttons similar to a VCR. With Video for Notes, stored video can be accessed, distributed, and managed as another data type within Notes.

Using Video for Notes

Video for Notes will benefit users in all market segments. For example, an advertising agency can embed video clips in Notes to assist in the creative review and approval process. This multimedia interaction ensures that an ad campaign meets business objectives. Using Video for Notes allows all members of the creative process—copywriters, video producers, account executives, and the client—to collaborate more efficiently and effectively, no matter where they are located.

Users also can view video when disconnected from the network. For example, a traveling salesperson with product video footage or testimonials could download video to a laptop, then play the video at the customer site to illustrate a point, showcase a product, or deliver testimonials.

Video for Notes is Easy for Administrators

Video for Notes has a secure, flexible replication and distribution model that allows administrators to balance user demands easily. The Lotus Video for Notes site manager distributes video to Notes users by generating a single copy of the video at each location regardless of how many times the video appears in mail messages or other documents. This distribution mechanism minimizes long-term storage problems. Video for Notes also allows the administrator to gradually introduce video into an enterprise-wide network and make video available to the entire workgroup regardless of size or current network connections.

Video for Notes provides direct and immediate playback from video servers for high-capacity local area networks. When a local area network does not permit continuous playback, staged playback is available. Staged playback allows video to be downloaded automatically from a server to the desktop and then viewed. For workgroup members that do not have wide area networks, Video for Notes applications can be distributed through CD-ROM.

Groupware Video Applications

For Lotus' more than 8000 Business Partners, Video for Notes provides a robust application development environment for the development of video servers and video editing environments. It also offers system integrators, consultants, and independent software vendors new opportunities to integrate video into their products.

In addition to standard network file systems, Video for Notes initially supports two video server software products from Partners Novell and Starlight Networks. Lotus Business Partners—including DSSI, GroupQuest, and Hayes Computer Systems Inc.—are currently working with Video for Notes to build and integrate video applications as well as develop vertical market extensions.

System Requirements

Initially, the Lotus Video for Notes site manager is available on an OS/2-based Notes server. Lotus Video for Notes requires Lotus Notes Release 3.1.5 or above client software and Release 3.2 or above server software, running a 486 or Pentium processor.

Video Notes 1.1

Video Notes 1.1 supports additional file formats and includes a distribution feature that lets users offload CD-ROM video clips. It supports MPEG, Apple Computer Inc.'s QuickTime (MOV), and Microsoft Corp.'s Windows sound files (WAV). Support for CD-ROMs enables developers to quickly develop custom applications and enable users to save multiple video clips quickly.

Support for video servers includes IBM OS/2 LAN server Ultimedia, Windows NT, and First Virtual Corp.'s Media Server support. Video Notes 1.1 also adds expanded cache support for Notes FX (Field Exchange) technology. FX allows users to save data from external applications to Notes fields. Every time a user creates a video clip, its properties are saved in Notes databases.

Video Notes' Site Manager Server enables replication of video over LANs and WANs. Users click on a video icon in the Notes user interface, and video clips are played back from the server.

Internet and Intranet Notes Products

Having set up publicly accessible World Wide Web sites, many organizations have found that Internet and Web protocols readily lend themselves to internal information publishing as well. This has given rise to corporate intranets that support the dissemination of policies and procedures documents, customer profiles, and other corporate information throughout the enterprise at a relatively low cost.

Intranets grow in a natural progression from simple one-way document publishing to more sophisticated corporate and departmental applications that support various business processes. Indeed, infrastructure planners see in the Intranet the potential to simplify and unify disparate systems into a single information architecture that can support the full breadth of business applications. The democratizing character of an Intranet stems from the standards on which it is based.

- *HTTP*—One of the barriers to the smooth flow of information throughout an organization is the presence of multiple operating systems on the desktop. Because low-cost browsers typically run on multiple platforms, a company does not have to sacrifice the benefits of specific desktop platforms for the sake of a common information access tool.

- *HTML*—Further complicating the sharing and delivery of critical corporate data is the use of multiple data formats that require specialized software to read. Corporate data exists as word processing documents, spreadsheets, graphic presentations, tabular relational data, rich text news feeds, electronic mail messages, SGML document management repositories, and images and video, each with its own proprietary editor and data format. In an intranet, all data is rendered in the *lingua franca* of HTML. In this way, all users can continue to use their editors of choice, each with its own proprietary data format, without compromising the corporate data format standard.

- *Java/JavaScript*—Although it is premature at this time to characterize Java/JavaScript as the universal scripting language of the intranet, there is a movement among development tool vendors to extend their own programming languages to interoperate with Java. This allows developers to continue to use their programming tool of choice while respecting an industry-wide standard.

Lotus Notes: The Enterprise-Scale Intranet Server

Clearly, the intranet solves several conundrums that have plagued infrastructure planners since the dawn of distributed systems. Of course, operating systems, data format, development languages, and network protocols represent only portion of an overall information architecture. Indeed, the most complex—and therefore the most costly—infrastructural issues are left unaddressed by intranets at this time: system management services, directory services, distribution services, and access control and security services. Moreover, as intranets grow in scope and scale, it is not clear how well simple Web servers will meet performance and reliability requirements. This leaves infrastructure planners balancing the unprecedented benefits of intranet harmonization on one side against a potentially costly and unmanageable mix of proprietary and unintegrated partial solutions on the other.

The Lotus Notes server is ideally suited as a scalable, mature, and robust server for corporate intranets. By virtue of its similar architecture to the Web, its integral messaging services, and its integration with other infrastructural components, Notes stands generations ahead of other Intranet servers.

The Notes server is fully compatible with intranet standards (HTTP, HTML, TCP/IP, and Java) and therefore fully supports Web browsers as well as Notes clients. The Notes server is fully integrated with the all corporate data resources: RDBMS data, OLE 2 objects, video and image and document management repositories, and back-end transaction systems. Notes bidirectional replication allows an organization to mirror its intranet servers for optimized performance and reduced network costs. Users in multiple locations can make changes directly against a local server, with all changes automatically synchronized, requiring no manual intervention by the Webmaster or system administrator. Notes server controls access to all documents and data at the field level. This allows an administrator to hide or publish specific documents or sections of documents according to client type, password, or user ID.

Notes includes a full suite of collaborative application services, from straightforward discussion databases to messaging and directory services and scripting capabilities, to create a high return on investment business process applications that transform an intranet from a basic publishing medium to a powerful platform for any line of business.

Notes servers include a full set of management and administrative tools to bring order to a corporate Intranet.

The Notes Client: Maximum Intranet Potential

When using a Notes server at the center of an intranet, infrastructure planners and users have the flexibility of choosing whatever client best suits their needs. Of course, the greatest benefit can be realized when using a Notes client in conjunction with the Notes server. The benefits include:

- Mobility
  - Users can make laptop replicas of Notes databases and applications, including pages from Web sites of competitors and customers.
  - Single user experience for client/server messaging, browsing, discussions, workflow applications, and Web publishing.
  - Full client authentication. Notes users have unparalleled secure access to Notes servers by virtue of Notes' RSA-based public key/private key certificates.
- Network independence
  - One of the only drawbacks of an intranet is that it demands standardization on TCP/IP over existing network protocols.
  - By using Notes clients connected to Notes servers, infrastructure planners can continue to leverage their currently deployed

networks without sacrificing Intranet commonality and functionality.

Clearly, Notes is perfectly suited to the emerging intranet architecture. As a server for the intranet, Notes lowers the cost of management and adds significant value to Web documents. As a client, Notes is unmatched in its support for mobile users, its client/server messaging, and its robust security.

InterNotes Products

There are two InterNotes products from Lotus Development Corp.: the InterNotes Web Publisher and InterNotes News. The InterNotes Web Publisher lets Notes users publish Notes applications to the Internet and eliminates the time-consuming problem associated with managing corporate documents published on Web sites. InterNotes News allows notes users to participate in Usenet "newsgroups" from the Notes environment.

INTERNOTES WEB PUBLISHER.   The InterNotes Web Publisher converts Notes databases and documents into the HTML format used for publishing on the World Wide Web (WWW), the Internet's fast growing application. The InterNotes Web publisher creates HTML pages of Notes views that automatically populate and manage changes to an HTTP Web server, such as Netscape Netsite.

Popular Web browsers, such as the Netscape Navigator and NCSA Mosaic, gain access to documents stored in Notes databases. Notes and InterNotes are a solution to automate the tedious process of maintaining and updating hyperlinked Web documents. Because of Notes' application development capabilities, items can be programmed to show a "New" sign or be present on the Web site for only a specified period of time.

USING THE INTERNOTES WEB PUBLISHER   IBM Global Network has had several customers who have Notes databases that they need to make available to people (their customers or employees) who typically use Web browsers and don't want to switch to the Notes client. The InterNotes Web Publisher allows you to easily have the same information in Notes databases and HTML format. One IBM Global Network customer, for example, said they had 2000 Web browser users who didn't have the Notes client installed but needed access to Notes databases. The Web Publisher provides a nice solution.

"A QUICK TOUR OF LOTUS NOTES" DISPLAYED BY INTERNOTES "A Quick Tour of Lotus Notes" is a database that is part of the Lotus Notes install code and is one that most installations make available to their Notes users. Figure 14.2 shows the "About" document for this database as displayed under the Notes client.

The "A Quick Tour of Lotus Notes" database was then converted to HTML format and placed on a World Wide Web server. The same "About" document is shown as it appears under a Web Browser (the IBM WebExplorer) in Figure 14.3. So far, so good. However, how does InterNotes handle the conversion of Notes functions such as "Views"?

THE NOTES DATABASE "VIEW" DISPLAYED UNDER INTERNOTES The "View" for the "A Quick Tour of Lotus Notes" database is shown in Figure 14.4. There are three basic choices in the view: Index, Print, and Table of Contents. Figure 14.4 shows that the "Table of Contents" was selected.

The InterNotes program uses hypertext to display the Lotus Notes view. This is shown in Figure 14.5. The "view choice" appears at the end of the "About" document. The hypertext for the view is displayed in the usual blue color, which indicates it is hypertext and that clicking on the text with your mouse pointer will bring up further details related to the hypertext.

Figure 14.2
"A Quick Tour of Lotus Notes" "About" document displayed under Notes.

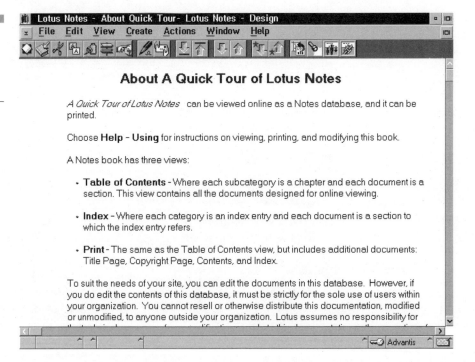

Figure 14.3
"A Quick Tour of
Lotus Notes" "About"
document displayed
under InterNotes.

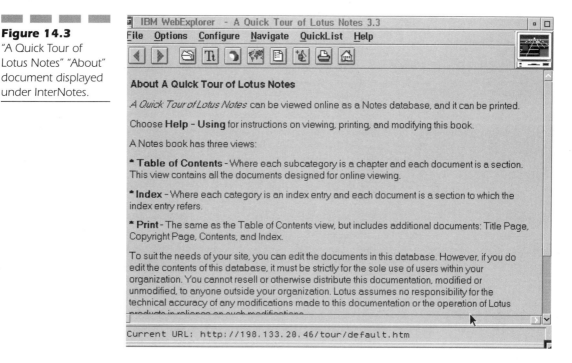

Figure 14.4
The "View" for "A
Quick Tour of Lotus
Notes" under Notes.

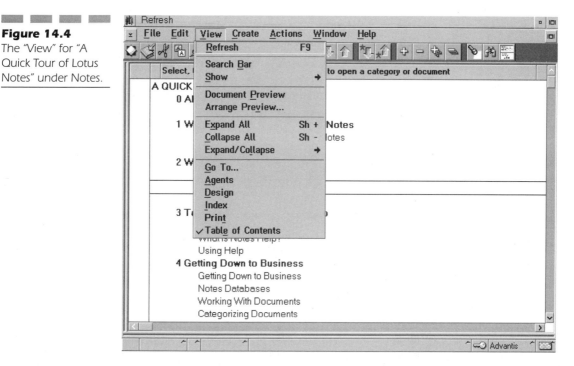

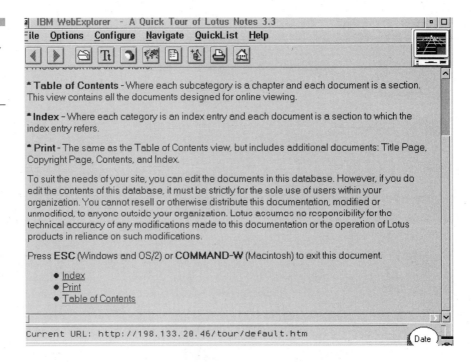

Naturally, InterNotes can't provide all of the Notes functions if you just want to use a Web browser. If it could, there would be no reason to use the Notes Client.

For example, the security function in Notes is not preserved once the database is converted to HTML. The database on the Web browser can be secured with passwords, but the Notes levels of security discussed in chapter 7 (e.g., access control based server, database, documents, sections, and fields) are not available.

Because both Lotus Notes and the Internet are becoming very popular, it is likely that products such as InterNotes will see wide use. With that wide use will come enhancements to the product in areas such as security and other Notes function not yet available.

INTERNOTES NEWS. InterNotes News is a Notes server application that exchanges Usenet news articles between Notes and UNIX servers using the Network News Transfer Protocol (NNTP). By bringing Usenet news articles into Notes discussion databases, users can leverage key notes capabilities, including full-text search, threaded views of discussions, and multiple indexed views of the articles. Because InterNotes News can be configured for specific newsgroups, administrators can provide access to only those newsgroups appropriate for business use.

Notes users can participate in Usenet news by posting a response from notes or by replying directly to the author using Internet Mail Transfer Protocol (SMTP) mail.

Lotus Notes: Newsstand

Lotus offers a content-hosting service called Lotus Notes: Newsstand for delivering business and industry publications to Notes users. Information providers—such as Patricia Seybold, Fidelity Mutual Funds, Dow Jones, etc.—replicate Notes databases to Lotus Notes: Newsstand. Newsstand then makes this content available to subscribers. This service has been available since July of 1995. It offers over 130 publications. Much of this information is industry specific, such as information from Best for the insurance industry and Patricia Seybold's Notes on Information Technology. Newsstand is available through Notes service providers.

Lotus Components

Lotus Components are scaled down applications designed to let users remain in Notes while using common office applications. The six applets included with the Lotus Components Starter Pack are Chart, Comment, Draw/Diagram, File Viewer, Project Scheduler, and Spreadsheet.

Why Components instead of Total Word Processor, Spreadsheet, etc.?

Lotus Components let Notes end users complete more of their daily tasks without leaving the Notes environment. Notes users select and activate task-specific applets and insert them directly into their Notes documents. There is no need for the user to load a separate application. Notes application developers can use Lotus Components to build more powerful Notes applications and deliver them more quickly because Lotus Components are fully programmable and customizable.

What's Available

Lotus originally began shipping Lotus Components in August of 1996. It described Lotus Components as "a set of fast, focused software applets (ActiveX controls) for Lotus Notes." These business productivity software applets were described as for use by both end users and application developers. At that time, the company also announced plans to deliver a new release of Lotus Components for interactive Internet applications in 1997.

Lotus Components is available for trial at no charge via the Lotus Components Web site at http://components.lotus.com. The Lotus Components Starter Pack includes six core components—Chart, Comment, Draw/Diagram, File Viewer, Project Scheduler, and Spreadsheet—and the Lotus Component Template Builder to customize the Lotus Components into business objects.

In 1997, Lotus plans to deliver Lotus Components business productivity applets for the Internet, enabling a new, more powerful class of interactive Web applications. Web developers will be able to use Lotus Components in building their applications. End users will then be able to interact with these Web applications by activating Lotus Components within their browsers.

Using Lotus Components in an interactive Internet application, for example, a brokerage firm could post an application to its Web site that allows its customers to query their investment portfolios. Customers could then use the Lotus Spreadsheet Component to analyze the portfolios and activate the Lotus Chart Component to chart their data.

Lotus Components are a key element in Lotus' Internet strategy. Together with the Domino Web application server, Lotus Components will provide a powerful platform for Internet and intranet applications. Lotus was the first major vendor to ship business productivity applets, and this brought the advantage of this valuable market experience and unique technology to developers as the Internet moves into its next generation: interactive Web applications. Lotus Components embrace and extend ActiveX and Java technology.

The first Lotus Component offering was a set of ActiveX controls, extended with unique technology from Lotus that enhances their usability, programmability, and data integration capabilities. The usability enhancements make Lotus Components easy to use, adopting similar techniques for accessing menus, formats, fonts, and other properties. Programmability extensions make each of the Lotus Components customizable for specific business needs by exposing the component properties and methods. The unique data integration features allow Lotus Components to exchange data with each other. Lotus also supports

JavaSoft's JavaBeans standards effort, which defines a set of component APIs that allow Java developers to create more powerful cross-platform applications for the Web. Lotus is fully embracing Java as a secure, cross-platform Internet technology, and they are working closely with JavaSoft to ensure that the JavaBeans design meets the needs of Lotus products and of our customers.

Lotus also provides the "Lotus Components Product Warehouse," a Notes Release 4 application that allows for distribution of Lotus Components and license monitoring within an organization. The version of Lotus Word Pro, which shipped as part of SmartSuite 97, is a container for Lotus Components.

System Requirements

In order to run Lotus Components Starter Pack 1.0, users will need an IBM PC or compatible (80486 or higher), Windows 95 or Windows NT Workstation 3.51 or higher, Lotus Notes Release 4.1 or higher, a 32-bit client for Windows 95 or Windows NT, and 12MB free disk space (20MB for install).

French, Italian, German, Spanish, Japanese, and Brazilian Portuguese language versions of Lotus Components are available.

Enterprise Conferencing with VM and Lotus Notes

In the corporate environment recently, there has always been a need to share information among people who are geographically apart. Some 20 years ago, a new idea arose that was to have interactive (electronic) discussions called "forums" in which anyone within the corporation could participate.

The IBM solution was to provide a facility in which users could "append" to a forum and read other users' appends. An append is much like a memo where it can initiate a discussion on a new subject or respond to another append. The "subject" of the append associates one append with another. A forum is defined to be discussion about a single, general topic such as Aptiva computers, and within the forum, there are appends discussing multiple subjects related to the general topic. All appends to a forum are maintained in a forum file in chronological order by append date and time.

Because all of IBM was performing its daily office work (i.e., e-mail and calendaring) on VM, the conferencing facility was implemented on VM. A service machine that ran continually was set up to be the processor of appends (i.e., one would send the append to this service machine, which would, based on the forum specified, add the append to the appropriate forum file). IBM's facility is called Toolsrun and was developed by IBM employees.

Toolsrun is a facility that allows conferencing disks to be set up. A conferencing disk is a VM minidisk. Conferencing disks generally contain forums that are somehow related. For example, one conferencing disk contains forums that are all discussions about PC topics. Another conferencing disk contains discussions about VM topics. Each conferencing disk is controlled by a Toolsrun service machine that processes appends for all the forums under its control. One service machine can control multiple conferencing disks.

Because network communications can be slow across long distances, in addition to these master Toolsrun service machines, shadows of conferencing disks were also created. Although users still have to append to the master service machine, with shadows, users can now access appends locally. A shadow is an exact duplicate of a conferencing disk. The master Tools run server is responsible for "replicating" its forums to the shadows at periodic intervals.

VM conferencing became very popular within IBM. In 1996, a popular Toolsrun server typically processed up to 1000 appends per minute, and a single conferencing disk contained 400 forums. So that DASD needs for these forums did not continuously increase, *pruning* of forums occurs at regular intervals. Pruning is the process of removing a subset of the earliest appends in a forum and archiving them on some other DASD. The pruned appends are available in case one needs to delve into a past discussion or retrieve a tip.

A sample append is shown in Figure 14.6. Let's assume this append is part of a forum called OS2WARP FORUM that is controlled by the IBMPC conferencing disk. "GMT" means Greenwich Mean Time, which is used to standardize time so that the order in which appends are received at the service machine is preserved. The "Ref" line indicates that this append is a response to another append.

Also included in a VM forum file is a header. The header is any number of records that describes such elements as the owner and purpose of the forum. It might also contain general information about how to obtain manuals or order service programs if the forum is for a store-bought product. The header would also contain a list of pruned appends and where they can be obtained. The first line of the append in Figure 14.6 is called the header of the append. Much of the informa-

Figure 14.6
A sample append to
the OS2WARP forum.

```
----- OS2WARP FORUM appended at 23:22:49 on 92/08/24 GMT (by THORNE at IBM)
Subject: I love the latest OS/2 WARP !!
Ref:      Append at 14:54:18 on 92/08/21 GMT (by MAREN at IBM)

Maren,  I agree that the new WARP has many enhancements that have made doing
office and development work much easier.  Have you checked out the Warp Center

tool bar?   It's a big improvement over the Launch Pad and the install
procedure is infinitely better.

Thorne Ventura
IBM Tools Reengineering
```

tion VMFM displays to users in the top of an append document is extracted from this header record.

On VM each line in a forum file is a fixed length record of 80 characters. Because VM is primarily a text-based system there is no provision for graphic inserts into forum files or font changing for emphasis. One is restricted to a single font and a subset of the 255 EBCDIC-character set (which presents a translation problem because the character set on a PC is ASCII; translation from EBCDIC to ASCII can cause unpredictable and undesirable results across VM to LAN gateways).

VM Conferencing from within Lotus Notes: General Concepts

As support and maintenance costs for VM systems increased, the need to find alternate solutions arose. At one point in IBM, there was some work done to duplicate in a PC LAN environment the Toolsrun server and conferencing disk concept. With the advent of Lotus Notes, which is a natural vehicle for conferencing with its concept of shared databases, one solution being implemented in IBM today is to "shadow" the VM forums into Lotus Notes databases. This shadowing allows end users to access VM forums from within the Lotus Notes environment without ever having to log on to a VM session. Because employees can spend many hours a day conferencing on VM, not having to log on to VM reduces Information Systems costs significantly (assuming that the cost of supporting VM services is considerably more expensive than the cost of supporting a Lotus Noes network). This section explains one such Lotus Notes implementation.

While implementing a Notes solution to conferencing reduces VM costs, within a corporate environment, it is not always possible to migrate everyone to a Notes environment simultaneously, especially because plans and priorities within corporate divisions are different. So that both employees who have migrated from VM to a Notes environment and those that have not migrated can take advantage of conferencing, shadowing VM forums to Notes databases works well. In order to use the Lotus Notes program to represent forums, the forum files must be "shadowed" into a PC environment. The forum to be shadowed is downloaded to a PC from VM and then converted into a Lotus Notes database.

VM FORUMS AS NOTES DATABASES. Representing a VM forum in a Notes database means copying the VM forum file's contents into a Notes database. Each append in the VM forum is represented by its own Notes document. Ideally all the information contained in the original append is preserved in the Notes document, such as the append author, subject and time, and most importantly, the text of the append. Because Notes provides the ability to customize the presentation of text in documents (especially with the use of tables and rich text fields), the possibilities for representing append information in a Notes document are almost infinite. With its concepts of views, Notes provides the ability to present the append information in various logical groupings, such as by subject alphabetically, by author, by append date, and hierarchically by subject such that responses to appends are grouped together showing the flow of conversation.

In order to shadow a forum into a database, a Notes database template (skeleton) is used. This template contains the design of the database including all forms and views that a user will use to view and respond to appends. When a forum is shadowed to a Notes database for the first time, this template is used to set up the database. The template might contain the following elements:

- Forms
  - Forum append for new subject
  - Forum header
  - Forum append to respond to another append
- Views
  - Appends in date order (ascending and descending)
  - Appends ordered by author
  - Appends ordered alphabetically by subject
  - Hierarchical views showing discussion threads

You might include buttons on some forms (in Notes Release 3) or action buttons (in Notes Release 4):

- Respond to displayed append (create an append that is a response)
- Modify displayed append
- Send a note to the append author
- Send the append to someone

CONVERTING VM FORUMS TO NOTES DATABASES: STANDALONE UTILITY AS KERNEL. Once the Notes template is defined, a program can be created that converts a downloaded VM forum file into a Notes database. The append shown in Figure 14.6 might be formatted into a document as shown in Figure 14.7.

Although Notes has the capability of reformatting paragraphs to fit within the left and right margins, the append text in Figure 14.7 is not reformatted. Reformatting is not desirable because sections containing, for example, code examples should remain intact. However, because screen resolution in part determines how many characters of the active Notes font will fit on a line, and because VM is always 80 characters per line, if the active font allows less than 80 characters per line, Notes will reformat the append text to fit on the line *and* preserve the line-break character from the original append. As a result, each line of VM append text in a Notes document can appear to have *two* line breaks, producing undesirable formatting.

The little graphical box next to the "References:" line is called a *doclink*. Doclinks are a powerful Notes feature that work well for following threads of a forum subject. A doclink is a pointer to another Notes document. Double-clicking on a doclink displays the document pointed to. A doclink in the "Responses:" field is a pointer to an append that the viewed append is a response to. A doclink in the "Responses:" field is a pointer to an append that is a response to the append being viewed. With doclinks in the "References:" field, you can trace the conversation backwards (i.e., in the opposite order in which the conversation

Figure 14.7
The VM forum append (shown in Figure 14.6) converted to a Notes document.

OS2WARP FORUM

```
Author:                        Received at Tools server:
THORNE at IBM                   08-24-92 23:22:49 GMT


Last modified by:              Last modification date:


Subject: I love the latest
OS/2 WARP !!
References: MAREN at
STLVM6 on 08-21-92
14:15:18 PM GMT
Responses:

Maren,  I agree that the new WARP has many enhancements that have made doing
office and development work much easier.  Have you checked out the Warp Center

tool bar?   It's a big improvement over the Launch Pad and the install
procedure is infinitely better.

Thorne Ventura
IBM Tools Reengineering
```

occurred). With doclinks in the "Responses:" filed, you can follow the conversation in the order it occurred.

THE OVERALL CONCEPT.   Now that we know the pieces of conferencing on VM and in Lotus Notes, how do we put it all together so that users have a facility with which they can monitor VM forums from within Lotus Notes, keeping track of appends they haven't read? Because Toolsrun already collects appends and makes them available in forum files, all that is needed is a continuous process that downloads these forums files to a PC and converts them into Notes databases. A Notes server is the perfect vehicle. We also need a database to control *which* forums should be converted into databases (unless we unconditionally convert all forums into databases, which could use a lot of PC DASD, especially when Notes full-text indexing is active). This database could also provide other administrator tasks. This database is called the Forum Manager. The entire facility we'll call VM Forum Manager, or VMFM for short.

A typical process to implement the entire facility might be:

1. Download all VM forum files of interest.
2. Read a view in the Forum Manager that lists all the forums to be converted to Notes databases.
3. For each forum to convert, add any new appends alter modified appends.
4. Repeat at step 1.

Note that step 3 states to *add* new appends and *modify* altered appends, which means appends altered on VM (users might modify their VM appends, for example, if they want to add more information or state that a question was already answered outside of the forum). One important aspect to this process step is that we do not delete the forum database or overwrite appends (unless modified) because then appends would be marked as unread the next time a user opened the database. (Whenever a document is changed, Notes marks the document as unread. Read and unread marks are maintained for each user who opens the database.)

We want users to be able to keep track of which appends they already read and which they haven't read, so each time they open the database, they can see what's new in the discussion. In order to maintain read and unread marks, the conversion program must keep track of the latest chronological append and not alter (unless altered on VM) any appends chronologically before the latest append. Also, note that, if append B is a response to append A, when the conversion program puts a doclink in append A to show that it has a response, the adding of the doclink (which is a modification to the document) becomes marked as unread.

Administering a VMFM Server

Now that we know the overall process of VMFM, let's look at some other aspects of maintaining it.

INTERNAL VS. EXTERNAL ACCESS. Our Notes server with all the forum databases is only going to be useful if users can access it. Notes has the concept of a domain (implemented originally for VMFM using flat certificates in Release 3 of Notes) that is a logical grouping of Notes servers that provides protection from external access. In order to access a server within a domain, a user must have in his ID file the certificate for that domain (the certificate is given to users by the domain administrator). In the Notes Release 4 version of VMFM, flat certificates were replaced by hierarchical certificates. Users must be certified to access the servers defined to a particular level of the hierarchy in order to access the databases on those servers.

Depending on network topology, Notes servers can be defined on a network's name server so that all servers are known to all users who have access to that name server. Name servers are maintained by network administrators. If a Notes server is not defined on a name server or the name server is not available, users must define the Notes server in a file on their workstations or specify the Notes server's address when opening a database on that server.

In the case of VMFM, users outside our domain (VMFM is administered using flat certificates) are told that they must access the Notes server via TCP/IP (NetBIOS connections are not supported), and they are told what two TCP/IP addresses are required in order to access the VMFM server: the TCP/IP address of the server itself and the TCP/IP address of the domain. Because it is unlikely that other IBM sites' administrators will define the VMFM server on their name servers, users would have to define the VMFM server in the TCP/IP configuration on their workstations.

HARDWARE CONFIGURATIONS/REQUIREMENTS. On the VMFM server, both the download and conversion processes execute on a single Notes server. To ensure maximum performance, there are minimum hardware and configuration requirements. Their requirements will vary depending on the number of forums (and their size) shadowed into Notes databases. A minimum recommendation to shadow 150 forums, each with about 500 to 1000 appends in them, is 32MB of RAM, a 1GB hard drive, and a Pentium processor. If full-text indexing is provided for these databases, Notes' maintenance of the indexes can cause hard drive usage to grow quickly (full-text indexing is explained later).

CONVERSION PROGRAM PERFORMANCE. Another performance consideration is the language used to code the conversion program. The conversion program, if it is to provide maximum functionality, performs a lot of CPU-intensive processing. Writing a conversion program in a compilable language (as opposed to an interpreted language), such as C, will significantly improve conversion time. A sample VM forum file containing 10,000 records took 2 minutes to convert all appends into Notes documents using a C program, while it took 10 minutes using an interpreted language such as REXX. This 10 minutes can be reduced by not updating the full-text index in the Notes database every time new appends are added.

MONITORING USAGE. Each Notes server has a log database that contains very useful information for monitoring accesses, reads, and writes that occur. A transactional log is maintained in the Miscellaneous Events view. The number of days that log documents are saved is controlled in the server's NOTES.INI file. This log is useful for collecting names of VMFM users so that any changes regarding TCP/IP address (in case the server is moved) can be broadcast to each user.

Maintenance

Maintenance of the VMFM server includes such things as pruning, compacting, and full-text indexing databases.

PRUNING DATABASES. It was mentioned earlier that forum files need to be pruned in order to conserve costly VM DASD. Although PC disk space is far less costly than VM disk space, if the shadowed database is to be a mirror of the VM forum, pruning must also be performed on the Notes database.

VMFM uses yet another database, a forums-to-prune database, to track those VM appends that have been pruned (this new database could be merged into another administrative view in the Forum Manager database). As part of the conversion process, the conversion program searches the forum file for the header of the first chronological (i.e., earliest) append and extracts the append's date and time. It then extracts the date and time of the first chronological append document and compares the dates and times. If the document date and time is earlier than the forum file date and time, a document indicating that this forum database needs to pruned is written to the forums-to-prune database. An administrator can view this database and run a pruning macro that will delete the appropriate documents. An improvement to this process would be to run a background macro nightly that automatically prunes databases.

COMPACTING.   It is important to conserve hard drive space on the Notes server. The server's Compact function should be enabled and run nightly to remove wasted space in all forum databases.

FULL-TEXT INDEXING.   VMFM provides the ability for users to search for appends that contain topics of interest without having to read the subject of every append. For example, in a forum about programming, you might want to see if any other users have found an efficient way to code a binary search. Notes' full-text indexing provides this ability. A user can enter in the search window the string "binary." If any appends contain this string, all of them will be displayed in a list to the user. However, Notes uses a lot of PC disk space to maintain full-text indices.

NOTES UPGRADE CONSIDERATIONS.   In an environment with many users, it is not uncommon for multiple levels of Lotus Notes to be supported, especially while the company is in the process of migrating users from one release to the next. Because it is likely that later releases will contain new functions not supported in earlier releases, it is important for Notes application designers not to use new functions before all users are migrated. Otherwise, users on the older release will encounter errors trying to access the databases. In VMFM, many functions are contained in Notes macros embedded in the template for all databases, so the least common denominator (i.e., the earliest release in use) should be used to design any Notes application functions.

Alternate Solutions to VMFM

There are several alternative solutions when implementing a VMFM type solution.

ALL FORUMS IN ONE DATABASE.   One alternate solution is to convert all the forum files into a single database where each database is its own Notes category and all the appends for a forum are documents in that category. The advantage of this setup is that it is easier to locate forums because they're all in a single database. The disadvantage is that this single database becomes huge. As a result, performance during searches and other application functions suffers because of the myriad of documents that have to be processed.

PROCESSING VM APPEND SUBSCRIPTIONS VIA MAIL-IN DATABASE. The Toolsrun server on VM provides users with the capability to "subscribe" to forums such that any new appends processed by the server are automatically sent to the subscribers. If you were to subscribe to all forums on a conferencing disk on behalf of the Notes server, a Notes mail-in database could be set up on the server to receive these subscriptions and add them to the appropriate forum databases. This solution would significantly reduce the overhead of having conversion and download programs. In fact, there is no need for a download program, and the conversion occurs one append at a time.

IMPLEMENTING A VM TOOLSRUN SERVER IN A LAN ENVIRONMENT. The entire Toolsrun facility could be coded to execute in a LAN environment instead of on VM. This solution can be very complex and will not be discussed here.

SEPARATING THE NOTES AND DOWNLOAD PROCESSES.
Another method that you could use to improve the turnaround of VM forums into Notes databases is to have the downloading task executing on one PC and the conversion to Notes databases executing on another PC. This way, the download and conversion tasks could execute simultaneously instead of serially (i.e., conversion cannot start until download completes). All that needs to be decided is how the conversion program on the Notes server would access the forum files on the other PC. Because this configuration has not been tried in VMFM, it is not known if the savings from concurrent processing are outweighed by the time it takes for the conversion program to access the forum files across the network.

VMFM Details

The following sections cover some of the detailed considerations we used in designing VMFM.

RECOGNIZING NOTES MEMOS AS APPENDS. When the code to implement the VM Toolsrun servers was originally written, only files in a certain format would be recognized as forum appends. Hence, the Toolsrun system included a facility with which to create and send appends to the desired conferencing disk server. This facility formatted the append so that the Toolsrun server would recognize that it was a forum append. When Lotus Notes came along, it was necessary to provide some mechanism to allow these new types of files to be recognized.

When a Notes memo is sent to a VM server, the memo must first be transmitted through a gateway between the Notes server and the VM network. Because there are no requirements for standardization of these gateways, how the Notes memo is transmitted to the VM network is left to network administrators. As a result, the file that a VM server receives could be in any format. In IBM, the mechanism chosen to transmit Notes memos to VM is to translate the memo into an Office Vision memo because many VM users process their mail using Office Vision.

Because a Toolsrun server would not recognize an Office Vision memo, some facility had to be invented. Software was created to allow a Toolsrun server to recognize a "magic line" in any memo. Once this software was installed on the Toolsrun server, any file that contained the magic line would be accepted as an append. Essentially the add-on software searches for the magic line and processes the file as an append if found. When a user appends from VMFM, this magic line is inserted into the append. The magic line must contain the Toolsrun verb APPEND, the conferencing disk that is to process the append, and the name of the forum to append.

CONTROLLING THE OVERALL SHADOWING PROCESS. On the VMFM server, to start the forum shadowing process, an administrator simply starts execution of the main program. In the case of VMFM, that program downloads VM forum files. In order to improve performance, this facility, which we call Benedict, keeps track of the last date and time that downloads were performed for each conferencing disk so that only those forums are downloaded that have been updated since the last time the download process last ran. Benedict downloads all forums for all conferencing disks being shadowed before calling the process to convert forums to databases.

When the conversion process is called, a program first reads the Forum Manager database (mentioned earlier), which contains a view that lists all the forums to be converted. The beauty of having such a view is that a VMFM administrator can easily choose to enable/disable conversion by clicking a button that toggles a yes/no field in the document representing the forum. When the conversion program executes, it reads all the documents in this view and checks to see which documents (forums) are enabled for conversion and converts only those forum files.

Now that the conversion process knows which forums to process, the actual conversion program is called to process each individual forum file. To determine the date and time of the last append processed, VMFM maintains an internal document in each forum database, not viewable by users, that contains the header of the last append processed the previous time VMFM processed that forum. The conversion program reads this header and locates the append in the downloaded VM forum file that contains this header. Now VMFM has a pointer to the last forum file append processed. VMFM starts its processing this time at the next append, adding each new append to the Notes database.

The Lotus Notes Toolkit contains APIs (Application Programming Interfaces) that can be used in C programs to access data in Notes databases. IBM has developed a REXX interface to these APIs that is more convenient for those programmers who prefer REXX to C.

The conversion program extracts the information it needs from the VM append to create the Notes append document, uses one of the forms in the template database mentioned earlier, and writes the document to the database. If the document is a response to another append, a doclink is added to link the two documents together. Also a doclink is added to the referenced append to indicate that it has a response.

Once all new appends are added to the database, the conversion program searches for modified appends in the forum file. In order to track modified appends, the conversion program lists the modifications that it has processed for an append in a hidden field in the document. Every time an append is modified, a new modification record is added to this field. So, in order for the conversion program to determine if a new modification has occurred, it compares the number of modifications in the forum file append (which are listed one per line) to the number of modifications listed in the append document's field. If the number of modifications listed in the append from the forum file is greater than the number listed in the database field, the append in the database must be modified to match the append in the forum file.

OVERRIDING GATEWAY ADDRESSING. Because of the inconsistency mentioned previously of transferring Notes memos to the VM network, VM addressing formats can differ from site to site. For example, given a VM user ID of THORNE and a VM node ID of AUSVMR, when you send a Notes memo to VM at one site, the format of the addressee might be "THORNE@AUSVMR," where at another site, the format might be "AUSVMR.THORNE@VM." VMFM overcomes this diversity of formats by providing a mechanism for users to specify the format of a VM address. Because a VM address is comprised of a user ID and a node ID, VMFM allows the user to specify the user ID as "%U" and the node ID as "%N." Then, given the two previous examples of VM addressing, one user would specify "%U@%N" and the other would specify "%N.%U@VM". VMFM substitutes the user ID and node ID in the appropriate positions of the format string.

NOTES MAIL TEMPLATE USAGE FOR APPENDS. When a user wants to start a new subject or respond to an existing append, VMFM provides a button in the form (in Notes release 3) for this purpose (an action bar button in Notes release 4). This button essentially displays a Notes memo formatted with the appropriate fields for creating an append. This button invokes a Notes macro that extracts from the viewed append the information needed to insert the magic line into the memo and provides an area for the user to enter the text of his append. The macro also needs to determine to which VM Toolsrun server to send the append memo. The addresses for all VM conferencing disk servers is another view in the Forum Manager database. The macro extracts the VM server name from the viewed append and uses that as the key into the VM servers view to extract the VM address of the server. This address is then written to the "To." field in the new append.

VMFM also provides the ability for users to create a signature that will always be added to the end of the append text. This feature is also implemented via a button.

CONFIDENTIAL FORUMS. Up to this point, we have assumed that all forum data can be shared among all employees. In Notes terminology, this means that the ACL, Access Control List, for a forum database lists the default access as Author (i.e., anyone can create appends and modify those appends they created). In a corporate environment, some information, such as work on unannounced products, might be deemed confidential and require that employees have a "need to know." Because the standard VMFM database template provides no way to restrict such information, a separate template database must be created whose ACL can be encoded with the names of only those users who are allowed to view the data contained in the forum.

This ACL must be modifiable as new people need access to the confidential data and others no longer have a need to view it. A VMFM administrator can always modify the ACL, but it makes sense to allow the VM forum administrator (usually the owner of the VM forum) to alter the ACL. Altering an ACL in Lotus Notes requires that a user have Manager access to the database.

Special code must be added to the conversion program so that it knows it is processing one of these confidential forums. When it encounters such a forum, the conversion program must also be able to determine which database template to use.

If confidential forums are supported on a VMFM server, no users must be able to access the server directly via TCP/IP's FTP or any other method. Replication of confidential databases must also be disabled.

AFTERTHOUGHTS. While VM Forum Manager is an excellent vehicle to introduce new users to Lotus Notes and to working in a LAN environment, it is only a step toward reducing the need to use VM systems for daily applications. The next step in the evolution of conferencing in a LAN environment would be to remove the forum files and Toolsrun servers from VM and establish a corporate-wide Notes domain in which all conferencing takes advantage of the more powerful data sharing, workflow tracking, and graphical interface capabilities in Lotus Notes. However, just as with the Internet, before network computing and LAN environments can realize their potential, the performance of networks must improve to support the increased amount of network traffic inherent when a large number of users are active simultaneously.

VM is an operating environment that provides much CPU horsepower and huge amounts of storage capacity. Despite much talk about the difficulties in migrating to and operating in a LAN environment, many companies continue to take advantage of the rich set of PC applications that can be shared in the less-expensive LAN environment. If a LAN environment is indeed more difficult to manage and its stability seems elusive, it is only because operating in a LAN environment and the new wave called "network computing" are relatively new to the computer world compared to the almost two decades that VM has had to become manageable and its administrators familiar with managing a large number of VM users. In time, LAN administrators will become equally adept at managing a large number of users and providing the distributed power of network computing.

15

Examples of Enterprise-Wide Use of Notes

Solutions are what it's all about.
Louis V. Gerstner, Lotusphere '97, January 27, 1997.

This chapter discusses some of the enterprise use of Notes. The first two examples are from IBM in Europe. They are based on the author's trip to Europe in April of 1995 to discuss the use of Notes by IBM Europe. Some of the other examples are based on worldwide Notes and Domino implementations by the IBM Global Network for IBM customers. Other examples are implementations for use by Advantis and IBM internally. The idea behind these examples is to show, with actual case studies, the diverse ways that Notes and Domino can be used in the enterprise, especially for projects that include workflow over the entire enterprise and among several enterprises.

In April of 1995, I had two meetings with IBM personnel in Sweden. First, I went to the IBM Forum building in Kista, outside of Stockholm. My contact there runs the IBMNORDIC Lotus Notes domain for IBM users in the Nordic countries and extensively deals with IBM customers on Lotus Notes requests. He says his customers are very interested in Internet access to Notes and the Internet in general. He mentioned that ISDN service to the Internet should be provided by the IBM Global Network because ISDN is inexpensive in Europe, other Internet service providers are already doing this, and the service would be a great way to access Lotus Notes. Because Sweden has the most deregulated telecommunications market in Europe, it would be a good place to start for new services.

Canon Notes Service in Sweden

Next I went to the IBM Nordic HQ building, also in Kista, and met with personnel from ISSC Sweden (IBM's outsourcing subsidiary, Integrated Systems Services Corporation) and the IBM Global Network in Sweden. They also said that Lotus Notes access via the Internet would be very popular in Sweden. IBM is currently providing a Lotus Notes replication service for Canon in Sweden. This service is typical of the type of Notes business for which there is a lot of demand in Europe. Through XPC dial, Canon replicates a Notes database to an ISSC hub Notes server. In turn, Canon resellers use XPC dial to access this database on the ISSC Notes server. The database contains Canon product information and details on shipments (e.g., when a certain Canon copier is due to ship to a reseller). IBM Sweden would like to extend this type of service to include Internet access and switched ISDN access.

IBM PC Company in Scotland

The IBM PC Plant in Greenock, Scotland, has a Notes requirement for Europe similar to the one being used by Canon in Sweden. The project is called GENIE. IBM Greenock would provide IBM prices, and product information to its PC vendors. For example, a PC vendor could use the IBM Notes database to find out when an IBM ThinkPad they've ordered is scheduled to ship. If PC vendors could access the IBM Notes database via a variety of methods (e.g., XPC, Internet, or ISDN) provided by the IBM Global Network, that would be ideal. The people from the IBM Global Network would be very interested in any details we could provide them on the process requirements for providing content on the NCC platform. Notes content with access by customers through a variety of methods (including the Internet) is the ideal way to start off with a Notes service offering. It appears there are plenty of opportunities for this type of service in Europe. The IBM Global Network people in Sweden were very interested in the Billing and Registration process, and I told them this was a high priority item and getting a lot of attention. They were also very interested in the details we provided them on the IVANS Notes project, because they have several Insurance customers who want Notes services.

Advantis Automated Line Ordering Process

The Advantis Company developed a Lotus Notes workflow application to manage orders and track implementation status of new customer connections to their commercial private multiprotocol (MPN) network called Internetworking 1.1. Customer orders for links are transmitted via a "shuttle" process from Tampa to a mail-in database in White Plains, NY, that is part of the Notes MPN Order Status application. This Notes database is replicated to servers at Tampa and Schaumburg.

Prior to the Lotus Notes MPN Order Status application, in order to obtain information on link status for Advantis customers, it was necessary to search around in several mainframe databases. Even then, to find the information you needed, you might have to call someone and have him look at the files in his desk. The following sections give details on how this Lotus Notes application helped "save the day" for Advantis. It is a good case study of how Notes can be used in the hands of empowered, knowledgeable users to dramatically improve workflow processes that are both dependent upon, and limited by, legacy mainframe applications that IS developers cannot easily replace or enhance. There must be countless situations and opportunities like this across the country.

Background on the Application

In October of 1995, there was a problem keeping up with enablement and technical support of the Advantis Internetworking 1.1 (I1.1). I1.1 had just reached an installed base of 1000 customer site routers and was growing at 10% to 15% each month. Marketing projected that orders would triple in 1996. An Advantis analyst was given a special staff assignment intended to focus on streamlining and improving the processes used to fulfill orders, especially involving the boarding of new customers.

At that time, Advantis was still dependent on a legacy order management application developed in the 1980s to support the ordering of SNA connections to the IBM Information Network. This legacy application had been minimally modified in the spring of 1993 to support orders for the new router-based Multiprotocol Network service called I1.1. Quick fixes like the VTAM Dummy Deck were employed to fool this SNA-oriented application into accommodating router orders, by treating a backbone router as an NCP and a site router as a control unit. Other gaps in the SNA process were filled by supplementing the legacy application with a collection of PROFS notes and VM shuttles like the original Router Configuration worksheet, or "R1 form." Long after new SNA orders effectively dried up, and I1.1 became the center of new order activity, Advantis was still dependent upon this fragmented and cumbersome stopgap process.

The Advantis development community acknowledged the limitations of the legacy application as a stopgap measure and offered a new strategic business application designed for the emerging router networks. Besides offering a modern, graphical user interface, NCMS, as it was called, was going to accommodate key requirements such as automation of a task involving the assignment of backbone slot/ports and associated IP address pairs, an integrated Router Configuration (R1) form, and some form of automated configuration support. I was told we could expect the R1 function in an NCMS release projected for February of 1995. However, the dates were forever shifting and slipping, and nothing at all had been delivered by October of 1995. Moreover, there was little reason to expect that we would see any help until well into 1996, after we had struggled through another 1500 to 2000 orders.

We were losing good technical people, and one of the factors they often cited, in connection to workload, was a frustration with what they saw as hopelessly archaic and cumbersome administrative processes that sapped technical time and energy. That contributed to their feeling that Advantis was not an innovative, leading-edge company destined for long-term leadership in a fiercely competitive and dynamic industry.

It was against this backdrop, in October of 1995, that the staff assignment was initiated that was intended to focus on I1.1 process improvement and automation. The staff assignee had, by that time, experimented with the use of Lotus Notes, in particular, as a workload/order management aid and was convinced of its potential to compensate for the shortcomings of the legacy application, streamline workflow, and generally enhance the productivity and optimism of the technicians involved. The staff person was well-suited to exploit these opportunities due to his intimate knowledge of the problems and requirements. His management expressed a desire to take the initiative, as opposed to waiting for Development to solve its problems.

The Lotus Notes Application

The staff assignee designed, developed, and implemented a comprehensive Notes-based application that has thoroughly transformed the way day-to-day work is done in the MPN Enablement Department.

From their perspective, the legacy application has been all but replaced by a Notes-based interface application that now encompasses all of the major steps in the Enablement process (presented in detail later in this chapter).

Each step has been made easier, faster, and less prone to human error.

Improvements include all of the functional requirements originally projected for delivery in NCMS during 1995, and now projected for delivery as part of the IGN Toolset in mid-1997. Many additional improvements never envisioned or requested of development have also been implemented.

The Notes application overcame a significant amount of natural resistance to change to win full voluntary acceptance by the end of the first quarter or 1996. Since that time, it has been the production vehicle behind more than 1200 new customer connections and will continue to carry the full production burden for MPN Enablement until at least the third quarter of 1997.

Acceptance by the MPN Enablement department and some of its key process partners has been totally voluntary, based entirely on the perceived merits of the application. Each new feature was presented as an optional alternative to previous methods, without benefit of management decree.

Design Review

The Network Design Group uses Notes views and forms to record the results of each deal that comes before a design review and to post the results online. Apart from the PROFS minutes, this represents the only record of design review actions. Unlike the PROFS minutes, it is an online database that allows all attendees and many other interested parties to quickly find and access design review information by date and/or account. It was developed in response to concerns that many people were trying to keep individual VM logs of the PROFS minutes and spending a lot of time combing through these logs to locate specific design review actions.

The views have the potential right now to serve as online agendas for future meetings as well as a searchable record of past actions. The developer is working with Design to make the underlying Notes form an online input vehicle for people who request time on the agenda. The current system of submitting requests and materials for copying and hardcopy distribution is, by all accounts, a highly inefficient and labor-intensive "administrative nightmare."

Capacity Management of Channelized-T 56Kbs Timeslots

The Advantis capacity management department in Gaithersburg, MD, relies solely on the Notes views and forms that were developed, at their request, to alleviate a heavy administrative burden and problems associated with the original vehicle, a flat file on the Capri Web site.

Gaithersburg uses a Notes view to monitor a semi-automated and dynamically changing online matrix showing the status of more than 1000 channelized timeslots in 10 backbone node cities. The matrix view shows how many timeslots are vacant, reserved (awaiting order placement), and assigned (on order).

They have 48 hours after a design review approves a channelized connection to provide the Sales Engineer (SE) with a channelized T1 circuit number and timeslot at the approved node for inclusion in the SE's circuit order request. A Notes form behind the view enables Gaithersburg to reserve the timeslot and generate a note to the SE in one step. The note automatically includes detailed information such as backbone router name and ID number, plus a 15-character AT&T circuit number and timeslot that previously represented opportunities for error when Gaithersburg had to manually key this into an ordinary PROFS note.

Another source of error was the need for Gaithersburg to wait for the order number to come back in a PROFS note and then manually post it to the flat file. This occurs automatically in Notes by means of a batch program (macro) developed to update records with order details, as well as the summary matrix view, on a daily basis.

Order-Entry and Synchronization with Legacy Applications

When the SEs receive the circuit number and timeslot, they enter this and other order information into a Notes shuttle form and send it to an order-entry department for re-entry into the legacy applications. Initially Network design manually created site order records in Notes following design review, and MPN Enablement manually entered the order number upon appearance of a task in the legacy application. This overhead was eliminated when the Notes application was mail-enabled and the staff assignee worked with developers to have VM/PROFS system notes sent from the legacy application each time an order was placed. Also developed was a macro that parses the ASCII system note and converts it into a fielded Notes order record. This keeps the Notes database in synch with the legacy application.

Notes adds value at this point by automatically extracting additional data, not available in the legacy application, from the Design review record in Notes. This includes Assignee information needed to post the order to customized Notes Work Queues and protocols approved at Design Review. Notes also computes the due date of Enablement tasks and the SE's prerequisite task to send a router configuration (R1) form to White Plains.

Customized Work Queue for MPN Enablement

One of the advantages of Notes over the legacy application is a set of Work Queues that are specifically designed for engineers in MPN Enablement. It lets them to see at a glance not only the complete list of orders assigned to them, but also due dates and status for the multiple enablement tasks associated with each order. It also shows the status and arrival date of the R1 Form that they need to complete their router configuration task. Subview options enable anyone to search through orders by assignee, account, order number, or customer request date.

The speed and convenience of these Notes views provided Enablement with a strong voluntary incentive to work from Notes as opposed to the legacy applications.

Serial IP Address Assignment and Data Input Task

Notes saved MPN Enablement from having to follow a complicated scenario for the 1384 task, especially for channelized-T orders. This is best illustrated by comparing the steps involved in that proposed scenario to the actual, current scenario in Notes.

Pre/Non-Notes Scenario

1. Monitor legacy application for 1384 tasks; page to comment field for T1 circuit number and timeslot.

2. Use the World Wide Web to access a flat file in Gaithersburg, and find the linename associated with that T1 circuit and timeslot.

3. Telnet to a Unix server, open another flat file for the channelized card, and assign a free pair of serial IP addresses by typing slot/port: interface, account, and location in remarks.

4. Access a VM application (RTRPROF) to check the SNMP Community name (existing accounts only).

5. Return to the legacy application, open the 1384 panel, and manually enter the six required items:
   - Linename
   - Serial IP address of the backbone port
   - Serial IP address of the customer router
   - Slot/port of the customer router
   - Hostname
   - SNMP community name

6. Telnet to the backbone router, and manually add the following (typical) lines to its config:

```
interface Serial3/1:15
description  AZ01 56000 AZ01DALL DALLAS, TX  American Automobile
Assoc.
ip address 129.36.148.113 255.255.255.252
no ip route-cache
bandwidth 56
shutdown
```

CURRENT NOTES SCENARIO

1. Monitor Notes WorkQ for 1384 tasks, open a 1384 form, and manually enter one or two items:
   - Hostname
   - SNMP community name (for new accounts only)
2. Telnet to the backbone router, and paste in a block of lines automatically generated for that purpose in the Notes 1384 form (see the previous six-line example).

Router Configuration Form and Task

The process calls for SEs with varying levels of technical knowledge to gather data from the customer and complete a router configuration worksheet, also called an R1 form. The form is then sent to the MPN Enablement group, where Network Engineers use it to create router configuration files, which are then sent electronically to a server in the Advantis warehouse. Configuration files are loaded into the router before it is shipped to a customer site for installation. The R1 form is intended to provide all of the information a skilled engineer would need to configure a customer site router, without guesswork or time-consuming inquiry. It is both complicated and key to an efficient router ordering process.

The R1 form was originally implemented as a formatted VM "shuttle" file. Re-implementing it in Lotus Notes opened the door to a wide array of process improvements and provided a powerful incentive for the SEs who fill it out and engineers who use it to switch from the legacy application to the new Notes interface. It illustrates an important lesson learned in the course of this case study: Any process that relies on the repetitive use of legacy e-mail systems, like PROFS, or the sending of VM files (especially formatted ones) can be improved significantly by switching these portions of the process to Lotus Notes. This is because traditional e-mail flows from one user to another. The sender and recipient might maintain a record of these private communications, but there is no public or online audit trail of who sent what to whom and when. Secondly, the information in traditional e-mail communications cannot be manipulated, even when formatted, because it is not usually in active fields, like it can be in a Lotus Notes form.

Once the R1 form was implemented in Lotus Notes, it became much easier to fill out and was so easy to manipulate that we were finally able to implement what amounted to a Notes macro-based configurator for

Cisco routers. On the copy of the form used by the engineers tasked with configuring the routers, a button entitled Preview Config was added. When activated, it triggers a macro that composes and displays a special form designed to look exactly like a Cisco router configuration file. This form contains static text for standard lines in the configuration, interspersed with a variety of "computed when composed" fields. These fields inherit values from corresponding fields in the underlying Notes R1 form and place them in the sequence and syntax required for configuration. Field formulas control much of the resulting display, including what happens when data is not found in the R1, indicating, for example, that a particular protocol, like IPX, is not applicable to the customer in question.

If the engineer is satisfied with a configuration he has "previewed" in this manner, he can click another button that exports the form data to his hard drive as an ASCII file. Ideally, that file is ready to run in a Cisco router and only needs to be shipped to the warehouse. At worst, it gives him a time and error-saving "base" configuration that he can then edit or embellish for routers with especially complex or unusual configuration requirements.

Automated Ping Test Notices

When a new leased line connection to the Advantis backbone network is installed and tested successfully, a Ping Test Notice is sent to a large number of interested parties. This seemingly simple step of the order process was one of the first to be automated via Lotus Notes, and it served to illustrate how Notes could be used to save time and improve accuracy.

Each leased line order is assigned by name to a specific individual, who configures the customer site router involved, but any member of the same department might handle the ping test, depending on who was available at that unpredictable moment. By informal agreement, the person conducting the test would take the time to compose a note and send it via PROFS (e-mail) to all other members of the department. Although Lotus Notes order forms were available to record information, such as IP address, contained in the note, there was as yet no similar agreement to update that record, because Notes was still viewed as an unofficial and optional database tool. The problem (and opportunity) was that the note took time to compose, and the important data it contained was not being captured in any convenient database, obliging each recipient to maintain his own reference log of these notes.

The staff assignee correctly saw this as a key opening for Notes. He added a button macro that included the Notes MailSend function to compose a note with data drawn from fields in the Note record and automatically sends it to a set distribution list. At the same time, it changed the order's status to "Installed" and captured the date.

Because it was far easier to fill in some fields and click a button than to compose a note, this enhancement was a sure winner. It standardized the appearance and completeness of the notices and spurred further use of the Notes database because it now contained valuable network topology data and reliable, up-to-the minute order status information.

Scheduled Customer Test Support Request Form

Notes offered a solution to another nagging coordination problem involving the scheduling of network engineers from the Enablement group to be available for the first customer test of production traffic over the new network connection. Previously this was "arranged" via PROFS notes or phone calls from the Sales Engineer/Account Rep. The notes or calls often arrived on short notice and lacked the details that would help an engineer prepare for the test. Again, as with any step of a process that relies solely on e-mail between individuals, there was no public record of what tests were scheduled to occur when, whether the requests were actually approved, or what the outcome was.

Another Notes mail-in form greatly standardized and improved this step of the process. It prompts the submitter for complete information and appears in a set of public database views by date, account, recipient, etc. It allows a two-way workflow dialogue between the submitter and the engineer whose time is being requested. Status of both the request and the actual test itself is recorded and accessible to anyone with the proper access. Estimated and actual test duration times are captured and used in process measurements and workload planning. The form was subsequently enhanced to include a button that helps the engineer compose and send a final "Customer Online Notice" once the testing is complete.

Automated Customer Online Notices

The final step in the customer enablement process is to send a "Customer Online Notice" upon successful conclusion of customer acceptance testing. This alerts all concerned, including the Advantis 24-hour network monitoring center and helpdesk, that a new customer connection is officially online. Prior to our use of Lotus Notes, the necessary network management information was entered into a VM-input panel, which sent the note and retained a VM flat file record for future reference. Even though this fairly voluminous information was all contained in Lotus Notes by this stage of the process, it unfortunately had to be re-entered in the VM panel because the network management center was not yet using Notes and there was no link between Notes and the VM-based tool.

This problem was solved by the creation of a Notes form that mirrors the VM Input Panel and a corresponding Notes view that identically matches the columnar structure of the VM flat-file record. After completing a customer test, the Enablement engineer opens an Online form in Notes for the order in question. It looks like the VM Input Panel but, by this stage in the process, is almost entirely prefilled. The engineer, who used to take an average of 8 minutes at this point to fill out the VM panel, now merely checks the form and clicks on a button that automatically uses the prefilled data to send the online notice and capture the date of this important change in order status. This takes only a minute, so there was an 8-to-1 reduction in technical time spent on an essentially administrative task. Needless to say, this was extremely well received and cemented support for the still "voluntary" use of Lotus Notes.

Once the online date is filled in, the Notes order record appears in the special view that mirrors the structure of the VM flat-file. At least once a day, in batch mode, the Notes File|Export command is used to create an ASCII file containing a flat- file record for each new online connection. Because the column widths of the special Notes "export view" conform exactly to the VM requirements, the resulting export file is easily uploaded and appended to the VM master file.

Worldwide Consulting Firm use of Lotus Notes

This example describes a Lotus Notes implementation by a major consulting firm with several billion dollars in annual revenues. With more than 30,000 consultants worldwide, the company faced the enormous problem of connecting far-flung users and enabling them to quickly communicate and make rapid decisions.

The company created what it calls the Notes Knowledge Exchange, which has more than 20,000 users and features 2000 databases, news, and electronic consultations. These Knowledge Exchange users are connected through Notes e-mail messaging.

According to the CIO of this major consulting firm, the Notes Knowledge Exchange enables the company to solve problems overnight that previously consumed as much as two weeks and thousands of dollars. The CIO went on to say that this service enables the company to keep up with the latest ideas and developments around the world.

What does robust network design bring to this operation (whether the network is from a Notes service provider or from the company itself)?

- A cost effective way to deploy this service globally to sparsely distributed users.

- The ability to easily connect to the company's customers and partners.

- The network administrative processes and tools that improve the ease and speed of deployment and administrative changes.

Figure 15.1 gives a graphical picture of the Notes service concept and the benefits from the use of Lotus Notes for this firm.

Worldwide Access to Company Information and Services

This example is based on a major pharmaceutical company that wanted to give employees, customers, doctors, hospitals, and HMOs access to

Figure 15.1
Consulting firm worldwide access to Lotus Notes.

"Problems that once took two weeks and thousands of dollars to figure out are solved overnight"
Chief Information Officer

$3.45 Billion
Consulting Firm

E-Mail

★ 30,000 consultants worldwide

Notes Knowledge Exchange

★ 2,000 data bases
★ News
★ Electronic consultations

E-Mail

"how we keep up with ideas and developments around the world"

★ 20,000 Knowledge Exchange users

information that resides on its secure, internal network. Because of its rapid emergence as a standard, the plan was to make the Internet a major part of the solution.

The Internet was also the ultimate distribution medium for the public dissemination of information such as financial information to investors and consumer information to end customers. Customers would often access this from consumer online services connected to the Internet, such as Prodigy and America On-Line. One question was: How could the company accomplish its goal and at the same time safeguard its internal network? Because this is a pharmaceutical company, security was a major concern.

The InterNotes Web Publisher provided the way to distribute public material over the Internet to the widest possible audience. Secure firewalls protected the company's network from access via the public network. Key pieces of this solution included a public network service provider, Lotus Notes and cc:Mail, InterNotes Web Publishing, and MQSeries, which monitors inter-application communications and ensures that messages are placed in a queue and sent to the target system with assured delivery.

Figure 15.2 depicts the network and services provided. In this example, the services provided extend beyond the Lotus Notes product, but Notes is the service that ties the system together.

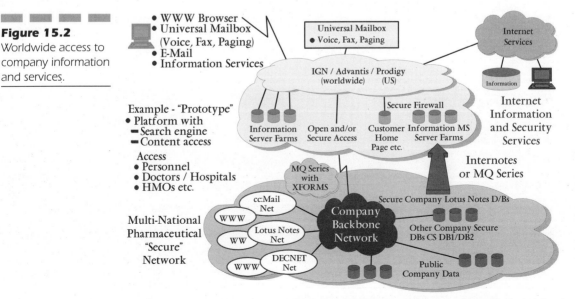

Figure 15.2
Worldwide access to company information and services.

Intraloan Banking Application using Domino and Lotus Notes Public Network

This example of the use of Lotus Notes and Domino for syndicated bank loans was originally presented at Lotusphere '97 by Rich Jenkins of IBM, Bill Conklin of Lotus, and Mark Adams of IntraLinks. It is presented in this chapter as an example of how Lotus Notes and Domino can be used to help banks collaborate over the Internet for syndicated loans. As discussed, the security features of Lotus Notes and Domino were paramount in the decision to use this technology.

IntraLinks is a small technology company that used Lotus Notes and Domino as the basis for developing a secure, inter-enterprise environment for creating, distributing, and managing highly confidential financial documents among participants in the global loan-syndication market. IntraLinks is backed by a Wall Street group that includes Arthur Scully (a former J.P. Morgan investment banker) and his brothers, John and David. The brothers are partners in Scully Brothers LLC., a New York-based investment and strategic advisory firm partnering with emerging technology companies.

IntraLoan, IntraLinks is a Lotus Notes-based application that targets the worldwide market for large corporate bank loans, about 2000 of which are funded jointly (or syndicated) each year. About 400 of the world's top investment banks and institutional investors are regular participants.

The Problem: Secure Transfer of Confidential Documents

Large syndicated corporate loans are classic inter-enterprise business transactions. Typically, the borrowing company retains one of the major investment banks as its agent or lead bank for the loan.

A highly detailed offering document describes the company's financial condition, the purpose of the loan, the proposed interest and repayment schedule, and terms and conditions for participating in the syndication. Interested investors then examine the offering, and a massive exchange of paperwork and highly confidential information ensues before the loan is syndicated.

Traditionally, because of security issues and operational concerns, banks have been reluctant to send syndication documents over inter-enterprise networks. Instead, they have relied on couriers, messengers, phones, and faxes—the so-called "sneakernet"—to exchange information. Though trustworthy, the sneakernet is also time-consuming and costly. Figure 15.3 shows the traditional workflow for these loans using the "sneakernet."

The Solution: Domino Delivers Financial Information Securely

Recent advances in networking technology presented IntraLinks with the opportunity to create a secure, reliable electronic environment for managing the exchange of documents among enterprises.

IntraLinks needed sophisticated groupware that would let lead banks manage the production and distribution of documents among multiple parties in multiple companies. The platform had to work with many

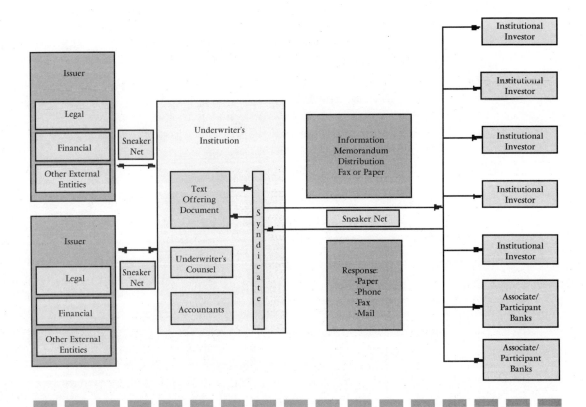

Figure 15.3

Traditional loan syndicate workflow using the "sneakernet."

other software and hardware platforms, and a scalable infrastructure was needed that was fully supported and absolutely secure. Lotus Notes and Lotus Domino met these requirements. Notes had the robust groupware capabilities and security features their customers needed.

IBM InterConnect's Domino servers allow them to offer IntraLoan, with all its inherent Notes features and security, over the Internet or corporate intranets. Customers can access and work with the information whether they are using Notes clients or Web browsers, such as Netscape Navigator or Microsoft Explorer.

IBM InterConnect for Lotus Notes provides all the network and server support. When you put it all together, IntraLinks can quickly and easily create a customized, dedicated, and secure network for each loan syndication. The loan syndication workflow using a Lotus Notes/Domino based system is shown is Figure 15.4.

The Results: Truly Collaborative Loan Syndication

IntraLoan eliminates much of the paperwork, not to mention legwork, associated with document management. The borrowing company and

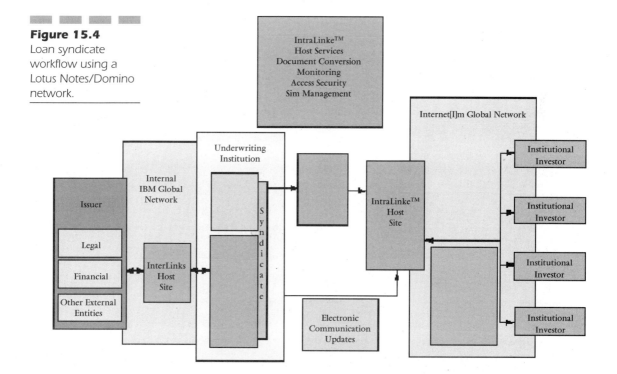

Figure 15.4
Loan syndicate workflow using a Lotus Notes/Domino network.

its investment bank use the Internet to access the IntraLoan application, hosted on Domino servers by the IBM InterConnect service. They create the offering document and associated materials using the communications and collaboration capabilities of Lotus Notes. The syndicate manager at the lead bank then distributes the document to other potential investors, authorizing controlled access to the Notes database. These firms, using either Notes clients or Web browsers, can review the loan document, ask questions, and request more information—all over the network. This IntraLoan Clearinghouse concept is shown in Figure 15.5.

The database management and security features of Notes and Domino ensure that confidential information is protected. The syndicate manager can establish a wide range of access privileges. Some reviewers can have access to the entire document, while others can be restricted to certain sections or even a single page.

Supporting documents in the IntraLoan database can also include a slide or video presentation. This online and real-time capability can

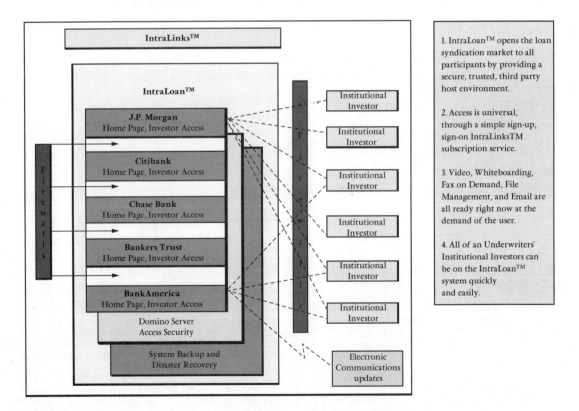

Figure 15.5

The "IntraLoan Clearinghouse" concept based on Lotus Notes/Domino technology.

eliminate some of the costly, time-consuming road shows that are often used to promote a loan syndication.

The Benefits: Cutting Costs by 30%

IntraLinks estimates that IntraLoan can reduce document creation and management costs by 30%. For example, document construction, which usually takes about 8 to 12 weeks, can be reduced to 4 to 6 weeks with IntraLoan. Plus, the banks avoid the costs associated with printing and physically distributing thousands of pages of documentation.

IntraLoan also drastically shortens the syndication cycle by making information instantly available to all prospective participants anywhere in the world. Customers can access and work with the IntraLoan application from their desktop browsers without having to purchase and install Notes servers or clients.

IBM InterConnect for Lotus Notes provides IntraLinks and its customers with a fully supported and scalable Notes and Domino infrastructure. With IBM's infrastructure and resources, IntraLinks can establish or remove dedicated intranets as loan syndications are offered or completed.

Following the deployment of IntraLoan, IntraLinks plans to extend the technology to the re-insurance and asset-backed securities markets, where many of the same institutions create and exchange financial documents.

An Enterprise-Wide Example: Customer Care

As with many companies, customer care is a top priority for the Advantis Company. In order to better care for their customers, Advantis developed an application based on Lotus Notes. This application is an enterprise integration application that uses LotusScript and middleware to interactively get customer data and network data from legacy databases on the mainframe and RS/6000 servers. This Customer Care Application is being rolled out to all of the IBM Global Services areas.

Previous Environment

Customer data for Advantis still exists in many different places. The information is in many different data stores, and different applications are used to work with this data. In some cases, the data is redundant across these systems. The applications are primarily host based and have very little automated integration. Programs have been written that aid in this integration and run outside of the applications but allow the exchange of information across some of the applications.

With all these systems and data stores with customer data, it is a time-consuming and sometimes difficult process to work with a complete and accurate set of Customer information. This has directly impacted customer satisfaction. Just think of it. . . . When a customer contacts us with a question or problem, and we have to ask the customer to give us information for services that we set up and bill them for, we know the data is in our systems, but it is difficult for our customer care agent to access this. Training for the Customer Care Agent is also time-consuming given that they must be educated on how to use all these different applications.

Customer problem calls used to be handled in what is usually termed *call-back* or *dispatch mode*. The call taker would log information from the caller in a new or existing problem ticket, and this ticket would be placed on a queue for someone to handle and call the customer back. To be able to "work" this ticket, with as much information about the caller as possible, the Customer Care Agent needed to know the caller's account and user ID. No good mechanism was in place to get this data without these two key pieces of information. The call would be handled, but any information that might have been learned during the course of this incident would be lost after the incident was closed out.

For example, if in the course of the resolving this problem, we learned the caller's name, phone number, fax number, e-mail address, what products they use, any unique aspects of how they use our services, etc. On a subsequent call, we would not be able to see this valuable information. However, even in the event that the caller does know the account and user ID, we are still in a situation where we ask the caller to give us, for example, their line name, TCP/IP address, etc.

Many times we were the ones that actually set these up to begin with. So why can't we tell the customer what we know about and verify which is the one related to the problem being called in? It is a subtlety, but this kind of exchange impacts customer satisfaction. Additionally, the existing host-based applications do not allow for the maintenance of other new pieces of data about the caller and their company, at least not without some significant coding efforts, if it is even possible at all. Customer configuration diagrams are not accessible at all from these platforms. There are countless other reasons why this method of working with our customers needed to be revamped. However, this above scenario gives a glimpse of the challenge.

The Solution

Several projects for Advantis Customer Care based around Lotus Notes with multiple release cycles were initiated, and significant parts are in production:

- A new Customer Care Organization with IBM Global Services Executive direction was put in place.
- A new Call Center was established (in Tampa, FL, initially) which included:
  - New phone switch and related software with extensive measurement and reporting features
  - New phones to all parts of the Tampa site
  - New PCs for the Customer Care Agents
- There was a conversion from call-back/dispatch mode to "live call" mode. Through choices selected from a Voice Response Unit (VRU) or from a dispatcher doing an immediate transfer, the caller is routed to a Customer Care Agent with the skills to handle the caller's request or problem. Call-back is only needed in the event that more detailed problem determination is needed.
- A new client/server application with improved data rationalization into an easily accessible data store was developed.

Key Technology and Products Used

- Lotus Notes R4 (OS/2 platform for clients, and AIX for servers)
- Lotus NotesPump (for regularly pumping some key data from DB2/6000 into Notes)
- Lotus Fax Server (for outbound faxes of Notes documents)
- DB2
  - DB2/CAE on the clients with TCP/IP connectivity to DB2/6000 and going through DRDA to DB2 on MVS
  - DB2/6000
  - DB2/MVS
- Early Cloud & Company's Middleware products called CallFlow and MDp.
  - CallFlow runs on the clients and on the RS/6000
  - MDp runs on the RS/6000 and on the MVS systems
- CICS (which is used by MDp and for the existing legacy applications on MVS)
- IBM Data Propagator Relational

With data being maintained in heterogeneous data stores (some DB2, some IMS, some VSAM, etc.) and across many different applications, it was difficult to access all the key customer data. The Advantis data team did extensive work in this area using tools like IBM's Data Propagator Relational and with some of their own home-grown code to bring all this data together into DB2 databases that are on either our AIX servers and/or on the mainframe.

The data has been rationalized from the many separate data stores into a key set of DB2 tables. Through the use of the LotusScript Data Object extension, this DB2 data is accessible and maintainable from the Notes User Interface. Some of the DB2 data is "pumped" into Notes on a regular basis using NotesPump.

Lastly, for data where real-time needs exist, the Early Cloud & Company middleware is used to work with the data from two of our key legacy systems on MVS. Figure 15.6 shows the different components of the IBM Customer Care Lotus Notes System.

As stated earlier, the initial targeted Customer Care Agents for this first release were for those who handle "live calls." (Account support personnel are also part of the CC organization, but they will be part of the next release's target users.)

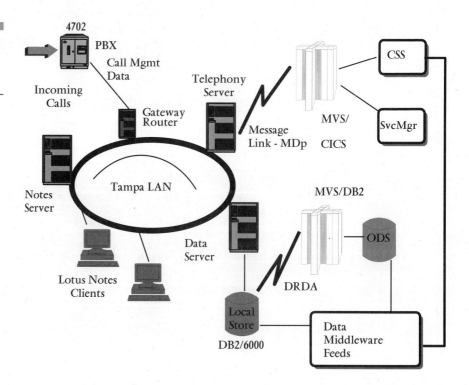

Figure 15.6
The IBM "Customer Care" Lotus Notes system.

These call takers use Notes as the user interface, and LotusScript is used extensively. It is used for accessing all our DB2 data and for maintaining some key pieces of DB2 data that were not added to the legacy system environment. Included here are the items mentioned earlier, like the ability to keep, maintain, and enable subsequent searches for information about the caller, regardless of whether or not they know their account and user ID. This data has been enhanced by allowing us to keep more information about the caller, like their fax number, e-mail address, and freeform notes about this person that might be helpful for any Customer Care Agent taking a call from this individual. Perhaps this is a key member of the account, a very technical person, or an administrative type person. Knowing this in advance might help the Customer Care Agent in handling the call.

The ability of LotusScript to call external C language functions is utilized to interface with Early Cloud & Company middleware that communicates to our legacy systems. A "C DLL" was written and is invoked from Notes and manages the real-time data messaging between CallFlow and Notes.

The question has come up about why the Early Cloud & Company middleware, and why not MQ, for example. Several factors influenced this decision. One, MQ is currently only installed on one of our legacy systems, and this is not the one that was the focus of the initial release. (It will be included in the scope of one of our future releases, though.) Next, the Early Cloud & Company product line has Computer Telephony Integration capabilities that go along with the new phone switch's software, as well as some of IBM's telephony related products. This, too, will be enabled in one of our later releases this year. However, most immediately, the messaging architecture they have is a good match for the two legacy systems that we needed real- time access to.

With our Notes-based application, the enhanced DB2 data, and real-time access to key legacy systems, the Customer Care Agent can easily get information about callers and their companies. This includes the ability to search for this information with a minimal amount of data. We can search by the person's name, the company name, the account, user ID, and, if calling back on an existing incident, the incident number. Wildcard capabilities have also been enabled for these searches. So now the Customer Care Agent will immediately be able to see all the key information about this customer without ever having to leave Notes. Very powerful! Very easy to use! Very easy to train new Customer Care Agents!

Once the caller is identified, if we know the account and user ID, we can also see a high-level summary of all the services enabled for both the individual caller and for their account, plus all the different connection types they have (e.g., leased line, dial, etc.). Then, through collapsible sections in Notes, a "drill-down" concept is used to get more detailed information about the connection or service. From within the Service section of the form, buttons are used to do the real-time data access to get very detailed information about several of our key services.

The Notes user interface has been set up with extensive use of "hide whens." The top part of the user interface stays the same throughout the entire handling of the call. However, depending on what pieces of information are requested next, the bottom three-fourths of the user interface changes. These choices are made from selections of Action bar items. Some Customer Care Agents prefer to use the mouse; others are fast on the keyboard and want to invoke these by keyboard. Action bar items make this very quick for them to use, either way.

Four Notes databases are used to support the Customer Care Agents: the main one that was outlined earlier and three others, which are described in the following paragraphs.

A Keywords database has been set up and is maintained primarily by a restricted group of Customer Care Agents. As you can probably ascertain by the title, the data in this database is used for our keyword lists in the main application.

A detailed Help database has also been set up. It, too, is primarily maintained by our users. Development seeded the database, but the users have customized it to give a Customer Care Agent perspective on using the new application. Additionally, development has used this database to document the error messages that the Customer Care Agent might see when using the application. Included in this documentation are workarounds, if any, plus what they should do in the event of getting this error. When they get an error from the main application, they have the capability to press a button and immediately link to the detailed help for this error. Plus an Action bar item exists on the main application that links the agent to context sensitive help. That is, we "know" what kind of data the Customer Care Agent was working with when the Help action was invoked, and they are taken directly to that part of the Help database.

Lastly, a Document Library database has been developed. The majority of documents in this database are fax documents that are mailed into Notes from several different VM based fax systems. The VM systems are currently the master source of these documents.

Some people that need access to these documents are not on Notes yet and still need them in VM. However, the Notes version brings all the faxes from the six or so different systems into one place. That Notes database is full-text searchable, faxable from Notes, and also "e-mail-able."

In the context of handling a call from a customer, the Customer Care Agent can quickly access, search, and transmit these faxable documents to the caller. Because we now keep the caller's fax number and e-mail address when a choice is made to fax, for example, instead of asking for the caller's fax number, we can verify the caller's fax number (a subtle way of letting him know we remember that you have given this to us before). Additionally we have provided temporary override capabilities in case the caller wants this fax sent either to another person and/or to a different fax number. At this point, a Lotus Fax Server solution is used to actually send the fax.

Next Steps

- More types of customer information accessible and maintainable from Notes (including graphics)
- Internet/intranet Web browser support
- CTI functionality (enabling a "screen pop" onto Notes with information already gathered about the customer just by knowing either the number they dialed, the number they are dialing from, and/or the VRU options selected, plus out-dial capabilities and data and voice transfer capabilities initiated from the Notes application)
- Support for the Systems Engineers that do Account Support functions
- More real-time data access to other non-Notes applications (this is where we're looking at MQ)
- International requirements for non-U.S. call centers.

At our peak, the Customer Care Notes team had five members. However, the majority of the time, we had three full-time developers. One focused on the Notes user interface functions and other pure Notes databases. Another focused on all the DB2 interactions. The third focused on the Early Cloud & Company middleware implementation from Notes. The project for the initial release took approximately six months for the Notes part, and included in that time was training for the team and rework needed as the solution changed to support only the live call Customer Care Agents. The end result has been a very robust Lotus Notes Customer Care application that has significantly improved response to customer queries.

16

What the Future Holds

Then came the Internet. Microsoft CEO Bill Gates called it the
"Internet tidal wave." After the initial hype by the popular press,
it soon became clear that the Internet's biggest immediate
impact would be on the corporate workplace. Internet technology makes it easy to publish and share information with colleagues, whether in the same office or in branches overseas.
Intranets, as these internal networks are called, seem to promise
all the capabilities of Lotus Notes, and they are easier to set up
and cheaper to maintain.

However, in fact, the Internet might have been what Notes
needed all along. Since it launched Notes in 1989, Lotus'
biggest challenge has been getting people to understand its
purpose. The Internet, the World Wide Web, and intranets give
people a window into the world of collaboration across a network. Eric Schmidt of Sun says: "Notes will offer more of the
pieces for intranets than any other solution."

Fortune magazine, July 8, 1996.

This book has described aspects of Lotus Notes and Domino network
design for the enterprise with the IBM Corporation used as an example
enterprise. In this chapter, the first section summarizes today's requirements for a fully functioned Lotus Notes system. The next sections discuss the future of Lotus Notes and Domino networks.

Today's Lotus Notes Network Needs

On an enterprise level, we have found the following:

- The need for robust network design that allows effective replication
  of Lotus Notes databases among the major site LANs in the
  enterprise. We have found TCP/IP to be an effective protocol for
  Lotus Notes over the WAN.

- A gateway design that allows easy routing to other e-mail systems,
  including mainframe-based systems.

- Easy dial access for the mobile user anywhere in the world.

- Strong network security, especially when the Notes network
  interfaces with the Internet and other companies via dial
  connections.

- The need for many other Notes services to support the enterprise
  LAN-based office, such as: Lotus Notes Fax Gateways, access to the
  Internet from Notes, dial access to Notes both from the Internet
  and via the internal LAN, Calendar options such as Lotus Notes,

Lotus Organizer or Time and Place, Lotus Notes Pager Service, Phone Notes, Video Notes, and a company-wide Notes Database Repository.

ATM Networks

The future of Lotus Notes for the enterprise will follow, to a large degree, the future of data networks. We will have ATM (Asynchronous Transfer Mode) to the desktop, at speeds of 155 megabits per second. Our concerns with overwhelming the network with Notes replication, Video Notes, or transferring large images will have largely disappeared.

ADSL and Cable Modems

Asymmetrical Digital Subscriber Line (ADSL) service is being offered by the telephone companies as a high-speed data link to the home. ADSL can be thought of as a competitor to cable modems because the downstream speeds for ADSL can go up to 6 Mbps. The upstream speeds are limited to about 600 Kbps, which is no problem for "surfing the 'Net" where almost all of the data traffic is from the Web server to the Web browser. For many Lotus Notes applications, the same holds true. The great majority of data traffic is from the Notes/Domino server to the Notes client.

Some current ADSL implementations support bidirectional 384 Kbit/sec links. US West, Inc. offered 1.55 Mbit/sec ADSL Internet access at the end of 1996 for $60 to $100 per month, and that will eventually drop to $35 (something like a faster replacement for basic ISDN).

Competition to ADSL for very high speed links to your home is coming from the cable-TV companies. A typical cable modem offers 10-Mbps data rates over existing cable-TV wiring. This modem includes an Ethernet connection which you attach to your PC. Currently both cable modems and ADSL are struggling for market acceptance.

Mobile Users

Following the trend of the last few years, the capabilities of mobile (laptop or notebook) computers will increase, and the cost will come down. Laptops will converge with their smaller brethren, the notebooks, and it will be easier to carry your computer with you. Perhaps even the very small Personal Digital Assistant (PDA), such as the Apple Newton, will have the capability to run sophisticated groupware such as Lotus Notes. Then you'll be able to carry your Lotus Notes client workstation in your pocket!

One very important trend that is happening now is the significant increase in modem speed for mobile computers. In 1994, the most common modem speed for existing laptops was 9.6 Kbps. This speed of 9600 bits per second has become the minimum modem speed deemed acceptable for computer applications like Lotus notes with their heavy use of graphics and other "rich text" capabilities. In the beginning of 1995, 14.4 Kbps was the modem standard for new purchases, and in mid-1995, 28.8 Kbps was the standard. Then the standard became 33.6 Kps and now we have 56 Kps modems. For laptops and notebooks, the miracle of PCMCIA "credit card" modem technology allows modem speeds at 56 Kbps with no significant modem weight or size to worry about. Even the price continues to drop rapidly. We've heard again and again how modem technology for transmission over regular telephone lines has reached its technical limit—first with the 9.6 Kbps modems, then with 14.4 Kbps, then 28.8 Kbps and now 33.6 to 56 Kbps. We'll see if this limit continues to be pushed forward.

Many of the PCMCIA credit-card-size modems allow attachment to a cellular phone. With today's technology, you can have a three pound notebook computer with a Pentium or PowerPC processor, a 56 Kbps PCMCIA modem with cellular phone capabilities, and Lotus Notes installed. This will give good groupware access performance for the most mobile of users. That's with today's technology. Tomorrow's access will be faster and cheaper, and groupware capabilities will continue to improve!

The best high-speed dial access should still be with ISDN. ISDN (Integrated Services Digital Network) provides two 64 Kbps data channels, called B Channels, and a 16 Kbps signaling channel, called a D Channel. The two B Channels can be put together with the right ISDN adapter to give 128 Kbps dial access for Lotus Notes. The only problem is that telephone companies in the U.S. are still struggling to complete the infrastructure to allow ISDN access everywhere, although this is happening quickly. In California and many other states, ISDN is readily available. ISDN is also readily available in Europe, and it's a great way to obtain high-speed dial access for Lotus Notes. When ISDN is readily available throughout the world, the Lotus Notes mobile user will have all the access he needs.

There are two additional technologies that hold promise for the future. One is PCS (Personal Communications Services). The U.S. Federal

Communications Commission has opened up additional bandwidth for wireless technology. PCS devices and services are starting to seriously compete with cellular phones in the area of wireless communications. By the end of 1997, more than one-half of the U.S. population should have PCS available to them. Prices are moderately less than cellular, averaging 25% to 30% less at lower usage levels and 15% to 20% less at higher usage levels. New concepts, such as simpler pricing plans, paying more for the phone, no contract, and free first incoming call minute have become trendsetters. Voice quality is better than analog, and there are fewer dropped calls. Actually, the major difference is not between cellular and PCS; it is between analog and digital.

We're also now seeing reactive and preemptive strikes by cellular operators. They're introducing "PCS-like" pricing plans, locking up distribution channels, and accelerating the digital roll-out. The migration to digital technology for all types of wireless communication will be especially helpful for all types of mobile PC use, including Lotus Notes.

The other essential communication technology of the future is software agents. For example, agents that "live" in the network can represent your preferences, so you can transparently connect to services using a variety of protocols and devices. Software agents hide specifics of the network from users, reducing the complexity of dealing with different services and providers. They can also route and filter messages, minimize traffic and connect time, convert data among different formats, and help users locate services on the network.

Notes Public Networks

Lotus has forged partnerships with several network service providers to deploy Notes as a network-enabled application. The first of these was with AT&T and was called AT&T Network Notes (AT&T Network Notes was discontinued in 1996). Since then, more than a dozen additional network service providers around the world have signed up for what Lotus is calling Notes Public Networks (NPN). These service providers are located in Europe and the Far East. This will spur international growth in the use of Lotus Notes. It is projected that Notes Public Networks will play a very important role in the deployment of Lotus Notes and Domino in the future. Each of the Notes Public Networks will, by design, interoperate with all of the other Notes Public Networks. This will give easy Notes accessibility world wide. Details on several of the NPNs, and the services they offer are given in chapter 1.

Lotus Notes and Java Technology

Java technology, as a part of the "Internet tidal wave," will have an immense effect on the future of Lotus Notes. Domino 5.0 servers and Notes 5.0 will have full support of Java in Lotus APIs and Java Beans. John Gage, of Sun Microsystems, has put it quite eloquently. He states that there are two or three things that are very different from the past, in that everybody's agreed upon a common environment and Web-based software and TCP/IP networking software that just runs on any computer, and that includes IBM, Apple, Sun, AT&T, Alcatel, and British Telecom—all the telephone companies—and all the chip companies and all of the cellular companies.

Gage explains that the swing to the common environment is a major shift that has never happened before. It brings old-style computers up-to-date, because they instantly run the same software as the hand-held computers. It changes everyone's business in the software markets enormously. Your market's no longer a Windows 95 machine or a Macintosh or a UNIX workstation, but every computing device. Using the Internet, you're able to build a new hierarchy of computing. Any computer linked into the network is part of a huge parallel computer. You can distribute some parts, and you can put intermediate memory a little closer. Today, processors, storage, and display have to be pretty close. The processor and memory speak at very high speed, so you're not going to pull them apart. However, disks are very slow—10,000 times slower than hardware memory. You don't even notice if you take a disk and move it from New York to Chicago. Because connecting things up costs very little, we can create a new form of computer architecture. The "network computer" will play an important roll in this new architecture.

There are three factors converging that make for watershed change. One element is the hardware side—the development of processors, memory, storage, and display—with all these things now getting cheap. They no longer belong only to the rich but can move out to hundreds of millions of people. That's the inevitable progress of hardware.

Couple that with the enormous development in networking. You can move enormous volumes of data—622 Mbps—across cheap, phone wire that costs 5¢ a foot. Fiber's being put in everywhere, and the wires are going everywhere, so there's ubiquity coupled with speed. We have these developments that allow all the machines to be connected to each other.

A third development is software that runs on all computers. You no longer ask, "Is it a Mac or a PC?" Everything from a wristwatch to computing elements in automobiles to the ATM to the telephone to the television can run common software. The change has come with Java. Suddenly there's a language that runs on everything. People can pick any computer in the world from the Nokia cell phone to the desktop computer, and write all the applications in Java. You write applications one time, and they just run.

The conjunction of these three forces—cheap platforms, ubiquitous and very fast network connections, and communications software running on all computers—allows an explosion of development that until recently has been fragmented.

A lot of people view the world of computer-based devices as being divided. However, in fact, they all are identical. They all have little display screens; they all have memory. They're all becoming smart devices. Those old industry divisions are evaporating. The "Java stuff" will remove the barriers. Companies will rewrite the basic parts of software you use every day so that it will run on every computer. Lotus Notes is a big part of this thrust with applications such as "Kona" components.

Lotus Notes and the World Wide Web

Web technology has been evolving at a very fast pace. Many of the Web technology enhancements are similar to the features included in Lotus Notes. In November of 1995, the *Wall Street Journal* carried a front page article questioning the added value of Lotus Notes because Web Technology was evolving at such a fast pace. The following sections give the IBM/Lotus response to that type of questioning. It is interesting to see what features Notes had in 1995 that the Web added in 1996 and 1997. Many Notes features are still not on the Web.

Both Lotus Notes and the Web were invented with the same purpose in mind: to support collaboration over local and wide area networks. Like other technologies, such as relational databases and desktop productivity tools that share some common attributes with groupware in general, the Web is sometimes viewed as redundant to, and therefore competitive with, Notes. However, just as those other technologies are appropriate to their own class of applications and have proven of greater value when integrated with Notes rather than used instead of it, a close look at the Web and its applications reveals its distinct and complementary nature to Notes as well.

World Wide Web Applications

The Web provides a global, public means of information publishing and consumption that is based on a "pull" model. The Web clients or "browsers" specify the URL or document name and request a copy of that HTML page, which is transmitted across the 'Net, then displayed on the client workstation. Those pages offer a document-centric, rich-media presentation of information, as well as a hypertext linking structure.

The World Wide Web is an ideal way to make marketing, corporate, and service and support information readily available for public access. In this way, the Web naturally complements traditional forms of communication, such as direct mail, advertisements in various media, "fax back" systems, toll-free telephone lines, and interactive voice response systems. Planned security and application development enhancements to the Web aim to make it a viable medium for transaction-based electronic commerce. In addition, a number of companies have deployed Web technology on private servers to act as an internal resource not available to the general public.

Lotus Notes Applications

Like the Web, Lotus Notes has been used by many organizations to serve as a publishing medium, focused more on information sharing within and among organizations. It has been Notes' role as platform for application development and deployment that has established it as a groupware standard. Notes' object model, messaging services, security facilities, desktop and database integration, and client /server architecture lend themselves to a class of applications not feasible on the Web. The characteristics of such application include one or all of the following:

- Interactivity
- Notification
- Triggered processes
- Customization
- Multiple levels of security
- Integration with other resources
- Mobile user support

INTERACTIVITY.   Business collaboration requires the active participation of team members, not just the one-way sharing of knowledge from one source to many readers. Activities such as brainstorming, idea sharing, and problem solving require that all participants can contribute their input.

NOTIFICATION.   Finding the right piece of information at the right time simply by browsing through a library of documents is an inefficient process. In many cases, users would benefit from having the information find them at just the right time. A system that combines information storage with integrated messaging services, such as Notes, augments the pull model of the Web with a push model of information distribution. In this way, when an author posts a new item to the system, he can also create an e-mail message that contains a pointer to that information and send it to the appropriate coworkers. Recipients simply click on the pointer to view the relevant document. To do this on the Web, a recipient must first go through a complicated multi-step process.

TRIGGERED PROCESSES. Similarly, Notes itself can monitor the state of data, initiating an action based on the state of particular fields. In many cases, the action might be the direct result of a lack of user activity (a sales person has not contacted a customer in 30 days, a contract to be approved is waiting for a specific person for over 24 hours) or an external condition (a client's credit rating has changed, a deadline is approaching). Such an application requires the ability to develop server-based agents.

CUSTOMIZATION. In many collaborative applications, individuals will find reason to want private, tailored views of data that support their specialized tasks. The capability to flexibly view information in different ways is fundamental to enabling all participants to collectively grasp the key elements of the business process and the context in which they exist. Notes forms and views provide each user with the ability to customize how information is presented to them without affecting the logic of an application.

MULTIPLE LEVELS OF SECURITY. Some business documents contain information appropriate for some readers but that should remain hidden from others. For example, a personnel review form might be routed through a series of team members who have the opportunity to make comments. It would not be appropriate for all members to be able to read the comments made by others or view the salary information, so those fields should be hidden from general view, while authorized readers have access to the entire form and its contents. While the Web is developing some measure of security and authentication, it will be some time before it can be applied to specific fields within a document.

INTEGRATION WITH OTHER RESOURCES. Not all information is native to a collaborative system. Some shared information resides in desktop productivity tools (e.g., spreadsheet tables) and host-based relational databases. Collaborative systems do not replace these systems, but rather must seamlessly integrate with them, so that as changes are made in a desktop document or a database, they are automatically reflected in the collaborative system.

MOBILE USER SUPPORT.   More and more users today need to perform their work away from the office. Users need to be able to continue to browse through data, compose edits and responses, and complete steps in a workflow process even while disconnected from the corporate or public network. Replication services allow users to make a copy of a database on a laptop computer, work with it while traveling, and later synchronize all changes with the server.

Notes and the Web: Distinct and Complementary

By its very nature, groupware overlaps with other information technologies: RDBMSs, desktop productivity tools, document management systems, etc. Of course, the commonalities between groupware and these other technologies do not render them competitive; on the contrary, groupware is essentially complementary to them. Each technology supports a distinct class of applications, while their integration with groupware leverages the strengths of both.

Likewise, the World Wide Web excels at a class of applications that involve the one-to-many publishing of public documents. The straightforward nature of "broadcast" applications lends them to the generic use of so-called "shareware" client and server software. Investments by third parties in the Web's forms-based metaphor are largely aimed at a set of transaction processing (i.e., electronic commerce) applications. Notes, on the other hand, excels in a broad class of business process applications that require ongoing action by some or all participants and in which the *state* of data as it moves through the process is as important as the data itself. The complex nature of these applications tend to require a highly sophisticated and customizable environment, such as Notes, which has demonstrated a remarkable return on investment when applied to mission-critical business processes.

Applications that include both types of requirements will naturally benefit from the integration of Notes and the Web, in which information most appropriately gathered and managed by Notes can be easily made available to the large population of Web users, and which, in turn, an event triggered by an activity on the Web can quickly become part of a nonpublic business process.

Beyond the Hype of the Internet and the World Wide Web

Lotus Notes and Domino are becoming inseparable from the Internet and the World Wide Web. What applies to the Internet and the Web will frequently apply to Notes and Domino. We have heard every extreme about the Internet and the World Wide Web, such as, "The Internet can provide any wonderful thing you can imagine" (except that it is often hard to find and takes a long time to get). Or, "the Internet is the root of all evil" (just like the television, the telephone, and the book before it), and "We should ban it." As usual, there is some truth in each view, and reality lies in-between.

What are some of the evil things said about the Internet? There is too much pornography, drugs, and gambling available on the Internet. People meet on the Internet and get lured to meet evil people in person and get physically harmed. Your credit card number can be stolen on the Internet. I do not condone any of these activities; however, one day, it struck me that, if I replace the phrase "on the Internet" with "over the phone," the same evil things can be said about the telephone. The point is that, after the hype dies down, the Internet is becoming another familiar communication and media channel like the television, the telephone, and the book. With the good comes the evil.

What are some of the good things about the Internet? In the future, you can do on the Internet essentially what you do in everyday life today. Some activities are available today. You can buy products (books, clothes, and computers), buy services (airline tickets, banking, stock trading, newspapers, entertainment, and sports events), receive education, interact with government agencies (motor vehicle registration, track bills in the U.S. Congress), and practice your religion. The selection is limited right now but is growing very rapidly. When you think about it, these activities are available by phone, fax, printed material, and mail. Again you can see that the Internet is becoming another familiar communication and media channel like the television, the telephone, and the book.

The Internet and the World Wide Web are still new, so the hyperbole and exaggerations might be expected. When the novelty wears off, the Internet and the Web will be as familiar as the telephone. Like the telephone, the Internet and the Web will be easy to use, reliable, and available with no delays.

The Web is rapidly evolving. Initially the Web was just for publishing material. People could browse your product catalog, course catalog, or any other printed material you would normally send to requesters. The second step was interactive Web sites. People could fill out forms to request more information, provide feedback, and order products and services.

The third step is personalization of Web sites, and this feature is spreading now. Odysseus buys lots of books about Lotus Notes and Domino from the Olympus bookstore Web site. (You know Odysseus likes to read whenever he travels.) The Olympus Web site remembers his buying pattern, sends Odysseus e-mail notices regarding new books about Lotus Notes and Domino, and he buys this one right away! Similarly, the Olympus Web site sends Odysseus a list of books by authors of books he has bought before. If he liked the first work by Homer, he will probably like the second work by Homer. Because Odysseus is an avid (and Ovid) reader and buys lots of books from the Olympus Web site, he automatically gets a discount. The more books he buys, the greater his discount (a frequent buyer program). The Olympus Web site appears personalized and customized for Odysseus and for every other individual user. This concept (sometimes referred to by the oxymoron "mass customization") is becoming more widespread on the Web.

An emerging trend on the Web is broadcasting. PointCast, and others, offer products that send information to your workstation without specific action or even an explicit request on your part. You configure your preferences and filters, and the information is sent directly to you from the Web. Once you entered your favorite sports team and stocks, you get updates only for those teams and stocks. You also might get unsolicited advertising for buying sports team clothing and stock investment newsletters.

Broadcasting is often called a "push" approach as opposed to a "pull" approach that characterizes "traditional" Web browsing. With "traditional" Web browsing, users have to know what they want from which Web site and "pull" the information from those Web sites. Broadcasting "pushes" information to you that you did not explicitly ask for. This is not a new concept, but borrowed from other media. We are all familiar with junk mail, junk faxes, and telemarketing over the phone. Now we have another "push" medium for marketing, sales, and advertising ("in your face!").

What does all of this have to do with Lotus Notes and Domino? As I stated in the beginning, Lotus Notes and Domino are becoming inseparable from the Internet and the World Wide Web. Lotus offers a generic Web server, Domino, but also a series of focused Domino Web servers including Domino.Merchant, Domino.Service, Domino.Broadcast, and LearningSpace. Domino.Merchant is preconfigured to provide a Web site like the Olympus bookstore Web site described earlier. Whereas the Domino.Merchant Web server is oriented for selling products, the Domino.Service Web server is oriented for selling services. Domino.Broadcast, as the name implies, allows your Web server to broadcast over the network. LearningSpace is a Web server for teaching courses over the network. Lotus products have been a part of the evolution of the Internet and the Web. In the future, Lotus will certainly continue to evolve and extend the Internet and the Web in unforeseen and innovative directions.

How profound will the evolution of the Internet and the Web be? Very profound, due to the rapid spread of ideas and information. (Warning! I might be offering my own hype and badly interpreting history in this section!) What can the Internet be compared to? When I was in school, I always heard that the printing press was one of the most important inventions of all times, but I did not quite understand why. I have done some recent readings about why the printing press was so profound, and here is what I learned. Prior to the invention of the Gutenberg printing press around 1450, books were hand copied by monks. The press produced more pages in one hour than a monk could produce in a week. This allowed a much more rapid spread in ideas and information.

Significantly, the first book printed was the Bible (the Gutenberg Bible). More people began to read the Bible and (in my simplified and, perhaps erroneous view of history) began to disagree with the interpretation of the Bible by the Catholic Church. This lead to the development of Protestantism. (Martin Luther posted his 95 Theses in 1517 to protest against the Catholic Church's methods.) Protestants were persecuted in Europe and were looking for safe havens. Meanwhile, with the aid of printed navigational tables, explorers were traveling around the world and settling in "new" lands. (Christopher Columbus landed in the "New World" in 1492). Many people looking for religious freedom settled in the Americas. These settlers later used printed farming guides and almanacs to help develop and grow their settlements.

Books on mining and metalworking led to the replacement of wooden machines by metal machines. Books on medicine led to healthier lives. Books on philosophy and political thinking led to upheavals in governments. The political, religious, and economic systems of the world changed forever. These changes were due, in part, to the spread of ideas and information made possible by the printing press.

Telecommunications has been spreading ideas and information more recently. Some people believe that the television, the telephone, the fax machine, e-mail, and Internet Web sites were partly responsible for recent political and economic changes in Eastern Europe, the Baltic, the former Soviet Union, and China. Television, cable TV, and radio news teams broadcast from around the world, around the clock. News organizations now broadcast from Internet Web sites to your workstation. The Internet will play an increasing important role of spreading ideas and information around the world around the clock. The increased speed of spreading ideas with the Internet can be compared with the increased speed that occurred with the invention of the printing press. (For some time, people have referred to this spread of ideas and information with computers and telecommunication networks as the "Information Revolution" to indicated that the changes are as profound as with the Industrial Revolution of steam and electric engines.)

The impact of the Internet on the world can be compared with the impact of the invention of the printing press. The evolution of the Internet that we have seen over the past few years has been very interesting. The emerging trends we foresee are quite exciting. The future that we cannot see or imagine is what will be truly astounding. Just as nobody could have predicted the profound impact of the printing press when it was invented, nobody today can predict what the profound impact of the Internet will be. We do not know what the impact will be, but we can be certain that whatever it is will be profound. We will just have to wait for the future to emerge clearly and to enjoy it in amazement.

Some Final Remarks

The future of Lotus Notes will be one of higher network and dial connection speeds. The explosive growth in the number of Notes users will continue to be fueled by Notes Public Networks. Lotus Notes and the World Wide Web will continue to complement and strengthen each other. Domino 5 servers and Notes 5 clients will have full support of Lava in Lotus APIs and Java Beans. Lotus Notes and Domino will continue to be a major part of the future of web technology along with Java and Network Computing.

APPENDIX A

Platform Hardware, Software, and Domino Strategy

This appendix describes the network platform hardware, software, and Domino strategy for Lotus Notes servers, workstations, and network components used by Advantis and the IBM Global Network.

Notes and Domino Server Platforms

Tactical: IBM 9595 Pentium Processors

The standard Lotus Notes Application and Mail Server is an IBM 9595-0QT Pentium Processor consisting of:

- 64MB ECC memory
- Two 1GB disk drives for system, change, and dump information
- Two 2GB disk drives for Notes and customer data

Tactical: IBM Server 720A SMP Processors

The IBM 720A SMP Server offers a scalable tactical platform. This platform is presently under evaluation.

Strategic: IBM SP2 Nodes

Multiple IBM SP2 Nodes running an AIX operating system with a combination of front-end OS/2 Servers for those applications not yet available for AIX.

The IBM SP2 (Scalable POWERparallel System 2) is a general-purpose scalable parallel system based on a distributed memory message-passing architecture. Generally available SP2 systems range from 2 to 128 nodes (or processing elements), although much larger systems of up to 512 nodes have been delivered and are successfully being used today. A parallel system of this size is referred to as a *massively parallel system*. For the design of the SP2, it was decided that, for systems of this type to succeed, they must be more general-purpose and less intimidating to use than they have been in the past.

The latest technology RISC System/6000 processors are used for SP2 nodes interconnected by a high-performance, multistage, packet-switched network for interprocessor communication. Each node contains its own copy of standard AIX operating system and other standard RISC System/6000 system software. A set of new software products designed specifically for the SP2 allows the parallel capabilities of the SP2 to be effectively exploited. The SP2 is the strategic platform for all of the applications (including Lotus Notes) available with the Network Centric Computing Platform (NCCP) service offering by IBM.

In April of 1996, the RS/6000 SP got some important new features. The new SP Switch, Adapter, and Expansion Frame:

- Improve reliability, availability, and serviceability.
  - Multi-node failures are minimized.
  - Single points of failure are eliminated by providing redundant fans and redundant power supplies.
  - Nodes can be added without disrupting operations.
- Improve performance.
  - Interprocessor communication bandwidth to user applications is increased by up to a factor of two based on node type.
- Protect your investment.
  - Upgrades to the new SP Switch are available.
  - These products continue the SP technology evolution.
- Expand short frame configurations.
  - The node drawer maximum has been increased from four to eight.
  - The SP Switch-8 is fully utilized.

IBM is enhancing the RISC System/6000^R Scalable POWERparallel™ Systems (SP) with new, higher performance switches and SP short model expansion frame capability. These enhancements benefit both scientific and commercial customers by providing improved switch reliability, significantly increased bandwidth, investment protection, and an extended low-cost pathway to high performance computing.

The Scalable POWERparallel Switch (SP Switch) is the next generation switch for the RS/6000™ SP systems. Building on the same architecture as the High-Performance Switch, it introduces improved reliability, availability, serviceability, and performance. Enhancements—such as the elimination of most hardware single points of failure, improved diagnostics and error management, and three times reduction in the component intrinsic failure rate—result in fewer switch incidents. In fact, failures affecting more than one node have been reduced by over an order of magnitude, and the switch maintenance price has been cut in half.

The new SP Switch and its adapter, when combined with Wide Nodes, deliver over 100MB per second peak unidirectional or bidirectional bandwidth measured at the user application level using the Message Passing Interface.

NOTE: *Actual application performance improvement will depend on the node type, node memory configurations, and ratio of communication to computation in the application.*

Existing systems can be upgraded to the new SP Switch, proving that the SP is an evolving system that incrementally introduces new technology and does not force a total system replacement and, therefore, preserves your investment. The SP Switch can be added to switchless systems, or the High-Performance Switch can be upgraded to the new switch. However, the two switch networks cannot coexist within the same system.

Furthermore, there are no application programming changes required when moving to the new switch if applications are written to standard interfaces such as MPI and TCP/IP.

Demonstrating the flexibility of the RS/6000 SP family, one short expansion frame is available on the previously announced SP short models that are switchless or are installed with the new eight-port SP Switch. The maximum number of nodes in these models remains the same; however, you can now have an additional four drawers capacity with the expansion frame. For models with the Scalable POWERparallel Switch-8 (SP Switch-8), this expansion capability will allow you to take full advantage of the low-cost switch. In addition, up to two system partitions are available in an SP system with the SP Switch-8 or a switchless short SP system.

Summary of Notes Server Platform Support

Operational support for the Lotus Notes Product Service consists of:

- Environmental management
- System change management
- System monitoring
- Alert management
- Problem management
- Daily system management

Environmental Management

Lotus Notes and Domino servers will be maintained in an environmentally stable and secure location. This location will maintain the proper

HVAC and UPS systems to house the servers. Shelving, cabling, and network connectivity components will comply with all Advantis environmental policies.

Notes and Domino Server Disk Architecture

Configuration

NON-RAID.   Physical disk drives one and two are 1GB in size and must be the same model. The SCSICOPY might not work if the disk drives are of different model types. Physical disk drives three and higher are 2GB in size.

RAID.   Physical disk drives one, two, and three are 1GB in size and must be the same model type. The SCSICOPY might not work if the disk drives are of different model types. Physical disk drives four and higher are 2GB in size. The use of mixed disk drive sizes in a RAID array will cause wasted disk space. The usable space in a RAID 5 array is determined by the number of disk drives in the array minus 1 times the size of the smallest disk drive in the array.

LABELING.   The use of ADSM as the disk data backup and restore utility requires that all partitions be labeled in the following manner:

```
computer-name_drive-letter
```

computer-name is the LAN Requester computer name (the same as the IP hostname). *drive-letter* is the letter assigned to the partition without the colon. For example, SCHNAV01_H would be the label of drive H: on system SCHNAV01.

Usage

Notes server disk usage is divided into the following eight components:

OPERATING SYSTEM.
- Partition C: of the first physical drive is the physical drive size minus 351MB and is formatted as HPFS.

■ This partition contains the OS/2 operating system, communications products, support products, and the Notes executable code.

CHANGE MANAGEMENT.

■ Partition D: of the first physical drive is 200MB in size and is formatted as HPFS.

■ This is an OS/2 bootable partition and is reserved for use by Change Management procedures.

■ Changes are copied to this partition as a staging area. The system is then booted from this partition at a scheduled time, and the changes are applied to the C: partition. The system is then booted from the C: partition.

■ This facility provides for full maintenance capabilities of the C: partition, including CHKDSK, without any user intervention being required.

DUMPS.

■ Partition E: of the first physical drive is 150MB in size and is formatted as FAT.

■ This partition is reserved for System Dumps.

■ System Dumps are created by either a Ring 0 trap or by the Ctrl—Alt—NumLock—NumLock key sequence, and no diskettes are required. The system is automatically rebooted after the dump has completed.

■ This facility provides for an automatic system restart after a critical system error, without any user intervention.

■ Dumps can be retrieved across the network without any user intervention being required.

BACK OFF AND CONTINGENCY.

■ Partitions F:, G:, and H: of the second physical drive are a mirror image of partitions C:, D:, and E: of the first physical drive.

■ These partitions are created via a SCSICOPY for non-RAID systems and via a drive swap for RAID systems after all software has been installed and customized.

■ These partitions should *not* be booted from via the Boot Manager.

■ In order to boot from these partitions, it is necessary to boot from the Reference Diskette (or System Partition for non-RAID systems) and change the boot sequence.

- This facility provides a Back Off capability. When changes are made to the C: partition by Change Management, those changes are not made to the F: partition until such time as the changes are determined to be stable and nondetrimental to the user community.

- In the event that a change is determined to be detrimental to the user community, it can be Backed Off by changing the boot sequence and booting the system from what was the F: drive.

NOTE: *It should be noted that, when the boot sequence is changed, drive letters from the two drives are swapped. Drive C: becomes drive F:, and drive F: becomes drive C:. The same swapping of drive letters takes place for the other drive letters on the first two physical drives. The drive letters of all other physical drives remain the same.*

NON-RAID DATA.

- Partitions I: and higher contain the NOTES\DATA directory.

- Partitions I: and higher contain Notes databases.

RAID DATA.

- On hardware systems that have RAID DASD, partition I: is a RAID 5 logical partition made up of a minimum of four Hot Swappable disk drives.

- One disk drive is used as a Hot Spare. In the event a disk drive fails, the data from the failing drive is automatically rebuilt on the Hot Spare drive without any user outage. The drive that failed can be replaced without taking the system down. This drive then becomes the Hot spare.

- All disk drives in the RAID Array must be of the same size. The usable space in a RAID 5 array is determined by the number of disk drives in the array minus 1 times the size of the smallest disk drive in the array.

- DASD capacity is reduced by two drives. One for the Hot Spare and one for Parity.

- Partition I: contains the NOTES\DATA directory and Notes databases.

CONTINGENCY NON-RAID.

- This facility provides contingency for the boot drive. In the event that the boot drive should fail, the boot sequence can be changed and the system can be booted from this drive with minimal down time.

CONTINGENCY RAID.

■ On hardware systems that have RAID DASD, the third physical drive is a continuous mirror of the first physical drive. No partition letters are assigned to this drive.

■ This facility provides for greater system reliability and availability. In the event that the first physical drive should fail, the third physical drive will automatically take over for the first physical drive with zero down time.

■ The failing drive can be replaced during off hours when there will be minimal or no user impact.

Domino Server Strategy

This section gives the Lotus Domino Server strategy developed for the Advantis company.

Platform: Windows NT 4.0

There are several reasons to use Windows NT (Win NT) 4.0. Domino is developed on Win NT, so the latest beta is available on Win NT first, then ported to other operating systems. Using Win NT allows developers to work with the most recent functions as soon as they are available. It is useful to have at least one Windows NT server to evaluate the latest beta code of Domino.

At the time of this writing, Win NT 4.0 has been available for several months with a recent fix pack. It is stable enough to begin development. Win NT 4.0 has features that Win NT 3.5.1 lack. Win NT has a good file system and logging security features. This is important for management of files used with Web sites.

Domino.Action

This is a Lotus product that provides templates to quickly set up and manage a typical Web site. By filling in a form in a Notes database, Domino.Action helps forward mail from "webmaster" to the Webmaster's personal Notes mail database, add the same footer to each page (e.g., a copyright statement required by the legal department), and automate other features that otherwise require manual maintenance.

Production Domino Server and Development Domino Server

A Development Domino server is needed to try out new concepts before deploying on the Production Domino server. The Development Domino server will be in the Test Notes Domain.

Intranet Webmaster to Manage Web Site Content and Design Web Hierarchy

A person or group of people must be designated to be responsible for the content of the intranet Web site. The Webmaster will manage the official intranet Web site w3.advantis.com home page. All other Web sites within Advantis will be linked to the w3.advantis.com Web site. All requests to link a Web site will be directed to the Webmaster. Any Advantis employee should ultimately be able to find any Advantis information from the w3.advantis.com Web site. This includes Human Resource information, Advantis code to download (e.g., the LIG dialer code), any IT strategy documents, and online help (for running the LIG dialer, configuring the Netscape browser, or using Notes!).

Internet Web Site

A Domino server will be placed on the Internet (specifically, on IBM's OpenNet piece of the Internet). The Internet and intranet Web servers will use Notes replication for updates. Domino 4.5 is socksified, so the replication can be done through the socks firewall. This method will be documented and will be approved using the IBM security process. The Internet Web site content will be a subset of the intranet Web site content. The Internet Web site will not contain any corporate confidential material. The intranet Webmaster shall be the Internet Webmaster also.

Server and Network Management

A technology support group will be designated to provide reliability, availability, and serviceability for the Domino server and the network connectivity. This group might or might not be in the same organization as the Webmaster. The Webmaster is responsible for the content of the Web server and not necessarily responsible for the hardware and underlying software.

Broadcasting to the Desktop: Domino.Broadcast and PointCast

Lotus Notes was designed for information to be primarily "pulled" by the end user. Documents are put in a database, and it is the user's responsibility to "pull" the information out of the database. Discussion databases and other forums are examples of "pull" type of databases. In some cases, it is desirable to "push" information to users. Corporate news bulletins, competitive news flashes, and network connectivity alerts are examples of more urgent information that needs to be "pushed" to the user without waiting for the user to go "pull" it when the user is available.

Domino.Broadcast, a Lotus product, can be used in conjunction with products like PointCast to broadcast information from a Notes database to the users desktop. The PointCast client is free from the PointCast Web site and was originally a screensaver with a personalized newsfeed from multiple public news sources. A private corporate channel can now be set up on the intranet to alert employees of important information. Users can respond to the alert by filling out a Web form and submitting it to a Notes database using the features of Domino.

Broadcasting to the desktop is an important complement to the "pull" type of databases offered by Lotus Notes. Domino.Broadcast and PointCast will need to be continually monitored to avoid overloading the network bandwidth.

APPENDIX B

LOTUS NOTES RELEASE 4 & 5

This appendix gives details on Lotus Notes Release 4 and 5. With the Lotus Notes Release 4 server, Lotus offered a scalable, reliable client/server platform for messaging, groupware, and application development. With the Release 4.5 of the code, the Lotus Notes Server was renamed the Domino 4.5 Server, in order to emphasize the significant amount of Web technology included with the server code. (The client software was still called the Lotus Notes client.)

The Domino server provides a distributed, replicated object store, a security service based on public key cryptography, and programmable agents. Notes clients provide a user interface for interacting with local and remote databases, for messaging, and for application development. Lotus has improved the robustness of the server to meet the needs of enterprise-wide and inter-enterprise applications.

Platform and Mail Support

SERVER SUPPORT.   Notes and Domino R4 and R5 continue to support the breadth of popular servers, including HP-UX, IBM AIX, Sun Solaris and SCO UNIX, as well as Microsoft NT, OS/2, and Novell NLM.

CLIENT SUPPORT.   Notes 4 and 5 clients include Windows 3.1, Windows 95, Windows NT, OS/2, Macintosh, and UNIX.

CC:MAIL USER INTERFACE.   Starting with R4 of the Notes code, the mail system conforms to the cc:Mail user interface specification. Users are presented with the cc:Mail client user interface and experience the same look and feel as the standalone cc:Mail product.

Support for the Enterprise

ENTERPRISE SCALABILITY.   Each release of Notes and Domino features greater scalability and performance. Many organizations choose to implement an enterprise layer of servers to manage large-scale applications. In addition to serving as a departmental server, Notes 4 and 5 are robust enough to allow administrators to deploy them for applications hub servers with no additional technology investment.

INTER-ENTERPRISE CONNECTIVITY. Lotus Notes Public Networks, which became widely available in 1996, allow both intra- and inter-company communication with minimal infrastructure investment. Lotus has also developed technology to allow Notes to act as a server on the World Wide Web on the Internet and to import information from the Internet into Notes databases, which can then be replicated internally.

NATIVE X.400 AND SMTP SUPPORT. Because X.400 and SMTP are such fundamental transports in enterprise and inter-enterprise messaging environments, the Lotus X.400 and SMTP MTAs can be installed directly on Notes R4 and Domino servers, in addition to installing them on Notes/cc:Mail Communication Servers. This feature provides departmental-level servers with inter-enterprise connectivity.

Detailed Changes for Releases 4.0 through 4.5

There are many changes for Release 4 (4.0 through 4.5) of Lotus Notes and Domino that significantly enhanced the product. These changes are given in this section.

Interactive Performance over Wide Area Networks

OBJECTIVE.
- Improve the interactive performance of Notes Clients that connect to Notes Servers via lower-speed wide area networks. This includes support for higher-speed modems, improved performance in various communication protocols, optimizations to reduce data sent over the WAN, and enhanced display logic to permit the display of an initial screen while the remaining data is still being transmitted.

ENHANCEMENTS.
- Modify prefetch logic for Views
- XPC performance
- Supports 28.8 Kbps modems
- Avoid delay for Index of database before access is allowed

- Select an individual document to replicate using the mouse
- Optimize option for TCP/IP when using dial-up PPP

Configurable Replication with Performance Enhancements

OBJECTIVE.

- Improve the performance of replication by allowing the Notes Data Base Administrator to configure how replication is performed (e.g., push/pull replication, prioritize and restrict databases to be replicated, stop replication if it exceeds a time limit, etc.).
- Enhance replication by reducing the data to be transferred (field-level replication) and maintaining more information about what needs to be replicated (event-driven replication based on document changes).

ENHANCEMENTS.

- Reduced time to determine what needs to be replicated
- Multiple replicators
- Push/pull replication option in Connection Document
- Replicate changes at the field level
- Database option in Connection Document
- Track replication events at the document level
- Replication priority by database
- Maximum time limit on the replication
- Event-driven replication

Scaling Notes on SMP Platforms

OBJECTIVE.

- Enhance notes Server code to take full advantage of leading SMP platforms by supporting better locking and queuing mechanism, optimization of code in heavily used critical sections, parallelization of heavily used server tasks (replicator), and better asynchronous I/O operations (write-behind).
- Initiate a performance benchmark that can measure performance and scalability and assist notes developers/administrators in locating performance bottlenecks.

ENHANCEMENTS.

- 500 simultaneous users
- Larger data structures for sessions
- Support for HP-UX, NT Advanced Server, and AIX
- Compatibility with future platforms
- Compatibility between versions
- Performance testing tools for Notes Servers

Reliability and Security Enhancements

OBJECTIVE.

- Improve the reliability and availability of Notes Servers by extensive testing under load and better logging of events leading to interrupts.
- Improve Notes security by preventing destructive writes to files (NOTES.INI), improved revocation of access rights, and limitation on retry of passwords.

ENHANCEMENTS.

- ACL or permission on pass through
- Revocation based on public key
- Mean time between reboots
- Improved availability of the server
- Quit with confirmation
- Limit failed attempts to enter a correct password
- Scan binary attachments (API)
- Prevent writes to NOTES.INI from Notes
- Disable Macro, Paste, and Mail-In on the server
- Option for immediate changes for the security groups
- Alternative to Catalog to observe ACL on listed databases (API)

Abstraction of Notes Servers

OBJECTIVE.

- Support for multiple Notes server instances running on a single, scalable SMP computer, so as to permit the hosting of many different Notes customers on a single machine.

- Support for Notes Server "clusters," in which multiple computers, interconnected by a LAN (or higher-speed interconnection), can host a single large Notes customer such that any Notes endpoint (Notes Client or Notes Server) can obtain service from any of the servers in the cluster (location of service is transparent to the user). The new notes connection protocol will provide load balancing redirection of connections to other servers in a Notes Cluster. The new connection protocol will also include nontransactional failover reconnection support, which means that the Notes Server Clusters are also a method of achieving very high availability.

ENHANCEMENTS.

- Ability to replicate unread marks on demand
- Company restricted mail addressing
- N&A Book restrictions
- Group names unique to the company, not the server
- Restrict database awareness by customer
- Hide server names during File|Open|Database
- Support for very large and small companies
- Resource utilization -DisWDatabase Space
- Unique N&A Books for each customer on a server
- Listed and unlisted names
- Enhanced ".DIR" links
- Accept/reject new load based on current performance

Remote Administration and Monitoring of Notes Servers

OBJECTIVE.

- Provide tools that reduce the cost of administering and monitoring a large number of Notes Servers and their associated Name and Address Books. Examples of administration include the tasks of creating, deleting, and changing user and group records and ACLs; configuring and limiting disk usage; establishing replication and mail connections; and migration from flat to hierarchical addressing. Examples of monitoring include presenting a graphical map of Notes Server interconnections, pinging Notes servers to report on status and detect failures; and tracking of multihop replication on a document basis to detect and correct replication errors. Much of this functionality will be delivered by Lotus' new NotesView product.

ENHANCEMENTS.
- Improved APIs for the migration of data
- Support for undocumented APIs (API)
- LotusScript support at the server level
- C and C++ API support on all platforms
- Automate and streamline user administration
- Allow Administrator to remotely create new replicas
- Broadcast a message to active users
- Notification when approaching disk quota
- Setting limits to databases
- Automated creation of a default user environment (API)
- API call for user recertification
- Automatic registration of an end user from a remote location
- Ability to track and debug multihop replication (along with NotesView V2)
- Improved N&A Book administration
- Migration utility: flat to hierarchical N&A Books (API)

NOTESVIEW V1.
- Notes server "Ping"
- Graphical system administration tool
- Remote SNMP network management capabilities

E-Mail Interconnectivity

OBJECTIVE.
- Support for high fidelity and extremely robust e-mail interoperability between e-mail services, including the various administration tools used to configure and monitor e-mail transfers.

ENHANCEMENTS.
- Notes/Internet Mail Exchange (LCM)
- Notes/X.400 Mail Exchange (LCM)

- X.400 Protocols (LCM)
- Mail Gateway: platform support
- Gateway activity reporting
- Enhanced management of mail routing
- Eliminated size limits to mail header fields

Support for Billable Services

OBJECTIVE.
- Record enough detailed information about Notes usage, so as to capture the value of the Notes based services that run on the carrier's Notes Servers. This information will be frequently flushed to reliable storage. An asynchronous, in-memory, message queue API will be provided to allow a billing application add-in server task to obtain the billing information from the Notes server.

ENHANCEMENTS.
- Record time in a database
- Record the size of a deposited document
- Fix a bug on retrieving reads and writes per database
- Provide event hooks
- Prevent billing "loopholes"
- Common recording infrastructure
- Recording for "background tasks"
- Recording for accuracy
- Record all events
- Globally unique document IDs
- Record each document accessed (API)
- Support recording of charge for accessing a document (API)
- Record disk space used by users and customers (API)
- Record agent activity on the server (API)
- New billing API (API)
- Service grade mechanism to extract billing data (API)
- Recording for mail (API)
- Record unique person identifiers (API)
- Record session IDs for dial-in service (API)

- Recording for replication (API)
- Record at start of and during session (API)
- Record document IDs of replicated documents (API)
- Record database replica IDs (API)
- Record the percentage of white space in a database (API)

Support for Routable Dial-in Protocols

OBJECTIVE.
- Support "routable dial-in" protocols (TCP/IP/PPP and SPX/IPX/PPP) in an efficient and easily configured manner. The Notes routable dial-in protocols will interoperate with all PPP-compliant gateways that interconnect dial-up lines to LANs.

ENHANCEMENTS.
- Support for PPP for TCP/IP dial-up users
- Support for PPP for SPX/IPX dial-up users
- Support of multiple TCP/IP addresses for Notes servers

Interoperability between Notes and X.500 and X.509 Directory/Security Products

Notes will be able to perform directory synchronization, and/or linking, between the Notes' Name and Address Book and other directories, such as Novell's NDS, via common X.500 protocols and APIs. Notes certificate formats will meet the X.509 standards so as to allow Notes to interoperate with other systems that also support X.509 certificates. Specific Notes security functions, such as user authentication, will permit extensions so that a user who has authenticated with a server, such as NDS, can also authenticate with a Notes server without having to supply a Notes password (for example, Notes could obtain the necessary password from the NDS directory in a secure manner).

ENHANCEMENTS.
- NDS directory service for Notes
- Use of DNS and NDS names
- Lotus Notes servers in the NDS NetWare Directory
- Single password and background authentication (API)

Detailed Changes for Release 4.6 and Release 5

Release 4.6 of Notes and Domino, which shipped in the third quarter of 1997, includes many of the features originally destined for Release 5. Therefore the details for these releases will be described together. Release 5 will be a totally cross-platform version of all of the Release 4.6 features and will additionally include support for the Internet Message Access Protocol 4 (IMAP) and Lightweight Directory Access protocol (LDAP), as well as Network News Transfer Protocol (NNTP). Release 4.6 is for Windows only.

End-user Clients

The Notes 4.6 and Notes 5 clients provide a work environment that is task oriented instead of application oriented. The enhancements to these two versions are covered in the following sections.

PORTFOLIOS. Portfolios allow you to represent a collection of Notes databases with a single database icon (on the Workspace) or tile (in the Folder pane). Users can easily navigate between multiple databases without having to return to the workspace or switch via the Window menu.

Portfolios allow users to group information by task, rather than by database. Personal Portfolios are available for organizing your related databases. For example, "My Stuff" might contain my mail file, personal address book, a personal journal in which I keep Notes and a personal Web navigator. As a second example, I could have Notes set up a local "portfolio" database added to my workspace with icons for my User's mail, Personal Journal, Personal Web Navigator, ToDo List, and Calendar.

Shared Portfolios reside on a Domino Server and can be used by many individuals. For example, Human Resources might provide new employees with a portfolio of databases on their desktop.

ACTIVE DOC SUPPORT. Active Doc Support allows you to use MS Word and Word Pro as alternate editors for Notes. A user can select which "editor" (Microsoft Word 95 and 97, Lotus WordPro 97, or Notes) launches when he creates a "Document Memo."

Type ahead addressing is provided. Notes Mail "Delivery Options" are available to the Word or WordPro Document. An MS Office Document Library template supports MS Office applications. Action Bar Buttons are passed to the Office applications.

Users then have office products with "security," "workflow," and other Notes services. Data from these office products is easily shared across the network.

INTEGRATION WITH MICROSOFT INTERNET EXPLORER. Users can specify in a Location document which browser to launch when a URL is clicked. The browser could be the Notes Server version of the Web Navigator, the Notes Personal Web Navigator, the Notes Web Browser Object (MS IE Active-X component), integrated Notes/Internet Explorer, Netscape Navigator (Netscape outside of Notes), Microsoft Internet Explorer (outside of Notes), or other browsers where Notes points to the .EXE of that browser. (Other browsers can be run outside of Notes.)

The Integrated Notes/Internet Explorer uses the MS Internet Explorer Active-X component to render HTML (including frames, recursive tables, animated GIFs, etc.) within the Notes environment. Other features include offline browsing/storage; full-text search of HTML pages; forwarding Web pages or just the URL; running agents to auto-surf or monitor pages; full fidelity HTML rendering; import of Netscape or IE Bookmarks into your Personal Navigator; and use of Internet Explorer when you view a cached page via Note's preview pane.

INTERNET MAIL CAPABILITY. POP3/SMTP Protocol Support provides a simple interface to a POP3/SMTP mail server, other than a Notes server, for mail retrieval and delivery. Users are able to retrieve messages from their Internet Service Provider (ISP) or any intranet server that supports POP3 retrieval and SMTP send.

Other features include a Universal Inbox for all of your mail where you can place corporate mail, personal mail from online services, home business mail, etc. into a single Inbox. You have one place to manage all of the information that is flowing in from all sources. All users have the same mail experience (e.g., they can send via Action button, etc.). Users have name look-up and groups supported via the Notes N&A Book. They can use the Replicator Page to replicate POP3 mail messages. They can create messages offline and replicate them back to the POP3 server.

OTHER IMPROVEMENTS. Other improvements include simplified Web installation and set-up, improved calendar and scheduling printing, UI improvements to help users re-use Mail file information, and the ability to have e-mail messages, to-dos, and calendar entries formatted in different ways.

Lotus Notes Designer for Domino

This feature enhances the Notes client capabilities as an application development tool for the Web. Overall, the Notes Designer improves the Domino Web application development process. Lotus Notes Designer is a design-only version of Notes.

NOTE: *Because Lotus Notes Designer for Domino is for design only, designers will use the Mail or Desktop clients for non-design work. The enhancements to Notes Designer for Domino are covered in the following sections.*

DOMINO WEB APPLICATION DESIGN.   Domino Web application design allows you to preview your Web page while in design. It makes it easy to embed views, folders, and navigator objects on a page. (It replaces 4.5 Domino Web Server $$ constructs, such as $$ViewBody.) Users have improved HTML handling, such as In-line HTML, the ability to enter HTML attributes for design elements via the programmer's pane "Event" menu, the ability to specify HTML attributes when designing forms or creating documents that override Domino defaults, a way to easily create HTML hotspots, and the ability to treat the contents of a view (including column text, view line text, etc.) as HTML.

The Navigator image mapping has been enhanced to allow you to save all navigator elements (text and graphics) together as an image map. Also you can import graphics into the navigator (versus copy and paste). The importing can be done as a graphic button or as background. The result is better color fidelity (versus the clipboard). Circular hotspots can be easily created.

Users can easily hide elements (e.g., views (look-up views)) for the Web or for Notes. Buttons on forms are now supported (like actions, any button not supported on the Web will automatically not display on a Web client). Anchor links can be placed in documents/Web pages, and you can create hotspot links to any part of a page. You can display alternate text while the browser loads deferred images. Usability enhancements include menu changes when in design mode and new documented templates that ship with the product.

PROGRAMMABILITY.   Programmability improvements allow you to embed Java applets in forms using the Create menu. You can run Java applets in the Notes client via support for the Java 1.1 Virtual Machine. Also you can create and run Java agents on the Domino server. You can also import existing applets into the Agent Builder and write Java agents that access and manipulate objects in the Notes back-end classes.

BUNDLED DEVELOPMENT TOOLS. The development tools bundled Notes Designer for Domino include the Lotus BeanMachine. This allows you to create multimedia 100% Java applets fast. Also you can preview the applet's appearance and functions as you create it using BeanMachine's built-in viewing capabilities or a Web browser of your choice.

You have a point-and-click interface for building applet logic. In addition, it is easy to add new beans (even existing applets) to BeanMachine's JavaBeans environment.

The Notes Global Designer is an add-in for Lotus Notes Designer that allows you to easily localize a Domino application. This designer supports users worldwide in their native languages efficiently and cost-effectively. Also Domino.Action 1.1 is included, which gives you the Web cutter sitemap applet and allows you to import a site to a Notes database.

Domino Server

The R4.6 and R5 servers are available on Win NT 3.51 and 4.0 (Intel and Alpha), as well as OS/2 and Unix (Sparc Solaris, AIX, HP-UX) and Solaris Intel Edition. The enhancements to these two versions are:

- A Server Setup database, with a new direct, brief, and easy-to-follow process, guides you through the server setup procedure. After server setup, the user can go directly to the Server Configuration database or perform configuration later.
- The Web-based Server Administration Tool is a Web-based administration database that can be accessed via a non-Notes browser, providing a subset of the administration functionality available through the full Notes client.
- Address Book populated from external source, which allows an administrator to have the Address book populated from an external source (when available). This allows automatic registration of users during server setup.
- Agent enhancements such as agent execution privileges control and the ability to run the agent on any server.
- Clickable form elements have been enhanced so that you can have clickable form elements with formulas behind them. Buttons, action hotspots, action bar buttons, and embedded navigator regions are now supported. This allows a form used through the Web server to behave more like a form in Notes.

- The SMTP MTA (Message Transfer Agent) is integrated and included with the core Domino server.

- IMAP and LDAP Protocol Support in the Domino Server allows users of IMAP client programs (such as Netscape Communicator) to access their mail files on a Domino server. The Domino Server supports LDAP V2 Protocol RFC1777 as regards to searching. The LDAP server supports a subset of LDAP Object classes for searching. This subset includes: Person Organizational Person, GroupOfNames, Organization, and OrganizationalUnit.

- MIME support maintains fidelity of incoming Internet mail while also giving users the option to take advantage of Notes document features. In Notes, the SMTP/MIME MTA stores incoming Internet mail within a message as an attachment, maintaining message fidelity. Users have this new feature as an option. IMAP and POP3 servers look for, and extract, the Notes attachment containing the original Internet mail message. The servers will then export the original Internet mail message to the user agent as opposed to converting CD records to Internet mail before exportation. IMAP server stores appended messages within a message as an attachment, maintaining message fidelity.

- News/NNTP support is a Notes Domino server add-in, which supports Internet standard News reader clients using the NNTP protocol. It also supports NNTP server-to-server "replication," or News feeds, to Internet standard News servers including Usenet News servers.

- Server-based Java agents include the Notes back-end classes Java adapter.

- Folder references in .NSF allow users to determine what folders, if any, a document is in, given only the document itself as a starting point.

THE LOTUS NOTES INTERNET COOKBOOK

This appendix contains the "Lotus Notes Internet Cookbook." The Cookbook is published electronically in the form of an FAQ (Frequently Asked Questions) document. It provides key information about how to take advantage of Notes' existing, built-in Internet support. It is included as an appendix to this book because it should prove useful to any Notes designer considering use of the Internet. The FAQ format is a good way to supplement chapter 8. The cookbook is included on the CD-ROM that comes with this book.

The Cookbook document was written by Barbara Mathers (bmathers@iris.com) and Dave Newbold (dnewbold@iris.com) of the Iris Associates subsidiary of the Lotus Development Corporation. It is included here with the permission of the document's authors and Iris Associates. The Cookbook is published online and is available through several electronic sources that are mentioned in the next two paragraphs. The softcopy for this appendix was obtained from the "Lotus Notes New Knowledge Base," which is available on the Notes server in the Lotus Notes Network (LNN). The Notes domain is "NOTES NET" to which over 13,000 Lotus Business Partners connect.

NOTE: *The online Cookbook was updated on the Internet (WWW.IRIS.COM or WWW.NOTES.NET) and via Notes (New Lotus Notes Knowledgebase) during 1996. Versions for both Release 3 and Release 4 of Lotus Notes are available. This appendix is based on the 1996 Release 4 instructions. However, the basic concepts for using Notes on the Internet are the same for both versions. This appendix does give instructions for accessing both versions of the cookbook.*

Sections from the Lotus Notes Internet Cookbook for Notes Release 4

If you are trying to figure out how to utilize the Internet from Notes, read on!

The Lotus Notes Internet Cookbook is a casual, working document that includes step-by-step information on how to extend Notes into the Internet. We wrote this to give you a shortcut guide to make the most of the connectivity that the Internet provides and the power of integration that Notes provides. With a complex topic such as Notes and the Internet, the secrets of making it work always seem buried in someone

else's mind. We hope to shed some light here. As with the previous version, we intend this to be a casual how-to document, not the definitive reference guide to Lotus Notes. We try to maintain the accuracy of this information as best we can, but we especially welcome your corrections and comments.

This version of the Cookbook describes how to connect to the Internet using Notes Release 4. For information on connecting to the Internet using Notes Release 3, see the previous version of the Cookbook. You can download both versions of the Cookbook from Notes:

```
welcome.nsf at home/notes/net  (205.159.212.10)
```

or from the Web:

```
ftp://www.notes.net/pub/faq/cbookv3.zip  (Notes Release 3)
ftp://www.notes.net/pub/faq/cbookv4.zip  (Notes Release 4)
```

Please send all your feedback to:

```
cookbook@iris.com
```

What Does the Internet Have to Do with Notes?

Just about everything. The Internet is a world-wide network of networks that connects millions of people together in a global work and recreation domain. Lotus Notes provides a way for users to organize, share, collaborate, and integrate information in a secure environment. With Notes Release 4, users are now seamlessly integrated into the two worlds. It is now easy to Web-enable your Notes daily life, from publishing information onto a Web site to sharing information pulled from the Internet—all within the Notes environment.

Lotus has made many advances in terms of integrating the standard protocols of the Internet into Notes, developing a common directory service (NotesNIC), as well as giving administrators greater flexibility in deploying Notes (such as anonymous access and authenticated Passthru servers.)

Aside from expanding the user's reach with greater external connectivity, the Internet offers organizations an inexpensive medium for Notes replication and WAN connectivity. Using the Internet as a communications solution can save you and your organization money if you have many point-to-point leased lines for communications to your branch locations or you make extensive use of dial-up phone lines for transferring large amounts of data. Having a single connection point for each location reduces data communications management time and reduces your connection fees.

The biggest savings will be realized by large and medium-sized organizations with an experienced and capable IS department. For small organizations, employing a direct Internet connection might not be a good investment due to the start-up costs and experience required. It is usually better to consider using a Notes Public Network Provider (such as AT&T, IBM Global Network, CompuServe, or WorldCom) to provide the mail routing and replication services you need.

Caution: As always, when connecting your Notes environment to the Internet, security becomes a crucial issue. Before making your actual Internet connection, be sure to read the section "Can I Create a Secure Environment with Notes" later in this appendix to learn how to create a secure Notes environment.

How Do I Set Up Notes to Connect to the Internet?

You can connect your Notes environment to the Internet in two ways:

- *Through an Internet Service Provider*—An Internet Service Provider (ISP) is a company that offers different types of network connections to the Internet. These connections range from simple dial-up connections to fully dedicated leased-line connections. With Notes, you can use any type of connection to hook into the Internet through an ISP.

■ *Through a Notes Public Network*—Notes for Public Networks (NPN) is a service that provides a secure, reliable computing platform for use on public telecom networks. End users subscribe to the NPN service through a network operator to establish inter-enterprise communications without incurring the cost of buying and maintaining the IS investment.

Connecting through an Internet Service Provider (ISP)

These are the setup steps you'll need to follow to set up your connection through an Internet Service Provider (ISP). The actual connection steps are in the section "How Can I Make a Connection over the Internet" later in this appendix.

1. Install the Notes Release 4 client or server software.

2. Obtain one of these two types of physical Internet connections from an ISP:

   ■ A leased-line connection that connects your LAN through a router to a leased phone line to the ISP

   ■ A dial-up modem connection that connects your LAN through a dial-up modem to the ISP

3. Install one of these types of network connections:

   ■ TCP/IP protocol software for leased-line connections

   ■ SLIP/PPP (dial-up versions of TCP/IP) for modem connections

4. Port. Test your Internet connection using a ping utility

HOW DO I INSTALL THE CORRECT NOTES CLIENT OR SERVER? Refer to the Notes R4 installation guides for information on installing the different Notes platforms.

HOW DO I OBTAIN A PHYSICAL CONNECTION TO AN ISP? Contact one of the many Internet Service Providers to find out about their rates and services. To locate an ISP in your area that suits your needs, visit the following Web pages:

■ http://thelist.com

■ http://www.cybertoday.com/cybertoday/ISPs/ispinfo.html

■ http://www.herbison.com/herbison/iap_meta_list.html

You can also post messages to the alt.internet.access Usenet newsgroup and ask for help on finding a provider in your area.

HOW DO I SET IT UP IN NOTES? To set up a TCP/IP, SLIP, or PPP connection to work with Notes:

1. Decide which network protocol you want to use and install it on your Notes server computer.

2. Set up and enable the network protocol on your existing Notes server(s).

3. Set up and enable the network protocol on your new Notes server(s).

4. Enable the network protocol on your Notes workstation(s).

WHICH NETWORK PROTOCOLS CAN I USE? You can use either TCP/IP (for a leased-line or LAN connection) or SLIP/PPP (for a dial-up modem connection). Both are considered to be "TCP/IP" LAN protocols for Notes. For the latest listing of tested and supported network protocols, see the Lotus Notes Network Configuration Guide.

Once you decide on a network protocol, install it by following the installation instructions included with that protocol.

HOW DO I SET UP AND ENABLE NETWORK PROTOCOLS ON AN EXISTING NOTES SERVER?

1. Stop the Notes server by typing QUIT or EXIT at the console.

2. Start the Notes client on the Notes server.

3. Select File|Tools|User Preferences, **click** Ports, **and then click** New.

4. In the New Port dialog box, enter the correct port name (for example, TCP) and driver (for example, TCP), and select any locations where you want the port to be enabled. Click OK.

5. Open the public Address Book for your server, and edit the Server document. In the Network Configuration section (see Figure C.1), add the port name that you specified and the name of the Notes

Figure C.1
Network
Configuration section
of server document
in the N&A Book.

Network Configuration

Port	Notes Network	Net Address	Enabled
TCP	TCP	122.00.100.41	ENABLED
		Band	DISABLED
		Band	DISABLED
		Band	DISABLED
		Band	DISABLED

named network that this Notes server belongs to (for example, TCP) in the Notes Network column. Enter the IP address of the computer in the Net Address column. Select Enabled in the Enabled column. Exit and save your changes.

6. Restart your Notes server to enable the new network protocol.

HOW DO I SET UP AND ENABLE NETWORK PROTOCOL SOFTWARE ON A NEW NOTES SERVER?

1. Run the Notes server setup program. Follow all the steps described in the appropriate installation guide for setting up the first Notes server. Be sure that you select the TCP driver in the Network type box that appears in the Additional Server Setup dialog box.

2. Open the public Address Book for your server and edit the Server document. In the Network Configuration section, add the port name and the name of the Notes named network that this Notes server belongs to (for example, TCP) in the Notes Network column. Enter the IP address of the computer in the Net Address column. Select Enabled in the Enabled column (see Figure C.2).

3. Start the Notes server to enable the new network protocol.

HOW DO I SET UP AND ENABLE NETWORK PROTOCOL SOFTWARE ON A NOTES WORKSTATION?

1. Choose File|Tools|User Preferences, click Ports, then click New.

2. In the New Port dialog box, enter the correct port name (for example, TCP) and driver (for example, TCP), and select any locations where you want the port to be enabled. Click OK.

3. In the Communication Ports list box, double click the TCP port to enable it.

4. Click OK.

HOW DO I TEST MY INTERNET CONNECTION?   After you establish your Internet connection, you should run a quick test to ensure that your connection works properly. It is best to run this test before you actually connect Notes to the Internet. The easiest way to

Figure C.2

Setting up the network protocol in the N&A Book.

▼ **Network Configuration**

Port	Notes Network	Net Address	Enabled
TCP	TCP	122.00.100.41	ENABLED
		Band	DISABLED
		Band	DISABLED
		Band	DISABLED
		Band	DISABLED

test your connection is to use a ping utility. (A ping utility allows you to ask another computer if it is running, and the protocol software can respond. Keep in mind that, even if you can ping successfully to the computer, the Notes server might not be running.)

When you ping another computer, make sure you attempt to ping a computer that is not in your immediate domain. If you can ping successfully into another domain, you will have verified that your router is working properly.

Connecting through a Notes Public Network (NPN)

Using a NPN service provider, you can use three types of networks:

- Internal networks on which the most proprietary inhouse information is kept.
- Secure NPN-like service running on public networks for closed user groups and confidential information.
- The Internet, best at this time for publicly available information, such as marketing materials.

These three networks can be used together. For example, you can use Lotus InterNotes Web Publisher in combination with Notes Public Network services to prepare information, in which some is confidential and some is public. Then, you can selectively replicate the confidential information to others on your NPN service through the NPN service provider and send the public information to the NPN clients as well as publishing the information through InterNotes Web Publisher to the Web.

HOW DO I SET UP MY CONNECTION? A service provider sets up and maintains the Notes Public Network. The service provider delivers full Internet connectivity and interoperability for Notes services and provides access to a Notes network without the costs of supporting private networks and Notes servers.

Through the NPN service provider's network, users can access all the features of Notes plus NPN enhancements: clustering, partitioned servers, and billing. Clustering enables servers to locate replica copies of a database if another server is down. Partitioned servers allow multiple Notes servers or customers to run on a single machine. Billing enables NPN customers to subsegment bills to their end-users, allowing each to pay for services used.

HOW DO I FIND OUT MORE ABOUT NPN?   For the latest listing and information on the NPN service providers, check the following Web site:

```
http://www.lotus.com/info/npn.htm
```

How Can I Make a Connection over the Internet?

Once you have completed the setup steps in the previous section to connect to the Internet through either an Internet Service Provider or a Notes Public Network Service Provider, you can go into Notes to establish the actual connection.

How Can I use a Notes Workstation to Access a Notes Server over the Internet?

You can connect a workstation to a server over the Internet to access its databases. (See Figure C.3.) If that server is a Passthru server, you can then use it to connect to other servers. (For more information on setting up Passthru on a server, see the Notes Release 4 Help.)

After you install and set up a connection to the Internet (see the previous section), you are ready to connect a Notes workstation to a Notes server over the Internet:

1. Find out the IP address of the destination Notes server with which you want to connect.

2. Create a Location document in your personal Address Book.

3. Create a Connection document in your personal Address Book.

Figure C.3
Accessing a Notes server from a Notes workstation over the Internet.

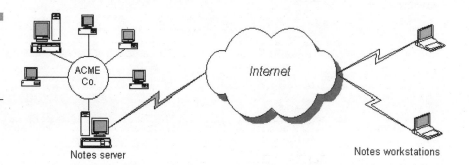

4. Establish the connection to the Internet outside of Notes.

5. Connect to the destination server and open databases.

OBTAINING THE IP ADDRESS OF THE NOTES SERVER. Before you can access a Notes server over the Internet, you need to determine the IP address and domain name of that Notes server. You should be able to obtain this information by asking the Administrator of that Notes server. With this information in hand, you can now work from within your Notes workstation to configure your connection.

CREATING A LOCATION DOCUMENT. Create a Location document so that, when you change to that Location, the TCP port will be enabled:

1. Open your Personal Address Book, and click Add Location.

2. Fill out the fields as per your needs using Local Area Network as the Location type.

3. Be sure the TCP port in the "Ports to use" field is checked. If there is no TCP port box, choose File|Tools|User Preferences, click Ports, and double-click the TCP port to enable it.

4. Exit and save the Location document.

CREATING A CONNECTION DOCUMENT. You need to create a Server Connection document so that you can open databases on the server. To create a Server Connection document:

1. Open your Personal Address Book, and choose Create|Server Connection (see Figure C.4).

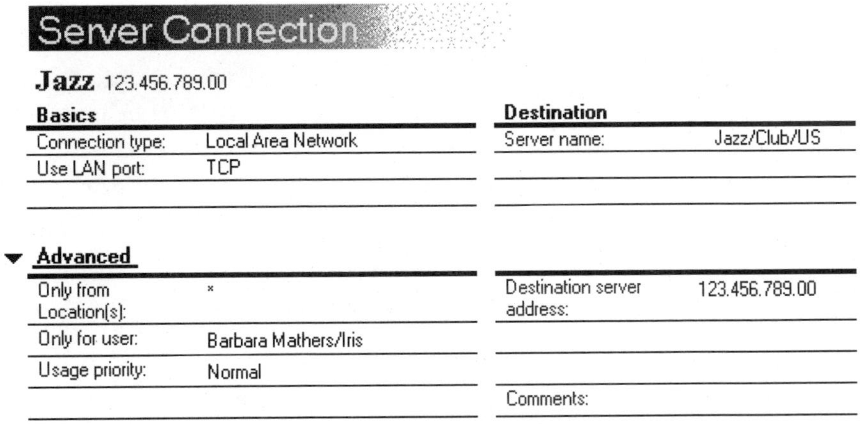

Figure C.4

Creating a Server Connection record.

Server Connection

Jazz 123.456.789.00

Basics		Destination	
Connection type:	Local Area Network	Server name:	Jazz/Club/US
Use LAN port:	TCP		

▼ **Advanced**

Only from Location(s):	×	Destination server address:	123.456.789.00
Only for user:	Barbara Mathers/Iris		
Usage priority:	Normal		
		Comments:	

2. Choose Local Area Network as your Connection Type.

3. Add the name of the destination server.

4. In the Use LAN port field, select TCP.

5. Open the Advanced section.

6. In the "Only from Location(s)" field, choose the location you just created.

7. Add the IP address of the destination server.

8. Exit and save your changes.

ESTABLISHING THE INTERNET CONNECTION OUTSIDE OF NOTES. Before you can open databases on a server over the Internet, you need to establish your Internet connection. If you are using a dial-up connection, dial the number of your ISP using your favorite connection software. (The type of connection software you use varies according to the operating system you are using. Contact your ISP if you need connection software.) If you have a leased-line connection, your Internet connection should already be established.

CONNECTING TO THE SERVER AND OPENING DATABASES. To connect to the Notes server and open databases:

1. Change to the location you just created (where your Connection document is valid and the TCP port is enabled).

2. Choose File|Database|Open.

3. Enter the name of your destination server to open it.

How Can I Use a Notes Server to Access Another Notes Server over the Internet?

You can connect a server to a server over the Internet so that you can perform normal server tasks, such as replication and mail routing. In this section, we'll walk you through the steps of setting up replication and mail routing from one server to another over the Internet (see Figure C.5). The specific steps you'll need to follow are:

1. Find out the IP address of the destination Notes server with which you want to connect.

2. Create a Connection document in your server's Public Address Book.

3. Establish the connection to the Internet outside of Notes.

Figure C.5

Connecting a Notes server to another Notes server over the Internet.

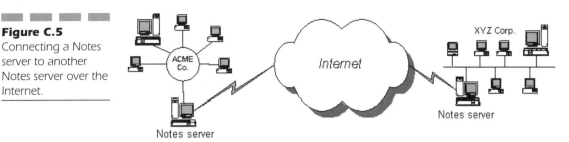

OBTAINING THE IP ADDRESS OF THE NOTES SERVER. Before you can access a Notes server over the Internet, you need to determine the IP address and domain name of that Notes server. You should be able to obtain this information by asking the Administrator of that Notes server. With this information in hand, you can now configure your connection.

CREATING A CONNECTION DOCUMENT. Replicating and mail routing to a Notes server over the Internet is done in the same way as over a LAN. You need to compose a Connection document in the Domain's Address Book. To create the Server Connection document:

1. Open your server's Public Address Book and click Add Connection (see Figure C.6).

2. Choose Local Area Network as your Connection Type.

3. Add the name and domain of the source and destination servers.

Figure C.6

Connection Record for Mail Routing and Replication over the Internet.

Server Connection

Band/Big/US to Jazz/Club/US

Basics

Connection Type:	Local Area Network	Usage priority:	Normal
Source server:	Band/Club/US	Destination server:	Jazz/Club/US
Source domain:	Club	Destination domain:	Club
Use the port(s):	TCP	Optional network address	123.455.000.11

Scheduled connection		**Routing and replication**	
Schedule:	ENABLED	Tasks:	Replication, Mail Routing
Call at times:	08:00 AM - 10:00 PM each day	Route at once if:	5 messages pending
Repeat interval of:	360 minutes	Routing cost:	5
Days of week:	Mon, Tue, Wed, Thu, Fri, Sat, Sun	Replicate databases of:	Low & Medium & High priority
		Replication Type:	Pull Push
		Files to Replicate:	(all if none specified)
		Replication Time Limit:	minutes

4. In the "Use the port(s)" field, select TCP.

5. Add the IP address of the destination server in the "Optional network address" field.

6. Choose the location where you want this Connection document to be valid.

7. Choose both Replication and Mail Routing as your tasks.

8. Enter the schedule of when you want the servers to connect.

9. Exit and save your changes.

NOTE: *Your Notes server needs to be in the same Notes domain as the Notes server you want to access. If it is not, your Notes server needs a certificate in common with the other Notes server. For more information on Notes certificates, see the Notes Release 4 Help.*

ESTABLISHING THE INTERNET CONNECTION OUTSIDE OF NOTES. Before connecting a server to another server on the Internet, you need to establish your Internet connection. If you are using a dial-up connection, dial the number of your ISP using your favorite connection software. (The type of connection software you use varies according to the operating system you are using. Contact your ISP if you need connection software.) If you have a leased-line connection, your Internet connection should already be established.

CONNECTING TO THE NOTES SERVER. To connect to the Notes server:

1. Establish your dial-up (modem) or direct (leased line) Internet connection through your ISP.

2. Your Notes server will connect to the other Notes server based on the scheduled times you set in your Connection document.

What Happens if I can't Connect?

This section covers some common problems you might encounter when trying to connect your Notes workstation or server to another server on the Internet.

PROBLEM 1. If you receive an error message from Notes about security, such as "You are not certified to access the remote server," your TCP/IP setup is working correctly, but you do not have proper authorization to connect to that Notes server. To solve this problem, contact the Notes Administrator to obtain the proper authorization. Also, see the next section for an alternate solution to connecting to servers when you don't have proper access.

PROBLEM 2. Messages such as "TCP/IP host unknown" or "Remote system not responding" usually mean there is a problem with the TCP/IP setup. If you were able to ping the remote host successfully with the IP address, the Notes server might not be running. If you use host names instead of the actual IP addresses in your Connection documents, you might be having a problem with name resolution.

To fix this problem, verify that your Domain Name System (DNS) can resolve the name to the IP address by checking that the IP address is entered in the hosts file. If you do not have a DNS, you need to add the entry to your local hosts file. The local hosts file is a file that contains a mapping of host names and IP addresses. The hosts file is usually located in the same directory as the protocol software. It has a format similar to:

```
IP Address        Domain Name        alias        comment
123.3.12.245      salt.usa.com       salt         #Salt server
123.3.12.678      pepper.usa.com     pepper       #Pepper server
```

PROBLEM 3. You might be blocked by a firewall server. To test this, try to ping the Notes server to see if you can access it. If you are able to ping but still cannot access the Notes server, try using telnet to connect to the Notes server on port 1352 (see your telnet documentation for details on how to do this). If you cannot connect with telnet, the firewall server might be blocking the TCP port, and you will have to contact the remote site Administrator to resolve the problem.

How Can I Access Servers if I don't have Proper Access?

Before a Notes workstation can access any database on a server, a process called authentication needs to take place. Three basic rules apply during the authentication process:

- Authentication is based on a challenge-response protocol using each participant's public key.
- Certificates are used to prove the binding between the public key and its owner's name.
- Authentication takes place in both directions: The workstation authenticates the server and the server authenticates the workstation.

During the two-way authentication:

■ If the workstation is not certified so that the server can check the workstation name-key binding, the workstation cannot be authenticated at the server. In this case, the only way to access the server without proper authentication is through a form of unauthenticated access called *anonymous access*. Anonymous access is an especially useful feature if you want to make your server databases available to users over the Internet who do not have a certificate or cross-certificate capable of being authenticated by your server. Once you set up your server to allow anonymous access, users can connect to your server over the Internet and gain access to your databases.

■ If the server is not certified so that the workstation can check the server name-key binding, the server cannot be authenticated at the workstation. In this case, the workstation can choose to access the server knowing that the server has not been authenticated. This form of access is called *unathenticated server access*.

HOW CAN I USE A NOTES WORKSTATION TO ACCESS A NOTES SERVER THAT CANNOT AUTHENTICATE ME? As a workstation user, you can access databases on servers even if you are not certified such that you can be authenticated by a particular server. The only consideration is whether the server administrator has allowed workstations to access the server in an anonymous way. If the server has been set up to allow anonymous access, gaining access is simple.

CONNECTING TO A SERVER USING ANONYMOUS ACCESS IN A NON-HIERARCHICAL ENVIRONMENT If you are in a Notes environment that uses nonhierarchical certificates, Notes displays the following messages in the status bar when you attempt to connect to a server that cannot properly authenticate you:

```
Server X cannot authenticate you because:
Your ID has not been certified by a certifier that is trusted by
the server.
You are now accessing that server anonymously.
```

CONNECTING TO A SERVER USING ANONYMOUS ACCESS IN A HIERAR-CHICAL ENVIRONMENT If you are in a Notes environment that uses hierarchical certificates, Notes displays the following message in the status bar when you attempt to connect to a server that cannot properly authenticate you:

```
Server X cannot authenticate you because:
The server's Address Book does not contain any cross-certificates
```

```
capable of authenticating you.
You are now accessing that server anonymously.
```

SETTING UP ANONYMOUS ACCESS ON A SERVER To set up anonymous access on your server so that workstations will be able to access your databases even if you cannot authenticate the workstations:

1. Protect the databases on your server by using Access Control Lists:
   - *Create an Anonymous entry*—Create an Anonymous entry, and assign the desired rights to that entry (typically, you give Anonymous users Reader access to databases).
   - *Edit the Default entry*—Assign the desired rights to the Default entry if you do not want to create an Anonymous entry. (If there is no specific Anonymous entry, an Anonymous user will be given the same rights as the Default entry.)

2. Provide access to your server from the user Anonymous. If you control access to your server through server access lists in the public Address Book for the server, add the user Anonymous in the Access Server field in the Restrictions section of the Server document. Shut down and restart the server for the changes to take effect.

3. Turn on the "anonymous access" feature on your server. In the public Address Book, choose Yes in the "Allow anonymous connections" field in the Security section of the Server document. Shut down and restart the server for the changes to take effect.

HOW CAN I USE A NOTES WORKSTATION TO ACCESS A NOTES SERVER THAT I CANNOT AUTHENTICATE? As a workstation user, you can access databases on servers even if you cannot authenticate a particular server. The only consideration is whether you are willing to trust that the server is really who it claims to be. (The server does not need to have anything set up to allow you to access it without being able to authenticate it.)

CONNECTING TO A SERVER USING UNAUTHENTICATED SERVER ACCESS IN A NONHIERARCHICAL ENVIRONMENT If you are in a Notes environment that uses nonhierarchical certificates, Notes displays the message shown in Figure C.7 when you attempt to connect to a server that your workstation cannot properly authenticate.

When you click Yes, you agree to access the server and its databases even though you cannot properly authenticate the server.

CONNECTING TO A SERVER USING UNATHENTICATED SERVER ACCESS IN A HIERARCHICAL ENVIRONMENT If you are in a Notes environ-

Figure C.7
The message
displayed in a
nonhierarchical
environment when
your workstation
can't properly
authenticate.

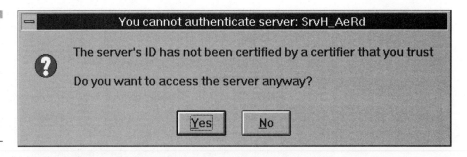

Figure C.8
The message
displayed in a
hierarchical
environment when
your workstation
can't properly
authenticate.

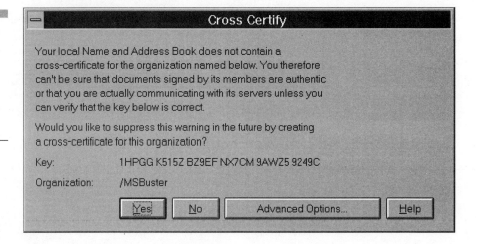

ment that uses hierarchical certificates, Notes displays the message shown in Figure C.8 when you attempt to connect to a server that your workstation cannot properly authenticate.

If you want to create a cross-certificate to access the server, choose Yes. The server will be authenticated. If you find out later that the server is bogus, you should remove the cross-certificate. To delete a cross-certificate, open your Address Book to the Certificates view, select the cross-certificate you want to delete, and press the Delete key.

If you want to access the server without creating a cross-certificate, choose No in the dialog box in Figure C.8, and Notes will display the message shown in Figure C.9.

When you click Yes, you agree to access to the server and its databases even though you cannot properly authenticate the server.

WHAT NOTES FEATURES ARE AVAILABLE WHEN I CONNECT TO SERVERS USING ANONYMOUS ACCESS AND UNATHENTI-CATED SERVER ACCESS? Unauthenticated server access (a client accessing a server that the client cannot authenticate) is only available

Figure C.9
The message displayed when you want to access a server without creating a cross-certificate.

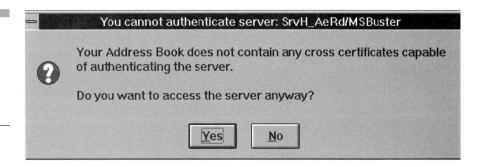

when you want to connect a workstation to a server; you cannot use unauthenticated server access to connect a server to another server. Anonymous access (when a server cannot authenticate a client) is available to all clients, provided the server is set up correctly.

These are the four workstation-to-server connections and what Notes features are available during each connection:

- *Properly certified workstation, properly certified server*—Workstation can open server databases, use mail router, replicate from desktop, and use background replication.

- *Properly certified workstation, unauthenticated server*—Workstation can open server databases.

- *Anonymous workstation, properly certified server*—Workstation can open server databases, use mail router, replicate from desktop, and use background replication (subject to ACL settings for the Anonymous entry).

- *Anonymous workstation, unauthenticated server (subject to ACL settings for the Anonymous entry)*—Workstation can open server databases.

How Can I Locate Other Notes Servers on the Internet?

There is a new Notes directory service now available that provides a way for Notes clients and servers to locate other Notes servers on the Internet. This service, NotesNIC, is an organized system for locating and communicating with Notes sites globally. The NotesNIC gives a Notes administrator the power to combine a standard Release 4 Notes server and the Internet economically and efficiently to allow Notes servers to replicate, route mail, and access databases around the world.

NotesNIC allows users to locate and communicate with Notes sites globally by:

- Using Notes Mail across the Internet (no need for gateways).
- Exchanging and updating information utilizing the power of scheduled replication.
- Enhancing anonymous access, allowing people to visit your Notes site and browse through a variety of Notes databases. (These could be databases ranging from Marketing/Sales material to information exchange databases like Notes Knowledge Base or user group discussion databases.)
- Utilizing a directory delivered to you via a standard Notes Name and Address Book making the set up of and managing of connections with other Notes sites around the world no different than administrating your own internal Notes domains.

For more information on NotesNIC, visit the home site for the NotesNIC on:

- Notes site: welcome.nsf at home/notes/net (205.159.212.10)
- Web site: http://www.notes.net

How Can I use Notes to Access Information on the World Wide Web?

The Internet is an exciting world filled with information. Once you get your Notes environment connected to the Internet, you'll be able to start searching for the information you need.

How Can I Access Information on the World Wide Web?

Lotus integrated Web access into Notes Release 4 with the InterNotes Web Navigator. Now, using Notes Release 4, you can automatically access information on the Web through a central database called the Web Navigator database. This database resides on a Notes server connected to the Internet, referred to as the InterNotes server. Each time a user retrieves a page off the Internet, the Web Navigator translates that page into a Notes document and stores it inside the Web Navigator database. Then, the next time anyone wants to read that Web page, it is available for instant access.

The InterNotes Web Navigator consists of the Web Navigator database and a server task that reside on the InterNotes server. The InterNotes server:

- Stores the Web Navigator database.
- Runs the WEB server task.
- Runs the TCP/IP network protocol.
- Maintains either a direct or proxy connection to the Internet.

Once up and running, the Web Navigator can easily be customized and managed. For example, the Web Navigator comes with powerful agents you can use to purge and refresh the Web pages stored in the Web Navigator database. In addition, you can write more agents to customize your database by using special LotusScript functions and Notes @functions designed for use with the Web Navigator database. As for security features, you get all the security of Notes plus the added feature of being able to control access to specific Web sites.

For complete documentation on setting up, administering, and customizing the Web Navigator, see the Lotus InterNotes Web Navigator Administrator's Guide (WEBADMIN.NSF). For complete documentation on using the Web Navigator, see the Lotus InterNotes Web Navigator User's Guide (WEBUSER.NSF).

How Can I Set up E-mail Access to the Internet?

Internet e-mail is the process of sending an electronic message through the Internet or through gateways and other networks. Internet e-mail is the most popular application on the Internet. Simple Mail Transfer Protocol (SMTP) is the protocol used to transfer e-mail between computers on the Internet. You need to be running or have access to a SMTP Gateway in order to send and receive e-mail between Notes and the Internet.

NOTE: *Most of the SMTP gateways run on Notes Release 3.x servers. Read the description for each SMTP gateway for specific information on supported platforms.*

To provide the ability to send and receive Internet e-mail from Notes, you can either:

- *Install a SMTP Gateway locally in your Notes environment*—Installing and configuring a local SMTP Gateway can be expensive and complicated. Before you begin, you will need a dedicated line to the Internet, the Lotus Notes SMTP Gateway, a Notes server computer, a TCP/IP stack, and an administrator who is familiar with all the different components.

■ *Use a SMTP Gateway remotely through a Public Notes Information Service*—If you do not meet all the requirements to set up a SMTP Gateway locally, you can access an SMTP Gateway remotely through a Notes Public Network service provider. Notes Public Network service providers provide most of the necessary hardware and software for your Notes users to send and receive Internet e-mail. All you need at your site is Notes and a modem.

WHAT SMTP GATEWAYS CAN I USE FOR NOTES? If you decide to set up a local SMTP Gateway, you can use one of the following gateways.

LOTUS SMTP GATEWAY V1.1 Lotus provides a SMTP Gateway that allows Notes mail users to communicate with Internet mail users. Because Notes and SMTP messages have different formats, messages sent from one system to the other must be converted to a format that can be read in the target mail system.

The SMTP Gateway software runs on a Notes Release 3.2 or 3.3 Server for OS/2. Also available is a version running on Sun Solaris 2.3 or 2.4. A Notes server with the SMTP Gateway software installed can handle SMTP mail for multiple Notes servers, including Notes servers on remote LANs. However, if you have a large, geographically distributed Notes system, it is more efficient to install the SMTP Gateway software on more than one server.

For specific steps on how to install and configure the SMTP Gateway, read the *Lotus Notes Mail Gateway for SMTP (OS/2) Release 1.0* online book or the *Lotus Notes Mail Gateway for Solaris 2.x Release 1.1* online book.

The current version of Lotus SMTP Gateway is Release 1.1, which costs $2500. You can call 800-343-5414 for more information.

What Happens to a Notes Document After Being sent Through the Lotus Notes SMTP Gateway? Briefly, the rich text document is converted into plain text and the formatting information is appended as a MIME attachment (and converted to a text representation using either BASE64 or UUencode, depending on the gateway configuration). MIME stands for Multipurpose Internet Mail Extensions and is an evolving standard for the format of a multi-part message with both text and binary objects using existing text-only SMTP mail servers. Any original attachments are appended as MIME attachments as well (again, encoded as text using BASE64 or UUencode).

This appears to the recipient as follows:

- If you are another Notes user, behind another Lotus SMTP Gateway using the same encoding technique, it will appear exactly like the original Notes document. The setting of the gateway configuration form field "Attach Notes Body" determines whether the plain text or rich text (if present) version of a document will comprise the Notes message body.

- If you are using a standard UNIX mail application, you will see the plain text of the e-mail followed with the text representation of the encoded rich text and attachments. If the originating gateway used the UUencode technique, you can manually convert (cut and paste to another file) the binary attachments using the UUdecode program and make use of them.

- If you are using a MIME-capable mail application, it should select the richest version of the document. That is, if a rich text version of a message body is present, the MIME mail application will display that version instead of the plain text version. Any original attachments are rendered as attachments. If the originating Lotus SMTP Gateway is using the UUencode translation technique, the attachments might not be converted or might be rendered as plain text. This is because UUencode, although a popular encoding technique, is not a MIME-standard Content Transfer Encoding type, and there are many variations in how this encoding is labeled.

Tip: Use a program called MPACK to decode and encode both BASE64 and UUencoded documents. MPACK is available at ftp://ftp.andrew.cmu.edu/pub/mpack.

What Happens to a MIME-encoded E-mail Sent Through the Lotus SMTP Gateway? The message is rendered as plain text within the Notes body field. MIME multi-part objects, both text and binary, are converted to attachments within the message. Due to limitations in the current gateway, nested multi-part objects are flattened and converted as a separate attachment.

SOFTSWITCH SoftSwitch provides two SMTP Gateway solutions for Notes users who want to exchange e-mail with foreign mail users: Lotus Messaging Switch (LMS) and SoftSwitch Central. Notes networks can connect to LMS and SoftSwitch Central with the new access unit for Notes, AU/Notes 2.0, which supports Notes Release 4.

Lotus Messaging Switch is a UNIX-based backbone switch. You can use Lotus Messaging Switch along with LMS' SMTP MTA to allow SMTP connectivity to the Internet and to interface with UNIX Sendmail.

SoftSwitch Central is an IBM host-based enterprise mail backbone switch. You can network SoftSwitch Central with any other SoftSwitch backbone products to form an electronic mail network backbone. To provide access to the Internet, you can use SoftSwitch Central in conjunction with the Central's SMTP Gateway.

The SoftSwitch line of products are well-suited for companies that have complex, heterogeneous environments with multiple e-mail systems with which to contend. For more information on SoftSwitch products, contact them at 610-640-9600.

CC:MAIL If you have cc:Mail users, you can install a SMTP Gateway for cc:Mail (available from Lotus) and then use a cc:Mail to Notes Gateway (available from both Lotus and third-party vendors). This solution allows Notes users to send e-mail through cc:Mail to the Internet.

LOTUS SMTP MTA The Lotus Notes SMTP/MIME MTA is an optional integrated Message Transfer Agent (MTA) that works with Notes Release 4. It is a scalable, high-performance component of the Notes Messaging Services. The key features of the Lotus SMTP MTA are:

- Is an MTA, not a gateway.
- Supports multiple SMTP and MIME Request for Comments (RFCs) and standards (STDs).
- Is easy to configure and manage.
- Leverages Notes built-in security.
- Supports MIME and UUencode.
- Supports native addressing to the Internet.
- Supports SMTP address conversion.
- Allows use of multiple Internet addresses for users.
- Supports multiple character sets for both MIME and non-MIME configurations.

The best way to become a candidate for the field test of the SMTP MTA is to connect to the Lotus Notes Network (LNN) and create a Beta Nomination Form. If you are not already connected to LNN, ask your Notes sales representative for more information.

WHAT SMTP SERVICES ARE AVAILABLE THROUGH NOTES PUBLIC NETWORK SERVICE PROVIDERS? Notes Public

Networks (NPN) service providers provide Notes users with SMTP services. An NPN service provider hosts Lotus Notes applications among other services, available through telephone companies and public network service providers. Through these services, NPNs can provide basic e-mail connectivity to allow you to send and receive Internet e-mail through a Lotus Notes connection.

Most NPN service providers offer SMTP services. Here are some of the NPN service providers: BT (British Telecom), CompuServe, Deutsche Telekom, IBM's Global Network, Nippon Telegraph and Telephone Corporation, NTT Data Communications Systems Corp., SNET (Southern New England Telephone Company), Telecom Italia, Telekom Malaysia, Telstra, Unisource, and US West. Visit the following Web site for further information on NPN service providers:

```
http://www.lotus.com/info/npn.htm
```

Both CompuServe and Interliant (formerly WorldCom) have SMTP services available now, based on Lotus Notes Release 3, but not as part of the NPN program. Read their individual descriptions for more information.

COMPUSERVE (FOR NOTES RELEASE 3) CompuServe provides a service called Enterprise Information Link for users of Lotus Notes. This service provides support for worldwide e-mail, magazine articles, news, financial information, private services, discussion databases, extensive file libraries, and Internet services. Users can communicate and share information in over 140 countries, with dial-up access via local numbers in over 400 cities. Access is provided directly through CompuServe's own network.

For electronic mail support, CompuServe provides server-based, standalone Lotus Notes users the ability to exchange electronic mail and files between people within their organization and the world.

INTERLIANT (FOR NOTES RELEASE 3) Interliant (formerly WorldCom) is a Public Notes Information Service from Wolf Communications Company. In addition to providing extensive news services and publications, Interliant connects Lotus Notes users to the Internet and to more than 100 other e-mail systems. Through Interliant, you can access many Internet services, including e-mail, Usenet newsgroups, the World Wide Web, FTP, SEC documents, and more than 5900 mailing lists. Interliant also provides multiple services to support sending e-mail to the Internet through Notes.

How Can I Access Usenet News?

Usenet newsgroups and a Notes discussion databases are similar. Both are used to collect and share information. Both provide a discussion mechanism similar to a BBS but allow for controlled replication to many sites. Both provide a way for navigating discussion threads, responding to specific issues, and organizing documents around general topics.

Usenet newsgroups are public, widely distributed, and mostly text-based. Notes discussions are typically private, have specific security controls (for access, replication and editing), support a consistent set of rich text, attachment and multimedia features, and usually are organized around one database per topic.

You access Usenet news in Notes through a Usenet-to-Notes Gateway. You can either subscribe to a Notes Public Network (NPN) service provider that provides Usenet news directly in Notes or deploy and manage your own on site gateway. Your choice depends on your capabilities and resources.

WHO PROVIDES NOTES-TO-USENET GATEWAYS? Notes-to-Usenet Gateways allow Notes users to participate in Usenet newsgroups almost transparently without additional training or support. Like other newsreader programs, Notes users can subscribe to newsgroups, read articles, post articles, and send e-mail replies (provided your site supports Internet e-mail) directly from the Notes environment. Experienced Notes users appreciate the power of Notes to organize, filter, search, and redistribute Usenet news, which often suffers from too much irrelevant content.

The Notes-to-Usenet Gateways are typically Notes server add-in applications that manage the connection to UNIX News Servers and do the translation between the Notes and USENET data formats. Unless you already have a News Server running at your location, your Internet Service Provider can provide the news feed at a moderate additional cost from their servers.

LOTUS INTERNOTES NEWS Lotus is offering a gateway that runs on OS/2 and Windows NT-based Notes servers. It utilizes a central configuration database and supports multiple gateways, News Servers and News databases.

- Notes server supported: OS/2 or Windows NT
- Protocol support: NNTP
- News Servers tested with: INN and CNews
- Contact: internotes@iris.com (e-mail for beta program)
- Info site: http://www.notes.net
- Version: 1.0 for Notes Release 3.x using pull replication (Free beta version for Notes R4 using pull or push replication)

- Price: $2500 (for Version 1.0) per server, unlimited users; free for beta version

JSOFT NOTES-TO-USENET GATEWAY JSoft makes two versions of their gateway: a server version for OS/2 Notes servers that utilizes a server add-in application to directly retrieve Usenet articles using the NNTP protocol and a Windows-based client version that translates a UNIX news spoolfile into a Notes database for offline reading and posting. C source code licenses are available.

- Notes server supported: OS/2 2.1 or later
- Protocol support: NNTP
- News Servers tested with: INN and CNews
- Special requirements: None
- Contact: Joseph Jesson at jsoft@mcs.com or call 708-356-6817
- Info site: http://www.mcs.com/~jsoft/home.html
- Version: 1.7
- Prices: $1458 (Usenet NNTP-to-Notes OS Version) and $320 (Usenet UNIX Shell Account to Windows)

WHO PROVIDES NOTES-TO-USENET CONVERSION SERVICES?

Notes Public Networks (NPN) service providers will provide Notes-to-Usenet conversion services. NPN service providers host Lotus Notes applications among other services, available through telephone companies and public network service providers. Through these Lotus Notes applications, NPNs can provide Notes-to-Usenet conversion services. You would typically set up a replication schedule with their News Servers to transfer news databases using dial-up or leased-line Internet connections. This might be a more expensive solution for large users, but it avoids the management and capital expenses of running your own gateway.

Most NPN service providers became available in 1996. Here are some of those NPN service providers: BT (British Telecom), CompuServe, Deutsche Telekom, IBM's Global Network, Nippon Telegraph and Telephone Corporation, NTT Data Communications Systems Corp., SNET (Southern New England Telephone Company), Telecom Italia, Telekom Malaysia, Telstra, Unisource, and US West. Visit the following Web site for further information on NPN service providers:

```
http://www.lotus.com/info/npn.htm
```

Both CompuServe and Interliant (formerly WorldCom) have Notes-to-Usenet conversion services available now, based on Lotus Notes Release 3, but not as part of the NPN program. Read their individual descriptions for more information.

COMPUSERVE (FOR NOTES RELEASE 3)

- Cost: $18.00 per hour with no setup or monthly fees
- Number of newgroups: Unlimited
- Type of replication: Dial-up, X.25, ISDN, and Internet planned
- Posting allowed: Yes
- Contact: 800-440-9604 or 617-524-0220
- Info site: http://www.compuserve.com/

INTERLIANT (FOR NOTES RELEASE 3)

- Cost: $50.00 per month per company plus replication charges; free one-week trial
- Number of newsgroups: Unlimited
- Type of replication: Dial-up or Internet
- Posting allowed: Yes
- Contact: 713-650-6522 or info@worldcom.com
- Info site: http://www.worldcom.com or ftp.worldcom.com
- BBS: 713-659-7119

How Can I Download Files Using FTP?

You can access any information on a FTP server by using one of these methods:

- Lotus InterNotes Web Navigator
- E-mail auto responder
- E-mail server

LOTUS INTERNOTES WEB NAVIGATOR. You can access FTP files using the Lotus InterNotes Web Navigator. You simply enter the URL of the FTP file, such as:

```
ftp://www.notes.net/pub/faq/cbook4.zip
```

and the Web Navigator downloads the file and attaches it to a page in the database. The Web Navigator saves the page and its attachment in the database that you can open by detaching it. To locate the FTP files later:

- Open the File Archive view, and locate the file by name or by URL.
- Open the All Documents view, and click the Type column to sort by file type.

NOTE: *If you are required to provide a username and password when accessing an FTP site (non-anonymous FTP), the Web Navigator stores the files in a private folder named after your username.*

E-MAIL AUTO RESPONDER. You can access FTP files by sending an e-mail message to an auto-responder program designed to retrieve files from a site-specific database. These auto-responder programs are known as *listservs* or *infobots*. Some types of listservs are: the Internic's RFC server, info servers used by commercial organizations, document servers (such as infodroid@wired.com for back issues of *Wired* magazine), and mailing list archive servers (such as LNOTES-L).

Internet-Drafts and other IETF material are available by mail server from ds.internic.net. To retrieve a file, e-mail a request to mailserv@ ds.internic.net (or your Notes SMTP gateway equivalent) with a subject of anything you want. In the body, put one or more commands of the form:

```
FILE /ietf/tao.txt
PATH jdoe@somedomain.edu
```

where PATH lists the e-mail address where the response should be sent. This will return an e-mail message with the document titled "The Tao of IETF to jdoe@somedomain.edu."

E-MAIL SERVER. You can also access FTP files by sending an e-mail message to a mail server, such as the UNC or DEC mail servers, to request files from any site. To get the ftpmail help document for commands on using ftpmail, send e-mail to ftpmail@sunsite.unc.edu or ftp-mail@decwrl.dec.com (or your Notes SMTP gateway equivalent) with "help" in the body of the message.

To send a request for the ftpmail servers, you would send something similar to this in the body of the message:

```
open <site> <username> <password>
cd <directory>
dir              # To obtain a directory listing (optional)
get <file>       # To retrieve a file
quit
```

If you want to send a request for a document that contains a list of FTP sites, you would send an e-mail message with something similar to this in the body of the message:

```
open rtfm.mit.edu jdoe@somedomain.edu
cd /pub/usenet/news.answers/ftp-list
get faq
quit
```

where *jdoe@somedomain.edu* lists the e-mail address where the response should be sent. You would get a document sent by e-mail back to you containing the Anonymous FTP FAQ written by Perry Rovers (where much of the information in this section of the Internet Cookbook has been derived). Please note that, if you use the DEC and UNC servers, they might take a while to return files due to their increasing popularity.

How Can I Access Gopher Menus?

You can access any information on a Gopher server by using the Lotus InterNotes Web Navigator. You simply enter the URL of the Gopher site, such as:

```
gopher://riceinfo.rice.edu
```

and the Web Navigator displays the Gopher menu. You can navigate through the menu and access any of the information by double-clicking the menu items.

How Can I Access Internet Mailing Lists?

An Internet mailing list is a collection of Internet addresses of people who are interested in a particular topic. You subscribe to these mailing lists, and then whenever someone sends an e-mail message to the mailing list, that message is resent to all the addresses on the list. Internet mailing lists are similar to groups in your Notes Address Book that you use to send e-mail to more than one person at a time.

If there are several people at your company interested in subscribing to Internet mailing lists, you can create a mail-in Notes database to collect and store all the e-mail messages sent from the mailing list. This way, your company only gets one copy of the e-mail message instead of many. Those interested in reading the messages can access them from the Notes mail-in database. In order to consolidate some of these mail messages into one central database that everyone can access, you make sure the mail-in database address is on the mailing list as opposed to everyone's address at your company.

How do I set up a Mail-in Database? A mail-in database is a Notes database of any design (typically, a discussion or mail design) with a Notes e-mail address to allow the Notes mail router to send e-mail messages to it. An excellent way to make use of a mail-in database is to use it as a repository for e-mail for Internet mailing lists. Mail-in databases are widely used in routing/workflow applications.

To set up a mail-in database:

1. First, create the mail-in database by choosing File|New Database. You can use any design for the database, but be sure that the default form supports the type of information that it will contain. Use the Notes Mail template if you want the database to have a mail format. (For example, you would use a Notes Mail template to create the repository for Internet e-mail for Internet mailing lists.)

2. Open the Notes Address Book for your domain, and create a Mail-In Database document.

3. Under the Basics section, fill in the Mail-in Name field: Enter the name you want to use as the address of the database. This is the actual Internet name that people will use when they send e-mail to the database. For example, if you enter LNOTES as the Mail-in Name, make sure that the name LNOTES@corp.com is on the Internet mailing list for Lotus Notes Information. Also, others in the company can send e-mail to this database by sending messages to LNOTES.

4. In the Location section, fill in the Server and Domain fields: Enter the name and domain name of the Notes server where you just created the mail-in database.

5. In the File field, enter the filename of the mail-in database. Save this document.

After this Mail-In Database document replicates around your domain, users will be able to send e-mail to the address you specified, and it will show up in your mail-in database.

For a complete listing of all the Internet mailing lists, copy it from one of these locations:

- ftp://rtfm.mit.edu/pub/usenet/news.answers/mail/mailing-lists
- Usenet newsgroups: news.announce.newusers or news.answers

HOW DO I FILTER OUT IRRELEVANT MATERIAL FROM INCOMING MAILING LISTS? You can filter out unwanted material from an incoming mailing list by creating agents that run whenever new documents have been created or modified. You can also add queries to the agent to perform a full-text search on any message coming into your database. Agents come in handy if you want to categorize and delete messages based on specified criteria in the body of the e-mail message. For example, if you do not want any e-mail messages that contain FAQs, you can create an agent that searches all message Subjects for the word FAQ and then deletes those messages.

How Can I Use Notes to Publish Information on the Web?

You can use Notes as an authoring and document management system for a Web site. Lotus and several other vendors have products that automatically translate Notes databases into collections of HTML documents. These products allow standard Web browsers, such as NCSA Mosaic or Netscape, to read documents stored in a Notes database.

Who Provides Notes-to-HTML Converters?

This is a brief list of the companies who provide Notes-to-HTML converters. Contact the companies directly for availability and more information.

InterNotes Web Publisher, Lotus Development Corporation. This Notes server add-in automatically converts multiple databases into HTML. Views, graphics, forms, doclinks, tables, and attached files are preserved during translation. InterNotes Web Publisher provides a simple, automated process for creating and managing interactive Web sites.

- Contact: Lotus Development Corporation, Sales Information
- Info site: http://www.internotes.lotus.com/
- Platform: Windows NT, OS/2, AIX, Solaris, and Solaris X86 versions
- Price: $2995 for Notes 3.*x*; free with Notes 4

NetFusion, Interliant (WorldCom). Interliant's NetFusion services include Web conversion based on the Walter Shelby Group's TILE and Lotus' InterNotes Web Publisher. Users replicate Notes databases to Interliant, which will convert them to HTML and distribute them to the

Internet through the Interliant server. Interliant will also set up separate Web servers for companies needing a higher degree of customization for an additional charge.

- Contact: WorldCom, 713-650-6522 or info@interliant.com
- Info site: http://www.worldcom.com/
- Platform: Lotus Notes
- Price: $2000 setup, $200 per month plus storage charges beyond 25MB

TILE, Walter Shelby Group. TILE is a software program that converts Notes databases into Internet accessible Web documents, Gopher files, and FTP directories. Views, forms, fonts, graphics, and hypertext links are preserved.

- Contact: Walter Shelby Group, 301-718-7840 or info@shelby.com
- Info site: http://www.shelby.com/
- Platform: Windows 3.1, Windows 95, Windows NT, OS/2, Solaris, and Macintosh.
- Price: $2995 per server

WebGate, WebWare Limited. WebGate provides an interactive way to extract information from a Lotus Notes database into the Web. WebGate supports browsing and searching of views and documents and supports forms and lets you define what combinations of fonts, colors, and pitch convert to the various HTML tags.

- Contact: info@webware.com, telephone +353 87 416627
- Info site: ftp.worldcom.com:/lnotes-l/internet/webgt01B.zip
- Platforms: SunOS, AIX, and OS/2
- Price: Free for noncommercial use; price for other uses on request

Can I Create a Secure Environment with Notes?

Internet security can be a complex topic involving many vulnerabilities. We suggest that you educate yourself on the basics of network security if you are responsible (even partially) for your organization's data security. For suggestions on books to read, see the section "Where can I learn More about Notes and the Internet" later in this appendix.

Security is more than just a network configuration exercise or using correct database ACLs; it needs to be viewed as a whole. Notes use of encryption technology and replication offers one of the most secure computing environments available today. The thoughtful design and configuration of a Notes implementation can provide more than adequate levels of security for commercial purposes when utilizing the public Internet.

Generally, attacks on your data system from the public network come from either direct network intrusions or indirect system attacks. Direct attacks try to exploit the vulnerabilities of the transport protocol suite (TCP/IP and UDP) and the software programs that enable the connectivity it provides. This vulnerability is best guarded by a solid router/firewall configuration and vigilant administration. These attacks typically are aimed at UNIX hosts and have been thwarted somewhat by running security checks with programs, such as SATAN, that expose vulnerabilities.

Indirect system attacks, such as viruses, use the higher level connectivity of messaging, transaction, and publishing systems to bring "stealth" software into your computing environment. They then either replicate and cause operational problems (such as flooding a hard disk or mail server) or are destructive. These are best guarded by user education, system containment, and again vigilant administration. Indirect attacks can be made against any system and are difficult to detect and eliminate.

Notes offers some powerful tools to help secure data. It is also a very efficient system for distributing viruses. The good news is that Notes offers excellent encryption tools that basically prevent unauthorized parties from stealing or altering confidential data. Encryption is the only technically sustainable way to protect data from disclosure or alteration at this time. For more details on encryption, see the next section.

The bad news is that Notes supports rich text fields, binary attachment support for documents, and a wide range of auto-activation features. This allows one to create destructive document buttons, specify actions (formula or LotusScript) on the document open event, OLE activation, and even DocLink auto-launch on document open. While Lotus does allow the user to shut off auto-activation (using the NoExternalApps=1 NOTES.INI variable), this prevents most of the new Release 4 templates from working. Again, cryptography comes to the rescue. These attacks can only be thwarted by educating users to understand auto-activation and demanding that the sender (or contributor) digitally sign the document. This ensures that the document has not been altered by third parties and was, indeed, created by a certified ID before you launch it.

What Encryption Tools Can be Used to Secure Data in Notes?

Notes provides security features that are rigorous enough to protect business information across the Internet yet are flexible enough so that authorized users can control access to private information. Done in a manner that works across vendors' operating systems, Notes utilizes public key technology licensed from RSA Data Security, Inc., which is the industry leader in cryptographic algorithms and services. To secure your information, Notes provides security features in several areas.

AUTHENTICATION. Security in Notes starts with client-server authentication, the two-way challenge-response dialogue for assuring the identity of both the user and server. Passwords for ID files prevent unauthorized access where malicious users have physical access to user IDs. Ensure that your users choose longer passwords that mix alpha and numeric characters (and are therefore more resistant to dictionary attacks). Release 4 allows the use of anonymous servers (for public publishing applications).

CHANNEL ENCRYPTION. If you are using the Internet as a transport network, turn on channel encryption on all your Internet accessible servers. You can find this in the Port configuration dialog (File|Tools|User Preferences|Ports). Select the option to Encrypt Network Data. This will encrypt all the data packets that are sent between the server and all Notes clients that communicate with it (clients do not have to turn it on as long as the server is configured.)

DOCUMENT ENCRYPTION. As a further protection against inadequate administration or physically insecure Notes servers, consider using document encryption keys in your database design. This should also be used, where possible, for sending e-mail as well. (It requires that you have access to the recipient's public key in your Name and Address book.)

DIGITAL SIGNATURES. This allows the user to trust who the document originator actually was, as well as the integrity of the document. As mentioned before, this is the best defense against auto-activation attacks, assuming that you trust the sender of a mail message or contributor to a database.

ACCESS CONTROL. This provides the ability to grant or deny access to shared databases, documents, views, folders, forms, and fields. Server access can also be controlled for individual users by either allowing or denying access to specific Notes servers within the organization. For example, deny access lists can be set up for all servers containing a list of terminated employees and/or contractors, while specific access lists could be set up for high-level security servers such as a legal department.

These cryptographic features of Notes ensure that all of the other control mechanisms (server and database ACLs, user roles, document section controls, etc.) work. Because no system is 100% secure, we recommend that you implement an organized backup schedule, with frequent archiving, to ensure your organization's data integrity.

Can I Use Notes with My Existing Firewall Server?

Security requirements vary from company to company. There is no single solution to cover all of the possible considerations. However, Lotus highly recommends that you practice both application- and network-level security when attaching your Notes network to the Internet. This includes the use of a single Notes server to "proxy" for all servers and clients on the secured local network, as well as filtering control over Notes traffic accessing the Internet. Be sure to take extreme care to properly configure all of the components of your firewall solution and maintain and monitor them as you would any critical resource.

In general, there are three types of firewall implementations:

- Packet filtering
- Application proxy
- Circuit-level or generic-application proxy

Read through the following sections to find out which implementation is best for your Notes environment.

PACKET FILTERING.　Packet filtering protection is found in most firewall systems, including the network routers themselves. An administrator can filter Notes traffic through the specific IP address of the host or even one step further by our well-known port of 1352. This port is the registered TCP socket for Lotus Notes. By filtering on this value in your packet filter process, you can maintain all of the administrative and management features that are native to your particular firewall system to control Notes traffic at the network level.

When packet filtering is combined with Notes Passthru as a proxy, the administrator can have single points of access control for both the network and application security levels of Notes.

Obviously, each corporate networking environment has its own set of security policies and components, but here are some basic examples of possible implementations using the services described earlier. Typically there are routers segmenting secure and unsecure LAN segments, but it is not in the scope of this section to define solutions that can entertain the security requirements outside of Notes connectivity.

In the following scenario, Notes traffic is filtered at the packet level on the router and/or firewall. It is up to the network administrator to determine whether or not the filter should allow inbound and/or outbound Notes traffic to pass through the firewall (see Figure C.10). Keep in mind that the port number of the requesting Notes node will be dynamic but not the destination port.

APPLICATION PROXY. Application proxy services need to first understand the application in some detail in order to provide this level of service. The application proxy will act as an intermediate node communicating "on behalf" of the requester. Most of the firewall packages support simple applications such as FTP, telnet, etc.; however, they cannot participate at the application level to understand a Notes network conversation.

Additionally, if the application proxy could understand Notes, it would then need to be able to decrypt the line conversation between the two Notes nodes if encryption was enabled on the port on one of the communicating Notes nodes. To provide a robust proxy service for Notes, Lotus recommends the use of Passthru. This Passthru process maintains all levels of Notes security, while allowing nodes of dissimilar protocols to communicate via a single Notes Server and access point. It is available on all Notes platforms in R4. R3 servers can be used in a Passthru connection only when they are the destination node.

The scenario shown in Figure C.11 adds Notes Passthru to the packet filtering implementation. In this scenario, Passthru can add the benefits of a proxy service for multiple Notes nodes as well as connectivity to the

Figure C.10
Packet filtering.

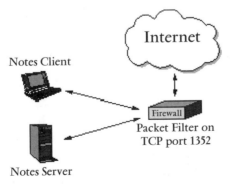

Figure C.11
Application proxy.

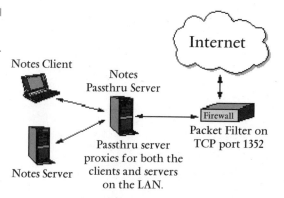

Internet for Notes clients and servers running protocols other than IP. By defining a single point of entry for Notes traffic to the packet filter, the administrator can centralize the administration of the application proxy— in this case, Passthru. Local traffic between the Notes clients and servers will only need the Passthru service if they have dissimilar protocols. It is not necessary to have multiple network interface cards in this scenario.

CIRCUIT-LEVEL OR GENERIC-APPLICATION PROXY. Circuit-level proxies (specifically Socks) work outside of the application layers of the protocol. These servers allow clients to pass through this centralized service and connect to whatever TCP port the clients specify. Socks servers also have the ability to authenticate the source address of connection requests and can block unauthorized clients from connecting out onto the Internet. Most TCP-based applications can become socksified by recompiling and linking them with a Socks client library. DLL-based vendor stacks have the additional benefit of being able to provide applications with Socks client capabilities without the need to recompile.

If passing through a Socks server is a requirement, Notes clients can be configured to do so by using TCP vendor stacks that support Socks transparently for all applications. An example of a TCP vendor that provides this capability is FTP Software's OnNet 2.0 product.

NOTE: At the time of this article, Lotus is working with IP stacks vendors to determine if there will be enough coverage of this support in the underlying stacks to provide a broad enough range of options for users to use Notes in Socks environments without having to move this support into the application. For more information on SOCKS, you might want to visit http://www.socks.nec.com.

The scenario shown in Figure C.12 depicts an outbound initiation only from the Notes client/server through a Socks server and then through a Packet Filter onto the Internet.

Figure C.12

Generic-application proxy with outbound initiation through Socks and Packet Filter.

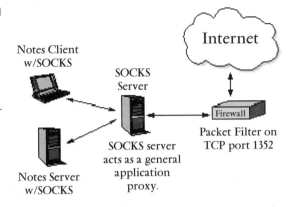

Figure C.13

Generic-application proxy with outbound initiation through Passthru, Socks, and Packet Filter.

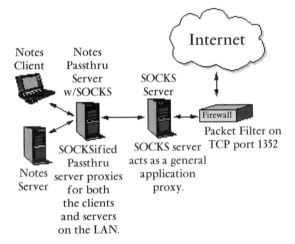

The scenario shown in Figure C.13 depicts an outbound initiation only from the Notes client/server through a Notes Passthru server (socksified), through a Socks server, and finally through a Packet Filter onto the Internet.

By leveraging Socks support that might be available in the underlying TCP/IP stack on the client or server, Notes can utilize centralized Socks services that might exist on the corporate network. It is important to remember that the IP stack must provide the SOCKS support in this configuration. The network administrator might also choose to use Packet Filters for the Socks ports defined. Communications between the local clients and servers can be configured to bypass the Socks server so that all local traffic does need to pass through the Socks server.

Important: Keep the following in mind when configuring both outbound and inbound connections through your proxy.

- Outbound connections:
  - Allow the Socks host access regardless of source socket. That is, filter by source address of the Socks server and allow any socket.
  - The destination address filter (on the Packet Filter) could be restricted to the host address, but require the 1352 socket.
  - For Notes, do not allow packets for any host other than the Socks Server when using a destination address containing the port 1352.

- Inbound connections: Because Socks is not used for connections originating from the Internet and whose source address is on the local net, inbound connections could be allowed to go straight to the target server, or perhaps a Notes Passthru server if configured depending on the customers requirements. Filters would probably look like the following:
  - If using a Notes Proxy (Passthru Server), then only allow packets with a destination address of the Notes Proxy Server on port 1352.
  - If not using a Notes Proxy Passthru Server, then allow the destination address to be any host as long as the port is 1352.
  - Optionally only allow inbound connections from trusted hosts using the 1352 port.

How Do I Use Notes as an Application-Level Firewall Server?

The section "Can I use Notes with My Existing Firewall Server?" described how you can incorporate the use of Notes with different firewall components. By combining Notes and firewall servers, you achieve both network-level and application-level forms of security for Notes. However, if you do not have access to firewall components, consider the following two scenarios as possible solutions to provide secure access to the Internet for your Notes clients and servers.

USING DUAL NETWORK ADAPTERS ON YOUR NOTES SERVER. The first scenario shows how to configure the Notes server to prevent unauthorized traffic from entering or leaving your LAN. This design involves putting two network adapters on the Notes server: one connected to the Internet (and configured for TCP/IP) and the other to the organization LAN (configured for your local protocol, such as

NetBIOS or SPX). See Figure C.14. TCP/IP could also be used for on the LAN side; however, you must turn off TCP/IP routing between adapters, remove any TCP utilities (such as FTP, Sendmail, NFS, and so on), and turn on all auditing and alarm utilities to track intrusion.

In addition, you should configure the Notes server with tight access controls for both the server access controls and the individual access controls. In Release 3 environments, this will only work to replicate data. However, in Release 4, the Passthru server can provide real-time access to destinations over the Internet if the administrator chooses to allow that.

This solution can be augmented by putting a router between the server and your corporate LAN that only allows traffic to the Notes server. While this first approach is used by many people, we can only recommend it if you use this type of a router and keep vigilant watch on the system.

USING DUAL MODEMS BETWEEN NOTES SERVERS. The second scenario shows how you can place the Notes server on an isolated LAN with Internet connectivity (Figure C.15). This Notes server then connects with the organization's Notes servers using a modem (or a null modem cable) to replicate Notes databases. The modem connection uses the X.PC protocol as a transport, is intermittent, and uses Notes authentication (and encryption, if enabled) to ensure security. This approach is simple, highly effective, and easy to implement, but it requires extra hardware. We recommend this approach if you have the required hardware and can tolerate the slower speeds of the X.PC connection.

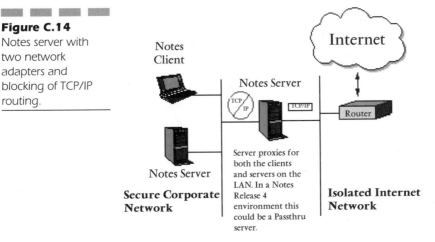

Figure C.14
Notes server with two network adapters and blocking of TCP/IP routing.

Figure C.15
Notes servers
connected with
modems to ensure
security.

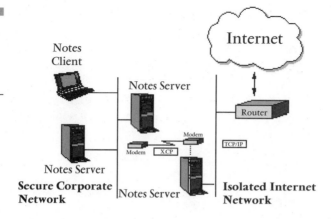

Where Can I Learn More About Notes and the Internet?

We've compiled a list of our favorite resources that cover both Notes and the Internet. Read on to learn where you can find this information.

Where Can I Find Out More About Notes?

There are many ways to find out more about Notes and participate in discussions with other Notes users and administrators. Read the following sections to find out how.

NOTES SITES. Visit the NotesNIC site for information on Notes: welcome.nsf at home/notes/net (205.159.212.10).

BOOKS FROM IBM AND LOTUS. Lotus Notes Release 4 documentation set includes:

■ Lotus Notes Administrator's Guide

■ Lotus Notes Network Configuration Guide

■ IBM TCP/IP for OS/2 Administrator's Guide

WORLD WIDE WEB SITES. Visit these Web sites for more information on Notes:

■ http://www.lotus.com

■ http://www.notes.net

- http://www.turnpike.net/metro/kyee/NotesFAQ.html
- http:///www.tile.net/smartnotes/
- http://www.cs.tu-berlin.de/~bene/lotus/Lotus_Mailingslist/index.html
- http://www-iwi.unisg.ch/delta/notes.html
- http://www.st.rim.or.jp/~snash/Notes/Notes.html
- ftp://ftp.worldcom.com/pub/lnotes-l/internet/ln.html
- http://www.application-partners.com/

USENET NEWSGROUPS. Read these Usenet newsgroups for discussions and information on Notes:

- comp.groupware
- comp.groupware.lotus-notes.misc
- comp.groupware.lotus-notes.admin
- comp.groupware.lotus-notes.apps
- comp.groupware.lotus-notes.programmer

INTERNET MAILING LISTS. Another way to keep current with the latest discussions on Notes and other topics is to subscribe to Internet mailing lists.

LNOTES-L MAILING LIST LNOTES-L is an Internet mailing list that contains postings by the general Notes user community. It is a mailing list created for the purpose of exchanging information among Notes users. LNOTES-L has over 2000 members, which puts you in contact with many other Notes users and provides you with access to many different Notes discussions. The Web site for the LNOTES-L mailing list is:

```
http://www.disaster.com/lnotesl.html
```

To send an e-mail message to the LNOTES-L mailing list, address your message to lnotes-l@medicine.wustl.edu. (Don't send your subscription request to this address.)

To subscribe to LNOTES-L, send a message to lnotes-l-request@medicine.wustl.edu. In the body of the message, enter SUBSCRIBE LNOTES-L address. *Replace* address with the e-mail address where the messages should be sent. You will be automatically subscribed to the mailing list.

To subscribe an address other than the one you are sending from, send a message to labatt@disaster.com. In the body of the letter, enter

SUBSCRIBE LNOTES-L address. **Replace** *address* with the address to send messages to. Approval generally occurs within a week.

LOTUS NOTES MAILING LIST ARCHIVE   LNOTES-L archives dating back to February 9, 1995 are available at the list site. Archives can be requested using the following format. To search the archives, send a message to listserv@medicine.wustl.edu with a body of:

```
SEARCH LNOTES-L search key
```

The request is case insensitive. For example, to search for messages with the word Solaris in them, send a message to the previous address with the following body:

```
SEARCH LNOTES-L solaris
```

You will receive a message back that contains the names of the files that contain the word "solaris" (or "Solaris" or "SOLaris" or whatever).
Then send a message to listserv@medicine.wustl.edu with a body of:

```
GET LNOTES-L filename
```

You will receive a message back that contains the messages archived in the filename. Some files might be split into multiple sections if the file grows over 20K.
For more information on the LNOTES-L archive, send e-mail to labatt@disaster.com or LNOTES-L-MGR@MEDICINE.WUSTL.EDU.

OTHER MAILING LISTS   There are a large number of mailing lists on the Internet. To help you locate a particular mailing list, there are many different Web sites that contain lists of mailing lists and search capabilities to help locate specific mailing lists. Some of these Web sites are:

- http://catalog.com/vivian/interest-group-search.html
- http://tile.net/listserv/
- http://www.neosoft.com/internet/paml/

Where Can I Find Out More About the Internet?

There are many resources for learning more about the Internet. Read these sections for some suggested resources.

PRINTED BOOKS. The following printed books are available from bookstores:

- *Network Security: Private Communication in a Public World* by Charlie Kaufman, Radia Perlman, and Mike Speciner
- *Firewalls and Internet Security: Repelling the Wily Hacker* by William R. Cheswick and Steven M. Bellovin
- *The Whole Internet User's Guide and Catalog* by Ed Krol
- *Managing Internet Information Services* by Cricket Liu, Jerry Peek, Russ Jones, Bryan Buus, and Adrian Nye

WORLD WIDE WEB SITES. Visit these Web sites for Internet-related information on the Web:

- http://www.w3.org/
- http://www.internic.net
- http://wwwcn.cern.ch/pdp/ns/ben/TCPHIST.html

Visit these Web sites for security-related information on the Web:

- http://www.ncsa.com/ncsafws.html
- http://www.v-one.com/pubs/fw-faq/faq.htm#head_whatis
- http://www.waterw.com/~manowar/vendor.html

BOOKS FROM VARIOUS INTERNET SITES

- *Zen and the Art of the Internet: A Beginner's Guide to the Internet* by Brendan Kehoe (http://www.cs.indiana.edu/docproject/zen-1.0_toc.html)
- *EFF's Guide to the Internet* by The Electronic Frontier Foundation (ftp://ftp.eff.org/pub/Publications/EFF_Net_Guide/bigdummy.txt and http://www.eff.org/pub/Net_info/EFF_Net_Guide/)

RFCS. Request for Comments (RFCs) are technical documents that define the different parts of the Internet. You can find RFCs on almost any Internet-related topic.

The following RFCs are useful starting points:

- Internet RFC 1118: *The Hitchhiker's Guide to the Internet*
- Internet RFC 1462: *What is the Internet?*
- Internet RFC 1630: *Universal Resource Identifiers in WWW*
- Internet RFC 1359: *Connecting to the Internet: What Connecting Institutions Should Anticipate*

- Internet RFC 1034: *Domain Names: Concepts and Facilities*
- Internet RFC 959: *FTP: File Transfer Protocol*
- Internet RFC 977: *NNTP: Network News Transfer Protocol*
- Internet RFC 1661: *PPP: Point to Point Protocol*
- Internet RFC 1055: *SLIP: Serial Line IP*
- Internet RFC 821: *SMTP: Simple Mail Transfer Protocol*

You can obtain RFCs from many different Internet sites, including:

- http://pubweb.nexor.co.uk/public/rfc/index/rfc.html
- http://www.cis.ohio-state.edu/hypertext/information/rfc.html
- ftp://nic.ddn.mil/rfc/
- ftp://ds.internic.net/rfc/

Tip: Some FTP sites require a username and password to access their site. Usually, you can log in as anonymous and use your e-mail address as the password.

APPENDIX D

MIGRATION BUTTON CODE

This appendix lists the Lotus Notes and Pascal code described in chapter 13 on migration from a single-level hierarchy for the ADVAN-TIS domain (/ADVANTIS) to an enterprise X.500 hierarchy (/Location/Advantis/US). This code is also provided in soft copy format on this book's CD-ROM.

Button (Conversion Utility)

```
REM "The following temp variables create the fully distinguished
name for the user's current mail server and for the user's new mail
server.";
REM;
CANONCURRMAILSERVER := @Name([Canonicalize]; CURRENTMAILSERVER +
"/ADVANTIS");
CANONNEWMAILSERVER := @Name([Canonicalize]; NEWMAILSERVER +
"/Schaumburg/Advantis/US");
REM;
REM "The following temp variables capture the user's mail file name
and the user's name.  These will be used to locate their current
mail file location and identify the document in the address book
that contains their ID.";
REM;
REM;
ADDRFILENAME    := "NAMES.NSF";
ADDRVIEW        := "($CrossCertByName)";
IDFILENAME      := CANONCURRMAILSERVER : "NEWUSERS.NSF";
IDCERTVIEW      := "Cross Certificates";
IDPEOPLEVIEW    := "People";
INIPROGRAMVIEW  := "Notes INI Executables";
MAILFILENAME    := @UpperCase(@Name([Abbreviate];
@Subset(@MailDbName; -1))) + ".NSF";
USERNAME        := @UpperCase(@Name([CN]; @UserName));
EXEDETACH       := @If(@Contains(@Platform; "OS/2"); "NOTESDOS.EXE";
"NOTESINI.EXE");
EXEDETACHPATH   := "C:\" + EXEDETACH;
EXEPROGRAM      := @If(@Contains(@Platform; "OS/2"); "CMD.EXE";
"C:\NOTESINI.EXE");
EXEPARMS        := @If(@Contains(@Platform; "OS/2"); "/C START /C
/WIN /MIN C:\NOTESDOS.EXE"; "");
REM;
REM "The following temp variables check to ensure that the
filenames are correct and do exist.  If their is an error it will
be trapped.
     ERRORA - Traps personal name and address book error";
REM;
REM;
ERRORA   := @If(@IsError(@DbColumn("" : "NoCache"; "" :
ADDRFILENAME; ADDRVIEW; 1)); "ERR"; "");
LOOKPHONE := @DbLookup("" : "NoCache"; IDFILENAME; IDPEOPLEVIEW;
```

```
USERNAME; "TieLinePhoneNumber");
ERRORB    := @If(@IsError(LOOKPHONE); "ERR"; LOOKPHONE);
REM;
REM;
PHONE         := @If(ERRORB = "ERR"; ""; "(" + ERRORB + ")");
ERRORAMESSAGE := PHONE + " received ERROR A:  It appears that the
Personal Name and Address Book doesnhave the right file name or the
right views.  Contact this person ASAP.";
ERRORBMESSAGE := PHONE + " received ERROR B:  It appears that they
don't have proper access to the 'Conversion Process' database on "
+
@Name([CN]; IDFILENAME) + " or that the person's name differs from
the new ID.  Contact this person ASAP.";
REM;
REM;
@SetEnvironment("CURRENTMAILSERVER"; CURRENTMAILSERVER);
@SetEnvironment("MAILFILENAME"; MAILFILENAME);
@SetEnvironment("CANONNEWMAILSERVER"; CANONNEWMAILSERVER);
@SetEnvironment("CANONCURRMAILSERVER"; CANONCURRMAILSERVER);
@SetEnvironment("ADDRFILENAME"; ADDRFILENAME);
@SetEnvironment("PHONE"; PHONE);
@SetEnvironment("INIPROGRAMOK")= "YES";
   @Do(
      @Prompt([OK]; "Program Completed"; "The Conversion Utility
has already been run successfully.  This document will be
closed.");
      @MailSend("Notes Announcements"; ""; ""; "**User tried to re-
run button macro for conversion.");
      @Command([FileCloseWindow])
   );
@If(ERRORA = "ERR";
   @Do(
      @Prompt([OK]; "An Error has Occurred-ERROR A"; "There has
been an error with the process.  A pager message has been sent to
support.  You will be contacted for assistance.  Click on OK to end
this process.");
      @MailSend("Support Pager"; ""; ""; ERRORAMESSAGE)
   );
@If(ERRORB = "ERR";
   @Do(
      @Prompt([OK]; "An Error has Occurred-ERROR B"; "There has
been an error with the process.  A pager message has been sent to
support.  You will be contacted for assistance.  Click on OK to end
this process.");
      @MailSend("Support Pager"; ""; ""; ERRORBMESSAGE)
   );
@Do(
  @MailSend("Notes Announcements"; ""; ""; @Name([CN]; @UserName) +
"\" + "1. Update Information has been STARTED");
   @Prompt([OK]; "Update Information"; "You need to be present
during this process to respond to process prompts.  This process
may take up to (10) minutes of your time." + @Char(13) + @Char(13)
+ "Please DO NOT press CTRL-C or CTRL-BREAK during this process.");
@Command([FileCloseWindow]);
@Command([FileCloseWindow]);
@Command([WindowMinimizeAll]);
@Command([FileOpenDatabase]; IDFILENAME; IDCERTVIEW);
@Command([EditSelectAll]);
@Command([EditCopy]);
@Command([FileCloseWindow]);
@Command([FileOpenDatabase]; ADDRFILENAME);
@Command([EditPaste]);
@Command([FileCloseWindow]);
```

```
@Command([FileOpenDatabase]; IDFILENAMe; IDPEOPLEVIEW; USERNAME);
@Command([Editdocument]; "0");
@Command([EditDetach]; "USERID"; "C:\NOTES.ID");
@Command([FileCloseWindow]);
@Command([OpenView]; INIPROGRAMVIEW; CURRENTMAILSERVER; "");
@Command([EditDocument]; "0");
@Command([EditDetach]; EXEDETACH; EXEDETACHPATH);
@Command([FileCloseWindow]);
@Command([ToolsRunMacro]; "PROMPT FOR UNREAD MARKS");
@Command([Execute]; EXEPROGRAM; EXEPARMS);
@Command([ToolsRunMacro]; "CHECK TO VERIFY INI FILE UPDATED")
)
)
)
);
```

Macro (Prompt for Unread Marks)

```
CURRENTMAILSERVER    := @Environment("CURRENTMAILSERVER");
MAILFILENAME         := @Environment("MAILFILENAME");
CANONNEWMAILSERVER   := @Environment("CANONNEWMAILSERVER");
CANONCURRMAILSERVER  := @Environment("CANONCURRMAILSERVER");
ADDRFILENAME         := @Environment("ADDRFILENAME");
REM;
REM;
@Prompt([OK; "Unread Mark Step"; "You may see more messages during
the following step.  If so, click on YES or OK to continue the
conversion process." + @Char(13) + @Char(13) + "This step may take
up to (6) minutes.");
@Command([RenameDatabase]; CURRENTMAILSERVER : MAILFILENAME;
CANONNEWMAILSERVER);
@Command([RenameDatabase]; CANONCURRMAILSERVER : MAILFILENAME;
CANONNEWMAILSERVER);
@Command([RenameDatabase]; CURRENTMAILSERVER : ADDRFILENAME;
CANONNEWMAILSERVER);
@Command([RenameDatabase]; CANONCURRMAILSERVER : ADDRFILENAME;
CANONNEWMAILSERVER);
SELECT @All
```

Macro (Check to Verify INI File Updated)

```
ERRORCMESSAGE := @Environment("PHONE") + " received ERROR C:  It
appears that the INI program didnrun properly.  Contact this person
ASAP.";
@SetEnvironment("TEMP"; "TEMP");
@If(@Environment("INIPROGRAMOK") != "YES";

  @Do(
    @Prompt([OK]; "An Error has Occurred-ERROR C"; "There has
been an error with the process.  A pager message has been sent to
support.  You will be contacted for assistance.  Click on OK to end
this process.");
    @MailSend("Support Pager"; ""; ""; ERRORCMESSAGE)
  );
```

```
@Do(
  @Prompt([OK]; "Your Password"; "Your new Lotus Notes ID password
is your PROFS ID in UPPERCASE.");
    @MailSend("Notes Announcements"; ""; ""; @Name([CN]; @UserName)
+ "\" + "2. Update Information has been COMPLETED"; ""; "");
    @Prompt([OK]; "Process Complete"; "The process is completed.
Thanks for your help.");
    @Prompt([OK]; "Lotus Notes Shutdown"; "Lotus Notes will now be
shutdown.");
    @Command([WindowMaximizeAll]);
    @Command([FileExit])
)
);
SELECT @All
```

■ ■ NOTESDOS.EXE

```
uses Dos;
var
  S: PathStr;
  erasef:file;
  oldf:file;
  newf:text;
  findf:text;
  count:integer;
  P: PathStr;
  D: DirStr;
  N: NameStr;
  E: ExtStr;
  i:integer;
  temp:string;
  dir:string;
  FromF, ToF: file;
  NumRead, NumWritten: Word;
  buf: array[1..2048] of Char;
  j:string;
begin
  S := FSearch('NOTES.INI',GetEnv('PATH'));
  if S = " then
    begin
      WriteLn('NOTES.INI not found.');
      writeln('Press Any Key to Continue.');
      readln(j);
      halt;
    end;
  P := FExpand(S);
  FSplit(P, D, N, E);
  Assign(oldf,P);
  {$I-}
  Rename(oldf,D + 'NOTES.OLD');
  {$I+}
  if IOResult <> 0 then
    begin
      WriteLn('Cannot rename NOTES.INI.');
      writeln('Press Any Key to Continue.');
      readln(j);
      halt;
    end;
  count := 0;
```

```
Assign(newf,D + 'NOTES.INI');
{$I-}
Rewrite(newf);
{$I+}
if IOResult <> 0 then
  begin
    WriteLn('Cannot write new NOTES.INI.');
    writeln('Press Any Key to Continue.');
    readln(j);
    halt;
  end;
Assign(findf,D + 'NOTES.OLD');
{$I-}
Reset(findf);
{$I+}
if IOResult <> 0 then
  begin
    WriteLn('Cannot open NOTES.OLD.');
    writeln('Press Any Key to Continue.');
    readln(j);
    halt;
  end;
dir := 'empty';
while not Eof(findf) do
  begin
    count := count + 1;
    readln(findf,temp);
    if POS ('Directory',temp) > 0 then
      if (count < 6) then
        dir := COPY(temp, 11,Length(temp) - 10);
    if (POS ('KeyFilename',temp) = 0) AND (POS('MailServer',temp)
= 0) then
        writeln(newf,temp);
  end;
if (dir = 'empty') then
  begin
    writeln('Cannot find data directory.');
    writeln('Press Any Key to Continue.');
    readln(j);
    halt;
  end;
writeln(newf,'KeyFilename=' + dir + '.id');

writeln(newf,'MailServer=CN=SCHNAV05/OU=Schaumburg/O=Advantis/C=US'
);
  Close(findf);
  Assign(FromF, 'C:.id');
  {$I-}
  Reset(FromF, 1);
  {$I+}
  If IOResult <> 0 then
    begin
      writeln('Cannot find C:.id.');
      writeln('Press Any Key to Continue.');
      readln(j);
      halt;
    end;
  Assign(ToF, dir + '.id');
  {$I-}
  Rewrite(ToF, 1);
  {$I+}
  If IOResult <> 0 then
    begin
```

```
      writeln('Cannot write new notes.id.');
      writeln('Press Any Key to Continue.');
      readln(j);
      halt;
    end;
  repeat
    BlockRead(FromF,buf,SizeOf(buf),NumRead);
    BlockWrite(ToF,buf,NumRead,NumWritten);
  until (NumRead = 0) or (NumWritten <> NumRead);
  Close(FromF);
  Close(ToF);
  writeln(newf,'$INIPROGRAMOK=YES');
  Close(newf);
  Assign(erasef, 'C:.EXE');
  {$I-}
  Reset(erasef);
  {$I+}
  if IOResult <> 0 then
    WriteLn('Cannot find NOTESDOS.EXE.')
  else
    begin
      {$I-}
      Erase(erasef);
      if IOResult <> 0 then
        writeln(IOResult)
      else
        Close(erasef);
    end;
end.
```

NOTESINI.EXE

```
uses WinDos,WinCrt,WinTypes,WinProcs,Strings;
var
  S: array [0..fsPathName] of Char;
  tempp,tempd,tempn,tempe:string;
  i:integer;
  erasef:file;
  oldf:file;
  newf:text;
  findf:text;
  count:integer;
  P: array [0 .. fsPathName] of Char;
  D: array [0 .. fsDirectory] of Char;
  N: array [0 .. fsFileName] of Char;
  E: array [0 .. fsExtension] of Char;
  temp:string;
  dir:string;
  FromF, ToF: file;
  NumRead, NumWritten: Word;
  buf: array[1..2048] of Char;
  j:string;
begin
  FileSearch(S,'NOTES.INI',GetEnvVar('PATH'));
  if S[0] = #0 then
    begin
      WriteLn('NOTES.INI not found.');
      writeln('Press Any Key to Continue.');
      readln(j);
```

```
          PostQuitMessage(0);
          halt;
       end;
    FileExpand(P,S);
    FileSplit(P, D, N, E);
    i:=0;
    while (p[i] <> #0) do
      begin
         if (p[i] <> #0) then tempp := tempp + p[i];
         if (d[i] <> #0) then tempd := tempd + d[i];
         if (n[i] <> #0) then tempn := tempn + n[i];
         if (e[i] <> #0) then tempe := tempe + e[i];
         i := i+1;
      end;
    Assign(oldf,tempp);
    {$I-}
    Rename(oldf,tempd + 'NOTES.OLD');
    {$I+}
    if IOResult <> 0 then
      begin
         WriteLn('Cannot rename NOTES.INI.');
         writeln('Press Any Key to Continue.');
         readln(j);
         PostQuitMessage(0);
         halt;
      end;
    count := 0;
    Assign(newf,tempd + 'NOTES.INI');
    {$I-}
    Rewrite(newf);
    {$I+}
    if IOResult <> 0 then
      begin
         WriteLn('Cannot write new NOTES.INI.');
         writeln('Press Any Key to Continue.');
         readln(j);
         PostQuitMessage(0);
         halt;
      end;
    Assign(findf,tempd + 'NOTES.OLD');
    {$I-}
    Reset(findf);
    {$I+}
    if IOResult <> 0 then
      begin
         WriteLn('Cannot open NOTES.OLD.');
         writeln('Press Any Key to Continue.');
         readln(j);
         PostQuitMessage(0);
         halt;
      end;
    dir := 'empty';
    while not Eof(findf) do
      begin
         count := count + 1;
         readln(findf,temp);
         if POS ('Directory',temp) > 0 then
           if (count < 6) then
             dir := COPY(temp, 11,Length(temp) - 10);
         if (POS ('KeyFilename',temp) = 0) AND (POS('MailServer',temp)
= 0) then
           writeln(newf,temp);
      end;
```

```
    if (dir = 'empty') then
      begin
        writeln('Cannot find data directory.');
        writeln('Press Any Key to Continue.');
        readln(j);
        PostQuitMessage(0);
        halt;
      end;
    writeln(newf,'KeyFilename=' + dir + '.id');

writeln(newf,'MailServer=CN=SCHNAV05/OU=Schaumburg/O=Advantis/C=US'
);
    Close(findf);
    Assign(FromF, 'C:.id');
    {$I-}
    Reset(FromF, 1);
    {$I+}
    If IOResult <> 0 then
      begin
        writeln('Cannot find C:.id.');
        writeln('Press Any Key to Continue.');
        readln(j);
        PostQuitMessage(0);
        halt;
      end;
    Assign(ToF, dir + '.id');
    {$I-}
    Rewrite(ToF, 1);
    {$I+}
    If IOResult <> 0 then
      begin
        writeln('Cannot write new notes.id.');
        writeln('Press Any Key to Continue.');
        readln(j);
        PostQuitMessage(0);
        halt;
      end;
    repeat
      BlockRead(FromF,buf,SizeOf(buf),NumRead);
      BlockWrite(ToF,buf,NumRead,NumWritten);
    until (NumRead = 0) or (NumWritten <> NumRead);
    Close(FromF);
    Close(ToF);
    writeln(newf,'$INIPROGRAMOK=YES');
    Close(newf);
    Assign(erasef, 'C:.EXE');
    {$I-}
    Reset(erasef);
    {$I+}
    if IOResult <> 0 then
      WriteLn('Cannot find NOTESINI.EXE.')
    else
      begin
        {$I-}
        Erase(erasef);
        if IOResult <> 0 then
          writeln(IOResult)
        else
          Close(erasef);
      end;
PostQuitMessage(0);
end.
```

LOTUS NOTES VS. MICROSOFT EXCHANGE Q&A

This appendix gives several comparisons of Exchange and Notes in the form of a question-and-answer (or comment-and-response) discussion. The information was originally a series of challenges from Exchange to which Lotus has responded. It is given here because most of the "comments" given are questions about the design of Lotus Notes that have been around from the beginning of the Lotus Notes "phenomenon."

Security System

EXCHANGE COMMENT. Exchange utilizes the security services of the NT operating system to provide a single point of security management. Even if an Exchange administrator leaves the company, the next administrator can remove the previous administrator's account and disable access to the system entirely and easily.

Notes is reliant on a dual-key security mechanism that places the "user ID" in a file that exists on the user's workstation. This file is protected by client-side password checking only and cannot be recovered if someone leaves the company or the ID is compromised. The immediate issue of concern is Notes' complete lack of Key Management.

In Exchange, if someone leaves an organization, all you need to do is delete the user from the NT security directory, which will remove their rights to access the network and messaging system at the same time.

NOTES RESPONSE. Exchange is dependent on NT for its security model. This, and other dependencies, inextricably tie Exchange to NT. At first glance, this model appears to leverage existing technology and streamline the administration process. In reality, this dependence excludes the use of any other operating system in an Exchange messaging infrastructure and brings customers back to the days of proprietary systems (a place they have spent the last decade getting out of). This is of particular concern when companies seek to share data with their customers, suppliers, and business partners. It is difficult enough for one company to standardize on a single vendor's back end, but nearly impossible for that company to force other enterprises to do the same.

Actually, reliance on the NT security model is quite limiting. NT security is user ID/password based. These types of systems have proven to be vulnerable to computer hackers. If a hacker can get access to a NT server—through the network, over a phone line, or even by walking into the owner's office—he/she can start guessing user ID/password combinations to gain access to the Exchange system. Such attacks, which are increasingly common these days, represent a real threat to systems based on this overly simple security model.

Notes, on the other hand, secures its data through sophisticated technologies based upon public key cryptography and digital certificates. In public key cryptography, there are two keys, and data encrypted with one key can only be decrypted with the other. No "passwords" ever go across a wire, only encrypted data. To gain access to a Notes server, one must possess a valid certificate (based upon the X.509 standard), which is authenticated through a set of carefully controlled steps.

Another example of NT security limitations is the fact that NT does not provide message encryption and digital signature support. In order to provide this "additional security," Exchange introduces a separate security mechanism provided by a "Key Management Server." While Microsoft faults Lotus for the use of private keys that require key management, KM Servers also use private and public keys. Unlike Notes, where dual key encryption is integral to the system, this encryption capability is an "added-on" feature.

This means that, with Exchange, administrators will be forced to manage two entirely separate security infrastructures. Moreover, when fully implemented, the set of features offered by these two radically different security architectures is in sum less than that offered by Notes. Notes' pervasive use of public key cryptography, implemented with a thorough understanding of the database architecture, allows, for example, individual fields or sets of fields within a single document to be encrypted or to be digitally signed. Neither of these features is available in Microsoft Exchange. An expense form or capital expenditure request requiring multiple approvers—who attach their digital signatures to approve—is an excellent example of an application that is limited in an Exchange environment due to the lack of this technology.

Even worse, Microsoft has consistently maintained that their next-generation operating system, "Cairo," uses an entirely different security architecture, called Kerberos. Kerberos, the result of an academic research project, is still based upon the user ID/password model and, while it goes to great lengths to make up for the weakness' inherent in such a model, is nonetheless vulnerable. However, certainly a greater concern for the administrator, regardless of its technical merits, is the knowledge that whatever Exchange does will be supplanted in a few years' time by something completely different that will result in a substantial disruption.

Another weakness of the current NT security model is that it is built around a managed security domain. The management of that security domain by an administrator can be very complicated in an inter-enterprise environment where interlocking security domains might be present.

The assumption is also being made in the Windows NT/Exchange model that security is tied to connected access to an enterprise. This means that disconnected or minimally connected users (occasional replication) are not addressed at all.

In Notes, the process of removing a user from the system is separate from the NOS administration by design. Because Notes infrastructures can be built on any number of operating systems, user administration should not be tied to any one of them. This does not mean that Notes administration must be difficult. In Notes, the process of removing a user from the system can be done by adding them to a company's "terminations" group.

User "Terminated List"

EXCHANGE COMMENT. Lotus claims that, by adding the terminated user's name to a "Barred" users list, they will be locked out of the system. In a large environment, this adds significantly to the administrative overhead. Also, checking this list adds to the time it takes to access the server. Server access is controlled by the DenyAccess= statement in an .INI file that, if the administrator forgets to set it, will allow terminated users access to the system.

NOTES RESPONSE. Contrary to this statement, the use of "terminations" groups does not add significantly to administrative overhead. Adding a user's name to an excluded group can automatically remove them from the system, no matter how distributed it is. Replication of this group will propagate the change throughout the organization automatically. In addition, if an administrator would like to streamline this process, Notes is a programmable environment. It is quite easy to programatically make this change at the same time an administrator removes a user from their NOS.

In Notes R4, this process is even further streamlined to ease user administration.

Regarding server access, this control is no longer implemented in the NOTES.INI file. This information has been stored in the N&A book since R3 of the product.

Notes Certifier ID

EXCHANGE COMMENT. If the Certifier ID in a Notes installation is compromised, anyone with it can create duplicate user IDs of valid users, which will allow them to impersonate these people on the network. They'll be able to access their files, even send messages bearing their names! The only way around this is recertification, which is a huge undertaking. This step would be required anytime an administrator leaves the organization!

NOTES RESPONSE. All security systems, from Notes to RACF (and including NT), place some responsibility on the system administrator. In Notes, this responsibility lies in the ownership of the certifier ID(s). Recognizing that organizations should be able to limit this responsibility, Lotus provides the ability to use hierarchical certificates to separate a Notes infrastructure into smaller organizational "units" that can be managed by different administrators. This significantly limits the impact that any one person can have on the system.

In Notes R4, work is being done to eliminate the single point of failure vulnerability cited previously.

Notes recertification, which is the most effective way to exclude an ex-administrator from the network, is not the burden that Microsoft would have customers believe. Recertification procedures are naturally integrated with Notes mail so that recertification can be done without physically "visiting" users. In Notes R4, we will streamline this process even further.

Physical Security

EXCHANGE COMMENT. Notes security is overly reliant on the physical security of Notes servers. Access to a server machine allows anyone to compromise Notes security. This makes it very difficult to deploy Notes in workgroup settings and in remote sites. In addition, if you use a Notes server also as a file server, anyone who maps a drive to the Notes data directory can bypass Notes security. This is an important issue now that customers are deploying Notes on NetWare servers.

NOTES RESPONSE. Notes servers, like NT Exchange servers, can be protected from physical access by "lockout" features provided by the operating systems that they are running on top of. While Lotus recommends that Notes servers be located behind closed doors, customers can also deploy them to workgroup or remote locations if one of these "lockout" features is employed. In addition, because Notes is cross-platform, customers are free to choose the operating system that best meets their security needs.

Lotus also includes a feature called "SET SECURE," which allows administrators to secure Notes server sessions without requiring password "lockout" of the whole machine. This feature is valuable, especially in situations where the machine Notes is running on is also serving printer functions.

Securing Notes directories from drive mappings is an implementation issue. Exchange data directories must also be protected from inappropriate file access. Like any product, if it is not implemented correctly, it will not perform as designed.

In Notes R4, Lotus further secures locally accessed data.

Replication and Security

EXCHANGE COMMENT. Given that only Notes servers enforce data security, another loophole is exposed. When applications are replicated to an end user's machine, they are no longer secure. A user can do whatever he/she wants to with the data.

Lotus will claim to have fixed this in Notes 4.0, but this fix is Notes client-based. If you use VB or ODBC tools, you can bypass this security.

NOTES RESPONSE. This statement is misleading. The Notes replication process preserves Notes security. Information that is not accessible to the user on the server will not be replicated to that user's laptop. Likewise, additions, deletions, and modifications that the user does not have the authority to make will not replicate back to the server. Therefore, even if a user has replicated an application to his/her laptop, he/she will not be able to bypass the security of the application.

In Notes R4, this is taken a step further as noted previously.

Contrary to Microsoft's claims, using VB or ODBC will not allow access to data that the user is not entitled to use. Access control is enforced in the client and not the client user interface.

Notes Scalability

EXCHANGE COMMENT. Notes was developed as a work group information sharing tool. Ray Ozzie and company designed it for 10 to

20 users, not 10,000 to 20,000. Lotus has spent the last 5 years trying to rectify these problems through major rewrites to the product between R2 and R3 and now R3 and R4.

NOTES RESPONSE. Contrary to this opinion, Notes was designed with a layered architecture, similar to most operating systems. It was designed this way to provide scalability and extensibility. Notes major releases have proven these capabilities of the architecture.

The premise of Microsoft's assertion is hypocritical and misleading (e.g., DOS was designed for 8086-class machines, and Microsoft has spent the last 15 years trying to work around resulting built-in constraints). Furthermore, if we've been successful in expanding the scope of Notes to address large-scale roll outs, the implicit assertion that the product is somehow flawed because it was originally designed to work well within the constraints of LAN infrastructures when it was first developed 10 years ago is irrelevant.

We are proud of the fact that we've been able to evolve our product to address larger contexts without requiring customers to exchange their entire infrastructure every few years, as future Microsoft communications products propose.

Notes DBMS Model

EXCHANGE COMMENT. Notes doesn't use a true DBMS model for data access and management. Each application exists in a separate file, including the directory (which has no special optimization for directory services). Mail boxes for all users are also individual files.

Because these databases are contained in individual files, a significant burden is placed on the OS running the Notes server. Each user connection requires management of file handles and file system locks. A traditional DBMS architecture would be much more efficient because it would be optimized for managing multi-user locking and internal data structure access.

Because each Notes database is self-contained, there is a significant duplication of resources on the server. If forms and views are used in more than one database, they are duplicated.

NOTES RESPONSE. The Notes storage facility is not modeled after a traditional DBMS because it was designed to handle different requirements. Notes databases are open object stores optimized to hold documents and design elements of differing structures. In addition, they are designed to be self-containing for distributed and disconnected use.

Traditional DBMS models simply cannot address these requirements. Dependence on central meta-data repositories with static schema and limited functions for composite data types, coupled with assumptions about "connected" use only, prevents their use in Notes-like environments.

Lotus does, however, believe that traditional database management systems are a vital part of the corporate world. We are committed to providing open connectivity to these information stores in order to meet our customers' business objectives. We believe that Notes and traditional DBMS environments are complementary.

The issue of file handles has simply never come up and is a diversionary tactic on Microsoft's part. In point of fact, it is a strength of the product that databases are in separate files. In Exchange, the Public Information Store consists of exactly one file with a purported maximum size limitation of 16GB. The PIS contains all public folders and Schedule+ data. If any part of the file header of this file is corrupted, all data is lost. Moreover, each public folder's size is clearly constrained by that of all the others, because the grand total cannot be larger the maximum size of the PIS. To our knowledge, an Exchange server can only contain one PIS.

In Notes, one can have as many databases on a server as there is disk capacity to support them.

In practice, duplication of design elements has never posed an issue in Notes. By design, Notes databases are self-defining in order to facilitate very flexible distributed use. Lotus Notes leads the industry in this regard. Notes developers don't have to worry about distributing their applications because it is done for them through replication.

Most databases have very different designs and would not benefit from a central "design" repository. Those databases that are more "standard" in design can use the "design template" feature in Notes. This feature allows many databases to inherit their design elements from a central design "template." In Notes, developers have built-in software distribution.

Notes Data Store

EXCHANGE COMMENT. Notes data is not stored efficiently. A good indication of this is how much compression you can get out of "zipping" a NSF file. Typically this is a 10:1 ratio. View indices are also stored directly in the NSF files and can grow to considerable size.

A good example of just how bloated these databases can become is to look at the N&A book of a medium-sized company. A MS Mail directory of 20,000 names is typically in the range of 8MB. The corresponding Notes Name and Address book is around 80MB! The enormous size of this database makes it very difficult for people to replicate to their laptops for offline use. Offline users must "guess" the spellings of recipient names.

NOTES RESPONSE. Keeping view indices in a database is a very positive feature of the product. This allows views to be presented to users very rapidly. Moreover, experience has shown that different users have differing preferences as to which view of a database they want to see. Indices are built on demand. Therefore, each database replica will only contain the indices that its users need. In addition, indices do not replicate, thereby reducing the resources needed to distribute Notes applications.

The MS Mail directory to Notes N&A book issue is comparing apples to oranges. Like the Notes N&A book, the Exchange directory maintains much more information than a MS Mail directory. In Notes, laptop users typically build a list of the recipients they usually send mail to (either from the corporate N&A book or from received memos). This list can be built manually (at the push of a button) or automatically, thus allowing laptop users access to the names of people that they most often send mail to with a very small impact on disk space requirements.

Notes User Mailbox

EXCHANGE COMMENT. Each mailbox is a separate Notes database. This means that when you send a memo to 100 people, the server has to open 100 files and write a copy to each. Each time it opens a database, it has to establish file handles and locks at the file system level. This again shows significant duplication of resources.

NOTES RESPONSE. When a user replicates her mail database, she needs to have her own copy of a message. No mail system, including MS Mail and Exchange, allows a user to take a message offline without requiring an additional copy be made of that message. Notes is the only messaging environment to date that allows users to replicate mail. Other mail systems allow users to download messages but not without removing them from server-based information stores. This places the burden of message management on the user instead of the mail system and hampers a single user's operation of multiple remote PCs.

In Notes R4, Lotus will support single copy storage of messages at the server level. This will permit very efficient storage of mail on the server while still allowing end users to create replica copies of their mail files to local machines.

Notes File Management

EXCHANGE COMMENT. File management of the database environment is a nightmare. ACLs are set on a per-database basis. Any global changes require the administrator to open each database separately. This is a very manual process.

NOTES RESPONSE. Most access control changes involve the addition or deletion of a particular user from an application. Through the use of group names in access control lists, most ACL changes are actually executed by adding or deleting a name from a group in the Name and Address book. This process can affect multiple database ACLs without requiring the user to open each database separately.

Because Notes is an open, programmable environment, third parties have been able to create management utilities that allow global ACL changes to be performed and managed without opening individual databases. In fact, many allow an administrator to make changes on a laptop, disconnected from the system entirely, and have them executed the next time the administrator replicates. Notes R4 will also address this issue.

Notes Replication

EXCHANGE COMMENT. Notes replication is much better in theory than in practice. This process breaks down in a large environment. Notes uses a table scan mechanism for replicating databases. A connection is established between two servers. The servers build a list of databases to be replicated between them. Each server opens the first database in the list. They check the last replicated date in each database to find the last synch point. The database is then scanned from top to bottom looking for records that have been updated since the last time replication took place. When a record is found, it is compared to the record on the other server, any update conflicts are resolved, and the records are brought into synch on each server. This process is a very time consuming, network consuming, and resource intensive.

NOTES RESPONSE. Peer-to-peer, selective, document-level replication is not a trivial process. In fact, no other vendor, including Microsoft, provides a similar capability. Despite Microsoft's claims, Notes replication is an efficient process that has been optimized over the last 10 years of development. Notes only replicates databases that are in common. Notes only replicates documents that have changed since the last replication. Notes only replicates documents that match a given selection formula. Notes only replicates the information desired (documents can be stripped of large attachments).

Thousands of companies use Notes replication today to replicate gigabytes of data over links of varying speeds and across many different platforms. Notes replication has proven to be a very efficient process, and one that continues to lead the industry.

Microsoft has yet to deliver peer-to-peer, selective, document-level replication. The replication architecture that is planned for Exchange has had problems in testing, has limited selection capabilities, and relies on messaging, which is unproven in large networks.

In Notes R4, replication will become even more efficient. Servers will no longer have to "scan" for replicas each time they connect, replication will be done at the field level versus the document level (so only changed field values will be replicated) and multiple replicators can operate simultaneously.

Notes Replication Impact on WAN

EXCHANGE COMMENT. Depending on the speed of the link between servers, replication can actually dominate the connection, disabling all other traffic. Many large Notes sites have been forced to upgrade all of their WAN links just to allow replication to take place.

NOTES RESPONSE. Replication traffic is efficient and controllable. Many corporations continue to use 9600 baud asynchronous lines to replicate tens of megabytes of information between dozens of sites. Laptop users use modems speeds as low as 2400 baud to keep their data in synch with Notes.

If for some reason, replication traffic becomes a burden, it can be separated to different network cards and/or different network protocols depending on an organization's needs. Also, Lotus' NotesView product provides complete monitoring and management of Notes servers, including the replication process. NotesView can notify administrators of any potential replication problems automatically.

Exchange has extremely high network bandwidth requirements. Microsoft recommends speeds of 1Mb/s and above within an Exchange site and at least 128Kb/s between sites.

Notes Replication via Spokes

EXCHANGE COMMENT.   In a complex environment with hub servers replicating to lots of "spokes," you can get to the point where no replication takes place at all! The replication "window" becomes smaller as you increase the number of servers that replicate. Replication is an all or nothing process. If the replication session exceeds the "window" time, then none of the records get synched. This means the process has to start all over again and usually fails all over again. In addition, because databases are only replicated one at a time, exceeding the replication "window" means that none of the other databases in common get replicated either. In an environment where you rely on replication to run your business, this is not a comforting thought!

NOTES RESPONSE.   This statement is misleading and inaccurate. Notes replication is not "all or nothing." If a replication session fails, all updated documents from the last successful replication session are indeed checked for replication, but only those that did not actually replicate are replicated. If another server is scheduled to replicate with the "hub" (in traditional "pull-pull" replication) while a previous replication session is still active, that session is placed in a queue until the previous session finishes. It is not skipped.

In addition, "push-pull" replication allows multiple "spoke" servers to replicate with the "hub" simultaneously. In this configuration, the replication "window" is not a significant factor because multiple servers can replicate at the same time.

Notes replication is extremely reliable and quite scalable. There are dozens of multi-thousand user Notes environments that have been replicating in complex, distributed environments with great success.

Microsoft has yet to deliver bidirectional replication. The replication architecture proposed in Exchange is based on an unproven store-and-forward foundation. In testing, it has shown problems due to the disconnected nature of message-based replication. Because the system cannot guarantee the delivery times of document updates, Microsoft will have trouble guaranteeing synchronically in large, multi-connection environments.

Summary

In summary, the comments and responses stated in this appendix represent two sides in the very competitive groupware arena. Both the Microsoft Exchange and the Lotus Notes/Domino products are evolving at a rapid rate, so the product functions discussed in the comments and responses might very well have changed with the latest releases of each product.

As with most competition, we the end users, will undoubtedly benefit. The strong competition among the major groupware products—such as Exchange, Notes, Groupwise, and SuiteSpot—has driven all of the product costs down significantly, and all of the groupware products are improving in features and performance at a rapid rate.

Competition among the groupware products has been a less publicized version of the "Web Browser Wars." I regularly use both the latest Netscape Navigator and Microsoft Internet Explorer browsers. That's easy to do as long as your PC has the available hard disk space. It's not easy to regularly use more than one groupware product. For one thing, groupware always includes e-mail and users want to consolidate, not expand, the number of e-mail addresses they have, also, as the groupware products embrace the Web, many of these products allow you to use the Web browser of your choice as part of your groupware setup. Lotus Notes, for example, allows you to use either Netscape Navigator or Microsoft Internet Explorer as your integrated Web browser.

APPENDIX F

CD-ROM USE

This appendix describes the contents and use of the programs and data files on the CD-ROM that comes with the book. There is a README file on the CD-ROM that gives additional details on how to use the databases and programs. The README is in two forms:

- README.NSF—A Notes database
- README.TXT—An ASCII text file

In addition to the README file and this appendix, all of the Notes databases have "About" documents, which will give you help in the use of the database.

The databases on the CD-ROM are divided into two directories: an "IBM" directory and a "LOTUS" directory. The following sections give the contents of each of these directories. An easy way to view the Notes databases is to place the CD in your CD-ROM drive and do a File|Database|Open from your Lotus Notes client. Then, in the "Filename" area of the Open Database window, type E:, *where* E is the drive letter for your CD-ROM. You can then add the database icons to your Notes workspace and explore. You can copy the databases to your hard drive (either via Notes or from a window under your operating system). Of course, the non-Notes files should be handled in the same way you handle other program and text files. Use your favorite editor or word processor to handle these.

IBM Lotus Notes and Domino Databases and Programs

The databases in this directory were developed to support the Lotus Notes users within IBM. The database filenames on the CD-ROM are given in parenthesis.

IBM GLOBAL DIAL HELP (DIALHLP.NSF). This is the dial help database that goes with the IBM Dial Mobility Kit. See "An International Dial Mobility Kit used by IBM Executives" in chapter 6.

IBM GNA APPLICATION REGISTRY (IBMAREG.NSF). The Enterprise Application Registry is available to help advertise, deploy, and control Lotus Notes applications across the IBM Corporation.

IBMUS MAIL TEMPLATE R4 (MAIL4US.NSF). This is a description with screen captures showing the features of the Mail Template used in the IBMUS domain by Notes R4 users. See "Use of Customized Mail Templates" in chapter 11.

IBMUS MAIL TEMPLATE R4.5 (MAIL45US.NSF). This is a description with screen captures showing the features of the Mail Template used in the IBMUS domain by Notes R4.5 users. See "Use of Customized Mail Templates" in chapter 11.

ICON DATABASE FROM IBMNOTES DATABASE REPOSITORY, "ICONZ, GET 'EM HERE!" (ICONZ.NSF). This icon database started out as an icon database supplied by Lotus, but it now has contributions from IBM Notes users over the past couple of years. See chapter 14 for a description of the IBMNOTES repository.

CLIP-ART DATABASE FROM IBMNOTES DATABASE REPOSITORY, "CLIP-ART, GET IT HERE!" (CLIPART.NSF). This database also started out as a clip-art database supplied by Lotus, but it now has contributions from IBM Notes users. See chapter 14 for a description of the IBMNOTES repository.

LOTUS NOTES PACKAGES "SHAREWARE" FROM IBMNOTES DATABASE REPOSITORY (LNPKGS.NSF). This database contains "shareware" from IBM Lotus Notes users. See chapter 14 for a description of the IBMNOTES repository.

"RECIPES" DATABASE FROM IBMNOTES DATABASE REPOSITORY (RECIPIES.NSF). This database can be considered a "fun" database. It contains food recipes contributed by IBM Lotus Notes users. See chapter 14 for a description of the IBMNOTES repository.

IBM PRESS RELEASES DATABASE (PRESSREL.NSF). This database is a Notes version of the press releases IBM makes available on its Web home page.

IBM ANNOUNCEMENT LETTERS DATABASE (ANNLTRS.NSF). This database is a Notes version of the Announcement letters IBM supplies to its customers.

IBM "IS THERE" CUSTOMER FORUM (ISTHERE.NSF) The "IS. There" Customer Forum is the Notes version of the discussion database

IBM makes available to its customers on the HONE VM system. This is a very general forum used when more specific IBM forums don't apply. See chapter 14 on "Enterprise Conferencing with VM and Lotus Notes".

IBM "MWAVE" CUSTOMER FORUM (MWAVE.NSF) The "MWAVE". Customer Forum is for questions on the IBM "MWAVE" modem. This is the Notes version of the customer forum available on the HONE VM system.

IBM GLOBAL NETWORK/ADVANTIS "BUTTON CODE" MIGRATION AID (BUTTON.ASC). This is "softcopy" of the migration code that is given in appendix D of this book.

IBM GLOBAL NETWORK/ADVANTIS IP TRACE ANALYSIS PROGRAM FOR OS/2 (IPTRACK.NSF). This bandwidth analysis tool is based on the IPTRACE utility that comes with TCP/IP for OS/2. See "Bandwidth Requirements Measurement and Analysis" in chapter 12 for further details. A description on the use of the Netmon tool for Windows NT workstation and servers is also included.

NOTESBENCH ANALYSIS REPORT (NTSBENCH.NSF). Notes Bench Disclosure Report for IBM PC Server 330 with Lotus Notes R4.11a for Windows NT 3.51.

SAMPLE FTP SESSIONS FOR LOTUS NOTES ADMINISTRATION (FTPSESS.NSF). This database contains sample FTP sessions for Lotus Notes administration.

Lotus (LNN) Databases

The Lotus Internet Cookbook database serves as an "example" of the Lotus Notes databases that are available to Lotus Business Partners on the Lotus Notes Network (LNN). Information on the Lotus Business Partner program and LNN is available on the Lotus Development Corp. home page at http://www.Lotus.com.

Originally we had intended to provide "samples" of several LNN databases such as the Lotus Notes Knowledge base, LotusScript Learning Center, Industry Solutions, Lotus Connects Online, Lotus Notes API Support Modem Survival Kit, and Lotus Notes and Domino Documents. However, we had difficulty obtaining the required "permissions" since the databases were developed only for Lotus Business Partners. We'll try again for the next book!

LOTUS NOTES INTERNET COOKBOOK FOR NOTES R3 AND R4 (CBOOKV4.NSF). The R4 version of this cookbook is included in appendix C of this book.

GLOSSARY

802.2 IEEE standard for logical link control.

802.3 IEEE standard for the CSMA/CD (Ethernet) network access method.

802.5 IEEE standard for the token-ring network access method.

Access Control A security feature that specifies the tasks that can be performed by each user of a notes database. Some users might have access to all activities, while others might be limited to specific operations.

ACL Access Control List. Data that controls access to a protected object. An access control list specifies the privilege attributes needed to access the object and the permissions that can be granted.

Action The core element of any Lotus Notes object because it determines what happens when users click the object. An Action adds automation to a form and speeds up repetitive tasks. Notes provides several simple actions that are easy to create and don't require any programming knowledge. Actions that use @function formulas require knowledge of LotusScript.

ActiveX Microsoft's answer to Java.

ADSM ADSTAR Distributed Storage Manager. ADSM is an IBM program product for enterprise-wide storage management for the network. It provides automatic backup and archive services to multi-vendor workstations, personal computers, and LAN file servers. Versions for AIX on RS/6000, VM and VMS on S/390, and OS/2 of PCs are available. The June, 1995 version of ADSM features an incremental backup capability for Lotus Notes databases.

Advantis A networking company jointly owned by IBM and the Sears Corporation. Advantis designs and manages the wide area voice and data networks for both its parent companies and also for commercial customers. In 1997, became part of IBM Global Services.

Agent A portion of a network management system that reports information about conditions and accepts commands to alter the state of one or more managed objects. Typical commands are GET, SET, CREATE, DELETE, and ACTION. Lotus Notes agents are macros that execute an Action or set of Actions. Lotus agents perform automatic tasks on multiple documents and consist of a document selection formula, a trigger, and one of more Actions.

AIX Advanced Interactive Executive. AIX is IBM's version of UNIX.

Alert A network management term for data sent from a network management agent to a manager indicating that some action is to be taken or a problem has occurred.

Alias A name, usually short and easy to remember, that is translated into another name, usually long and difficult to remember. Alias names are often used in TCP/IP or on the Internet.

All-In-One An office/mail system for use on DEC host machines.

ANYNET This IBM product allows network users to send data with different protocols. The original design of ANYNET was to allow TCP/IP or NetBIOS traffic over SNA networks. The product has been enhanced to allow SNA over TCP/IP networks, etc.

API Application Program Interface. A functional interface supplied by the operating system or by a separately orderable licensed program that allows an application program written in a high-level language to use specific data or functions of the operating system or the licensed program.

APPC Advanced Program to Program Communication. An API embodied in SNA U6.2, which supports synchronous program-to-program communications.

Applet The Java name for a small chunk of code that is brought down over the network from a server and executed on a client.

AppleTalk A networking protocol developed by Apple Computer for communication between Apple Computer products and other computers.

Application-centric Refers to a dependency on legacy mainframe applications that require extensive skill and time to change with business needs. There is also a growing demand for access to data independently from application programs in an object-orientation.

Application Layer The topmost layer in the OSI Reference Model, providing such communication services as electronic mail and file transfer.

APPN Advanced peer-to-peer network. The network protocol IBM has positioned to replace SNA.

Archie An Internet application for locating publicly available files.

Architecture A specification that determines how something is constructed, defining functional modularity as well as the protocols and interfaces which allow communication and cooperation among modules. For computer networks, the architecture gives this broader aspect, while the network design gives specific details as to which device goes where, etc.

ARP Address Resolution Protocol. A method used to translate an IP address into a MAC (Media Access Control) address.

ARPA Advanced Research Projects Agency. The research arm of the Department of Defense. ARPA funded the research of packet-switched networks that led to the Internet.

ATM Asynchronous Transfer Mode. A method of high-speed data transmission that can dynamically allocate bandwidth to multiple data streams of different types by breaking each stream into 53-byte cells (5-byte header, 48 bytes of data) of information. It is intended to support many different types of traffic on a single network (e.g., voice, video, and data). ATM is planned to be used as a basic data transport technique for high-speed WAN and LAN applications. Operating speeds are from 25 Mbps to 155 Mbps for individual connections today. Gigabit range transmissions are predicted with the next five years. In 1990, the CCITT selected ATM technology as the standard mechanism for support of Broadband Integrated Services Digital Networks (B-ISDN). One of ATM's most attractive features is the promise of seamless integration of LANs and WANs.

Authentication   Verifying that a data transmission from a person (or a process) is indeed from that person (or process). A password is a common method used for authentication.

Backbone   That portion of a network used to interconnect major sub-networks. For example, the cable used to connect the networks of two buildings on a site could be called a backbone.

Bandwidth   The amount of data that can be transmitted across a particular network. Token-ring has a 16 Mbps bandwidth, while Ethernet has a 10 Mbps bandwidth. Technically, bandwidth is defined as the difference, in Hertz (Hz), between the highest and lowest frequencies of a transmission channel.

Bps   Bits per second. The rate of data transmission across a network.

Bridge   A device that transparently interconnects two LANs that use the same logical link control protocol but might use different Media Access Control (MAC) protocols. This is performed in a manner transparent to higher levels on the protocol stack. Only data destined for the other LAN is transferred. Bridges can usually be made to filter packets, that is, to forward only certain traffic. Related devices are repeaters, which simply forward electrical signals from one cable to another, and full-fledged routers, which make routing decisions based on several criteria.

Broadband   This term refers to any transmission at a higher speed than T1 (1.54 Mbps). A broadband network can carry voice, data, and video signals simultaneously.

Broadband ISDN (B-ISDN)   Digital data transmission standard to handle high bandwidth applications, such as video. The architecture supports wide-area transmission at speeds of 51, 155, and 622 Mbps per channel. Eventually 2.4 Gbps will be supported. Fiber-optic cable is required.

Browser   See *Web browser.*

CCITT   International Consultative Committee for Telegraphy and Telephony. A unit of the International Telecommunications Union (ITU) of the United Nations. An organization with representatives from the PTTs of the world. CCITT produces technical standards, known as "Recommendations," for all internationally controlled aspects of analog and digital communications. As of January of 1995, the function of the CCITT was taken over by the ITU-T.

cc:Mail   An electronic mail system from Lotus Development Corporation.

Certificate   A unique electronic stamp stored in a user ID file. Certificates permit you to access specific Notes servers. Your user ID might have many certificates.

Certifier ID   A special Notes ID file used to certify or "stamp" other Notes IDs so that they can access the Notes servers under the control of the Certifier ID.

CGI   Common Gateway Interface. A scripting programming language used on the Internet.

Circuit   A direct data stream between two systems on a network.

Circuit switching   A communications method in which a dedicated communication path, on which all packets travel, is established between two hosts. The public telephone system is an example of a circuit-switched network.

Client   A computer or process that accesses the data, services, or resources of another computer or process on the network.

Client/Server   A method of computing in which client systems (usually intelligent PCs and workstations) run user applications. The applications use a combination of their resources and a portion of the storage and computing resources of one or more server systems (e.g., high-performance PCs) to perform useful work. The application portion on the client requests server resources by communicating over a network. The client/server model usually replaces a mainframe-centric setup.

Clustering   This technology allows up to six separate Notes servers to co-operate together to appear to the user as if they were a single Notes server. This is accomplished by a process known as real-time replication.

Collaborative Computing   A method of information sharing using computing. Collaborative computing technologies include groupware and meetingware products. These products are intended to improve personal interaction, group work, and discussion.

Components   The Lotus name for software applets that can be included with Lotus Notes R4 and R5 to embed application function such as spreadsheet, file viewer, chart, draw/diagram, comment, and project scheduler.

Connectionless service   A data transmission service in which each frame of data can be broken up into one or more independent packets, each containing both a source and destination address. The packets might be delivered out of sequence or dropped. Sometimes called *datagram service*. Examples: LANs, Internet IP, and ordinary postcards.

Connection-oriented The model of interconnection in which communication proceeds through three well-defined phases: connection establishment, data transfer, and connection release. Examples: X.25, Internet TCP, and ordinary telephone calls.

CSMA/CD Carrier Sense Multiple Access with Collision Detection. The media access control method employed by Ethernet networks.

Daemon An event-triggered program, equivalent in function to an agent.

DARPA Defense Advanced Research Projects Agency. See *ARPA*.

Database A group of Notes documents and their forms and views, stored under one name and opened through a database icon on your workspace. A database is always one and only one file, with the .NSF extension. Notes databases can range from a very small personal database to a very large complex enterprise-wide database.

Database Catalog A special Notes system database that contains information about databases stored on a single Notes server, a group of servers, or all the servers in a domain. The database catalog (CATALOG.NSF) resides on a server and can hold an inventory of databases.

Data Link Layer Layer 2 of most network architectures. Defines the method to transmit data between two network entities across a single physical connection or a series of bridged connections.

DCE (1) The Distributed Computing Environment, which is a definition of networking support programs as provided by the Open Software Foundations (OSF). It defines the interface from a workstation to a network and includes a naming service, a remote procedure call, and a distributed file system. Also called an Andrew File System or AFS. (2) Data Circuit-terminating Equipment, a device used in X.25 networks that accepts/starts calls.

DES Data Encryption Standard. A widely used, government-sponsored data encryption method that scrambles data into an unbreakable code for transmission over a public network. The encryption key has over 72 quadrillion combinations and is randomly chosen.

Design When used in the designation "Network Design," this refers to the details of putting together network components in order to have computers "talk" to each other. Network Architecture refers to a broader concept of how a computer network should be put together. See *Architecture*.

DHCP Dynamic Host Configuration Protocol. An IP-based protocol that dynamically assign an IP (Internet Protocol) address to a host upon request.

Dial-up Notes A Notes feature that allows you to access databases by calling Notes servers using a modem instead of using Notes on a

LAN. It includes special functionality to reduce the amount of data transferred and stored on the local workstation.

Directory   In DCE, the directory contains information about resources, services, objects, and users on the network. This makes it simple to find each of these things by using only the name and makes it possible for each component to be moved in the network as business or technology dictate without needing to change applications.

Distinguished Names   A method of ensuring that each Notes user is assigned a unique name that will not be confused with any other name in the system, no matter how large the Notes installation grows. When a hierarchical certifier registers a user, the name on the user ID is inherited from the distinguished name of the certifier. For example, if the certifier's name is Sales and he certifies John Doe, the fully distinguished name is John Doe.

Distributed database   A collection of several different data repositories that looks like a single database to the user.

DNS   Domain Name System. The distributed name/address mechanism used in the Internet. It is through the DNS that the domain name of every computer on the Internet gets mapped to its IP address so that each computer can send information back and forth on the Internet.

Document   In Notes, a document is the default form type that is assigned to a form. A document is created by using a form on the Compose menu. A Notes document can range from a short answer in a discussion database to a multi-page analysis of a particular problem. Documents consist of fields containing text, numbers, graphics, scanned images, or even voice messages.

Domain   A group of Notes servers with the same public Name and Address Book. Domains are used to define the scope of a Notes mail environment. Domain is also a common term used in computer networks to mean: (1) that part of a network in which the data processing resources are under common control and, (2) in TCP/IP, the naming system used in hierarchical networks.

Domino   The Web-enabled Lotus Notes Server. With Release 4.5 of the Notes product, the Lotus Notes Server was renamed the Domino 4.5 Server in order to emphasize the Web technology built into that version of the server.

Domino.Action   The Lotus Web site creation tool that allows you to select and create all the basic components of a Web site with "point-and-click" technology.

Domino.Broadcast   Works with the PointCast product to broadcast information to a group of users. This is a good way to communicate fast-breaking or mission-critical news to employees via Notes.

Domino.Connect   Allows Notes to integrate with enterprise applications. The enterprise applications could be transaction systems such as CICS, legacy DBMS such as DB2, and ERP systems such as SAP R/3 and PeopleSoft. NotesPump R2 and Notes SQL are two of the Domino.Connect products that allow this.

Domino.Merchant   Lotus product that allows creation of a Web site for electronic commerce. Essentially Domino.Merchant adds electronic commerce to Domino.Action

EDI   Electronic Data Interchange. Allows computer-to computer exchange of business documents, such as invoices, shipping orders, payments, etc., between different companies. Independent service organizations provide EDI services that enable users to interconnect with another organization's network regardless of type of equipment used.

EISA   Extended Industry Standard Architecture. A PC bus architecture that extended the ISA architecture and was the PC Vendor's answer to IBM's Micro Channel Architecture (MCA).

Encryption   The manipulation of data in order to prevent any but the intended recipient from reading that data. There are many types of data encryption that are the basis of network security.

Ethernet   A 10 Mbps standard for LANs, initially developed by Xerox, and later refined by Digital, Intel, and Xerox. All hosts are connected to a cable where they contend for network access using a Carrier Sense Multiple Access with Collision Detection (CSMA/CD) method. Adopted as an IEEE standard (802.3).

Extranet   A network connecting independent companies, such as customers to suppliers. This is a "private Internet", like an "Intranet", but used to connect to outside companies.

FAQ   Frequently Asked Question. A FAQ list is a file that contains commonly asked questions and their answers. On the Internet, a typical newsgroup has a FAQ that answers questions readers have about or relating to that particular newsgroup. There are also FAQs that focus on questions new users have about newsgroups in general.

Fax (Facsimile)   A machine that scans a paper form, converts the image to a coded stream of bits, and transmits it over the public telephone network.

FDDI   Fiber Distributed Data Interface. A high-speed (100 Mbps) protocol for fiber-optic LANs.

File Server   A server system that stores files created by application programs and makes them available for sharing by clients on the network.

Filtering   When applied to a network bridge or router, filtering refers to the recognition of the type of a data packet and redirecting it to another path on the network or rejecting it.

Firewall　When applied to a large site network, a firewall is a computer-controlled boundary that prevents unauthorized access to the network for transmitting and receiving data.

Forms　These determine how you enter information into a Notes database and how that information is displayed and printed. Forms for a Notes database are created by Notes application developers. The Notes user makes use of "Views" to choose the way he wants data displayed. Forms and Views are part of the basic design of every Notes database.

Forums　Another name for Discussion Databases or Bulletin Boards. IBM uses the name "Forum" for their discussion databases on their main-frames.

Frame Relay　A WAN service protocol for high-speed, long-distance digital data packet transmission. It is a simplification of the X.25 protocol on a faster channel. Therefore, it is ideal for interconnecting two LANs over a WAN and is replacing X.25 commercial services. It is designed to provide high-speed packet transmission, very low network delay, and efficient use of network bandwidth. Transmission speeds supported are from 64 Kbps to 1.544 Mbps (T1).

FTP　File Transfer Protocol. A protocol (and program) used with TCP/IP to send files over a network.

Full-duplex transmission　The transmission of data across a network in both directions at the same time.

Gateway　A device that acts as a connector between two physically separate networks. It interfaces to more than one network and can translate the packets of one network to another.

GB　Gigabyte. A billion bytes of computer or hard-disk memory.

Gopher　A popular Internet application that uses a menu system to search for and download information.

Groupware　A new classification of application software that allows users to conveniently share data across a telecommunications network.

GUI　Graphical User Interface. A pictorial way of representing to a user the capabilities of a system and the work being done on it.

GW　Gateway.

Half-duplex transmission　The transmission of data across a network in which only one side can send at a time.

HDLC　High-Level Data Link Control. A data link protocol for public networks that is an ISO standard.

Homepage　Generic term for the hypertext document users see when they first access a World Wide Web server.

Hop　A term used in routing. A path to a destination on a network is a series of hops, through routers, away from the origin.

Host　In the TCP/IP sense, a computer that allows users to communicate with other host computers on a network. Individual users communicate by using application programs, such as electronic mail and FTP. Also, used to refer to a large computer system, such as a mainframe.

HTML Hypertext Markup Language. This is the language used to write World Wide Web documents, or pages. It is a subset of ISO SGML.

HTTP Hypertext Transfer Protocol. This is the protocol used by the World Wide Web to transfer documents between clients and servers.

HTTP-NG HTTP-Next Generation. This proposed protocol includes caching, preemptive caching, and open sessions between browser and server.

Hypertext Text within a online document that contains links to text within other online documents; selecting a link automatically displays the second document.

Icon A graphical element that represents an object, command, or option on your computer screen. Notes uses several different types of icons, the most important of which are the icons on your Notes Desktop that represent databases.

IEEE Institute of Electrical and Electronic Engineers. A professional society and standards-making body.

IEFT Internet Engineering Task Force. Part of the IAB, a volunteer group of engineers responsible for Internet standards development.

IBM Global Network (IGN) The subsidiary of IBM that handles all wide area networking for both IBM internal customers and for commercial customers.

IMAP Internet Mail Access Protocol.

IMLG/2 The IBM Mail LAN Gateway/2 product, which runs on OS/2 and provides mail gateway service between OfficeVision/VM and Lotus Notes, among other applications.

INET Dialer This is the dial program for Internet access developed by IBM. The dialer provides SLIP and PPP TCP/IP access to the Internet. OS/2 WARP includes the INET Dialer as part of the BONUS PACK.

InterfloX An IBM software product that serves as a REXX API to Lotus Notes.

Internet A set of connected networks. The term Internet refers to the large and growing public-domain internetwork developed by DARPA that uses TCP/IP. It is shared by universities, corporations, and private individuals. To be on the Internet, you must have IP connectivity (i.e., be able to telnet to or ping other systems). Networks with only e-mail connectivity are not actually classified as being on the Internet.

Internet Address A number that identifies a host in an internet. It is a 32-bit address assigned to hosts using TCP/IP.

InterNetNews (INN) This is a news software program that allows a computer to operate as a Usenet News Server.

Internet Service Provider A commercial organization that provides you with services to connect to the Internet.

Internetworking   The connection of two or more networks.

InterNIC   Internet Network Information Center. Part of the central authority for the Internet. The InterNIC (sometimes called NIC for short) registers new Internet domain names and address ranges.

InterNotes Web Navigator   A Notes Release 4 feature that allows you to navigate through pages on the Web directly from your Notes environment. The Web Navigator is much more than a Web browser; it combines the features of a Web browser with the powerful capabilities of Notes.

Interoperability   The ability of unlike systems to work heterogeneously. Computers of different sizes and brands can communicate together, sharing resources, information, and software applications.

Intranet   An internal corporate network employing Internet and World Wide Web technologies, such as TCP/IP, HTTP, and HTML protocols.

IP   Internet Protocol. The network-layer protocol for the Internet protocol suite.

IPng   IP next generation. Under development by the IEFT IPng committee, IPng is intended to upgrade the current IP to greatly increase the number of networks and nodes beyond the current 4 billion addresses, plus other enhancements to support future high-speed internetworking requirements.

IPX   Novell's Internetwork Packet Exchange network communication protocol.

I/S   Information Systems. Usually the in-house information systems organization that provides network and systems help within a company.

ISA   Industry Standard Architecture. The PC bus architecture used on most PCs. ISA was extended with EISA, and now both are being replaced with the PCI bus architecture.

ISDN   Integrated Services Digital Network. A standard architecture, specified by CCITT, for integrating different types of data streams on a single network, such as data, video, and voice. It is just recently being offered by the telephone carriers of the world.

ISO   International Organization for Standardization.

ISP   Internet Service Provider. Companies such as Netcom, PSINet, or IBM that connect customers to the Internet.

ISSC   Integrated Systems Solutions Corporation. The IBM subsidiary that provides information system services to external and internal IBM customers.

I/T   Information Technology.

ITU-T   International Telecommunications Union. See CCITT.

Java   Sun Microsystem's object-oriented language developed for programming on the Internet. It's been stated that "Java offers the promise that the network will become the computer."

Java Beans   The applet components developed with Java. Kona is the Lotus code name for its Java Bean Components.

Kbps   Kilo (thousands) of bits per second.

Kerberos   The security system of the MIT Athena project. It is based on symmetric key cryptography.

LAA   Locally Administered Address. This LAN adapter address can override the Universally Administered Address (UAA) that is "burned in" to each adapter. The LAA can be coded to be a "meaningful" number to help LAN administrators track a problem. Compare to UAA.

LAN   Local Area Network. The interconnection of several personal computers and other hardware such as printers. Designed originally as a means of sharing hardware and software among PCs, it is now used as a general means of communications between PCs.

LAN Distance   This IBM Product provides dial access to a LAN. Because the product simulates a LAN Bridge, the user has all the features of Token Ring access to a LAN. This connection will support many protocols, including NetBIOS, TCP/IP, and SNA.

LAN Hop   This IBM Product provides dial access to a LAN for TCP/IP access via a SLIP connection. LAN Hop users are assigned TCP/IP addresses based on the IP subnet of the LAN Hop Gateway they dial into. This TCP/IP access differs from LAN Distance TCP/IP access in that LAN Distance TCP/IP users can only dial to a specific LAN location because they have a fixed TCP/IP address, while LAN Hop users could dial any of many LAN locations because their TCP/IP address is assigned for every session.

LAN adapter   Moves data to and from a personal computer's memory to transmit and receive data over LAN cable.

LAN segment   See *segment*.

LIG   Local Interface Gateway. The dial service offered by the IBM Global Network that allows local access with a wide variety of supported protocols.

Logical Link Control (LLC)   The upper portion of the data link layer, as defined in IEEE 802.2. The LLC sublayer presents a uniform interface to the user of the datalink service, usually the network layer. Beneath the LLC sublayer is the MAC sublayer.

Login   A UNIX term for gaining access to a UNIX processor. The term "logon" is usually used to describe the process for gaining access other LAN servers or mainframes.

Lotus Messaging Switch (LMS)   A scalable backbone switch for interconnecting different messaging system.

Lotus Notes   A workgroup application for sharing information and building business applications with integrated e-mail.

LotusScript   This is the object-oriented basic language used by Lotus Notes application developers. It is used to perform more complicated tasks than can be done using the Notes formula language.

LU   Logical Unit. A port through which an end user accesses a SNA network in order to communicate with another end user or system.

LU 6.2 Logical Unit 6.2   A peer-to-peer (system-to-system) communications protocol that supports program interoperability developed by IBM as part of SNA. Also called APPC.

MAC address   The hardware address of a device connected to a shared media.

Mail gateway   A machine that connects two or more electronic mail systems (especially dissimilar mail systems on two different networks) and transfers messages between them. The IBM Mail LAN Gateway/2 (IMLG/2) and Notes SMTP Gateway are examples.

MAIL.BOX   The database on every Notes server that serves as the temporary "storage" for Notes "store and forward" mail routing.

Mainframe   A large computer system, such as an IBM System/370 or System/390 architecture computing system running either the MVS or VM operating system. See *Host*.

MAN   Metropolitan Area Network. A LAN-like network that covers larger geographic distances (up to 50 km), possibly crossing public rights-of-way.

MB   Megabyte. A million bytes of computer or hard-disk memory.

MCA   Micro Channel Architecture. IBM's PC bus architecture.

Mbps   Mega (millions) of bits per second.

MHS   Message Handling System. A server-based application used to process electronic mail messages. MHS was developed by Novell for its NetWare product.

MIB   Management Information Base. A rigorously defined database for network management information. The MIB is the conceptual repository of management information within an open system. It consists of the set of managed objects, together with their attributes. Standard, minimal MIBs have been defined, and vendors often have private enterprise MIBs.

MIME   The Multipurpose Internet Mail Extensions. This protocol enables users to send multiple kinds of binary data (video, sound files, and so on) as attachments to an e-mail message.

Modem   A device that converts binary information into on-and-off analog tones that can be transmitted over analog telephone lines.

Mosaic   An application, with a GUI interface, that supports easy browsing of information on the Internet. In addition to standard binary and text files, Mosaic supports multimedia information. It was developed by the Software Development of the National Center for Supercomputing Applications (NCSA) at the University of Illinois.

MPN   Multi-Protocol Network. Also, the name given to the router-based T1 network used by IBM for its internal use. This IBM network supports TCP/IP, NetBIOS, IPX, and SNA protocols.

MQSeries   IBM middleware software that links message-oriented application.

MTA   Message Transfer Agent. The Lotus Domino MTAs are like gateways in that they translate messages, headers, and attachments between other message formats. However, unlike gateways, Domino MTAs are integrated with the Domino server. One such MTA is the Lotus SMTP/MIME MTA used for transferring Lotus Notes mail to and from the Internet.

MVS   IBM's Multiple Virtual System operating system for mainframe computers.

Name and Address Book (N&A Book or NAB)   The N&A Book is a special Notes system database. This database serves as the central facility to schedule replication between servers and manage security. The "Public Name & Address Book" is a database that contains the names and domain address of every Notes user, user group, and server in a domain. The "Personal Name & Address Book" is a database that contains the names and computer addresses of users and user groups for your personal use. It is added to your workstation when Notes is first installed. The personal N&A Book only contains information you add yourself. It can be considered to include your personal list of "nicknames" for the addresses of Notes users with whom you frequently communicate.

Name server   A server system used in a TCP/IP network that converts TCP/IP network addresses into names and vice versa.

NetBEUI   Microsoft's NetBIOS Extended User Interface. Client software that redirects application software requests for service across the network.

NetBIOS   Network Basic Input Output System. Client software that redirects application software requests for service across the network. Originally developed in 1984 as a high-level programming interface to the IBM PC Network, it quickly became a *de facto* session layer standard for the LAN industry. Most major LAN vendors support NetBIOS.

NetFinity   An IBM product used to help users with hardware and general systems management functions. NetFinity is comprised of NetFinity Services and NetFinity Manager.

Netscape Navigator   A commercial browser that provides an interface to the World Wide Web.

NetSP   An IBM firewall product.

NetView   An IBM program product used to monitor a network, manage it, and diagnose its problems.

Network   A computer network is a data communications system that interconnects computer systems. A network can be composed of any combination of LANs, MANs, or WANs.

Network address   The network portion of an IP address. For a class A network, the network address is the first byte of the IP address. For a class B network, the network address is the first two bytes of the IP address. For a class C network, the network address is the first three bytes of the IP address. In each case, the remainder is the host address. In the Internet, assigned network addresses are globally unique.

Network architecture   The logical structure and operating principles of a computer network. The network architecture consists of the set of layers and protocols that defines the interconnection of systems in the network.

Network-centric   This is the concept where the network becomes the computer.

Network Computing   The simplified name for Network-centric Computing.

Network design   The details of implementation when connecting systems in a network. These details include link size (bandwidth), alternate routing, routing control, design for network management, etc. See *network architecture*.

Network Information Center (NIC)   A group at SRI International, Menlo Park, CA, responsible for providing users with information about TCP/IP and the connected Internet.

Network layer   The OSI layer that is responsible for routing, switching, and subnetwork access across the network environment.

NFS   Network File System. A network service that allows a system to access data stored on a different system as if it were on its own local storage. Part of the ONC standards developed by Sun Microsystems.

Network Operating System (NOS)   A family of programs that runs on networked computers. Some programs provide the ability to share files, printers, and other devices across the network.

NIC   Network Information Center. See *InterNIC* and *NotesNIC*.

NOTES.INI   A critical configuration file that defines how your Notes system runs.

NotesNIC   Notes Network Information Center. An organized system for locating and communicating with Notes sites globally. NotesNIC is run by the Lotus Development Corporation.

NotesView   A Lotus management tool used for displaying real-time data on Notes databases.

.NSF extension   Stands for Notes Storage File, and describes regular databases created in Notes.

.NTF extension Stands for Notes Template File, and describes database templates created in Notes.

OC-n Optical Carrier-n. SONET hierarchy for fiber-optic transmission. (See Table G.1.)

Office Vision An IBM VM application that supports electronic mail, calendar, documents and file management. Also known as Office Vision/VM, OV, or OV/VM. Formerly called PROFs.

OLE Object Linking and Embedding. A technology that lets you share data between applications and is supported by Microsoft Windows and Macintosh operating systems. For example, OLE lets you embed or link data—such as a 1-2-3 chart, Word Pro document, or Freelance Graphics presentation—as an object in a Notes document.

OpenNet This is IBM's worldwide network that connects directly to the Internet and can be considered part of the Internet.

OS/2 Operating system for the IBM Personal System line of computers. OS/2 Extended Edition provides multitasking, a common communications interface, an integrated relational database, and query functions.

OSF Open Software Foundation. A not-for-profit organization that develops and delivers open technology to its members. There are sev-eral hundred members of OSF, including Apple, DEC, HP, Hitachi, IBM, ICL, Lotus, Microsoft, Motorola, Novell, and Xerox. OSF delivered five technologies: SOF/1, DCE, DME, Motif, and ANDF.

OSI Open Systems Interconnect. Standards developed by ISO for an open network environment. The OSI Reference Model is an abstract model defining a seven-layer architecture of functions, with protocol standards defined at each layer.

OSI Reference Model A seven-layer structure designed to describe computer network architectures and how data passes through them. Developed by ISO in 1978.

OV See *Office Vision.*

OV/400 The operating system for the IBM AS/400 midrange computer.

OV/VM See *Office Vision.*

Packet switching The transmission of data in small, discrete packages that are sent by different routes through a network and then reassembled at the receiving end.

PC Personal Computer

PCI Peripheral Component Interconnect local bus. A standard internal connection in a PC for data transfer to peripheral controller components. The PCI standard is beginning to replace the ISA, EISA, and MCA bus standards used with PCs over the past several years. PCI greatly increases bus bandwidth capabilities and adds greater flexibility when compared to the standards it is replacing.

PCMCIA Personal Computer Memory Card International Association bus. A standard external connection that allows peripherals adhering to the standard to be plugged in and used without further system modification. PCMCIA is the standard used for the "credit card" modems or other adapters used with laptop computers.

Ping Packet Internet groper. A program used to test reachability of destinations by sending them an ICMP echo request and waiting for a reply. The term is used as a verb: "Ping host X to see if it is up."

Platform A product, design, or architecture upon which applications, other products, or designs can be built.

PolyPM A product used for remote management of OS/2 systems.

POP3 Post Office Protocol 3. POP3 and IMAP are the Internet's standards for e-mail.

PPP Point-to-Point Protocol. Used for multiprotocol transmission of datagrams over serial (point-to-point) links (RFC 1171/1172). Can be used as the underlying protocol for TCP/IP transmission over telephone lines, for example. It is seen as the successor to SLIP. PPP, unlike SLIP, actually monitors and verifies Internet packets as they are sent.

PROFS Professional Office System. IBM's office system product for the mainframe.

Proxy Server A computer that provides access to information (such as URLs or network protocols) through a firewall server.

PTT Post, Telegraph, and Telephone. The name for the communication authority in many countries outside of North America. The PTT is simply a branch of the national government.

PU Physical Unit. A term used in IBM SNA networks.

RAID Redundant Array of Inexpensive Disks. Disk storage with a very high degree of reliability, etc.

Remote Procedure Call (RPC) The DCE communications mechanism that enables subprograms to execute on several servers in the network while an application is running. A message from the client application is sent to the remote server, requesting its services. The RPC can be represented by a program written in any computer language.

Replica A copy of a database that is updated by exchanging information with the original database, either on a regular schedule or as desired. Notes servers can connect to other Notes servers and periodically update all replicas of a database so that they become identical.

Replication The process of updating replicas of a database to make them identical.

Replicator The task running within the Lotus Notes product that handles replication. In Release 3 of Lotus Notes, only one replicator could be running. Release 4 allows multiple replicators.

REXX An IBM-developed programming language, available for several operating systems (including VM and OS/2), used to create applications involving high-level manipulation of operating system resources.

RFC Request for Comment. Documents created by Internet researchers on computer communication.

Rich text A term describing the capability of a document or a part of a document to contain text, graphics, scanned images, audio, and full motion video data. Lotus Notes allows rich text.

RISC Reduced Instruction Set Computer. A computer in which the processor's instruction set is limited to constant-length instructions that can usually be executed in a single clock-cycle.

RJ11 Recommended Jack 11. This standard is for the familiar telephone jack.

Router A device that is used to interconnect networks and intelligently route data traffic based on the transmission protocol employed. A router is preferred in TCP/IP internetworks because of its ability to support the complex networks in which the protocol is typically used. Individual LANs or groups of LANs can be treated as a logical subnetwork by a router within a larger, more complex network. Routers are devices that terminate data link or logical link protocols, making it possible for routers to match protocols from LAN to WAN.

RSA Data encryption standard that is used by the Lotus Notes product for all encryption requirements. Because of export restrictions by the Federal government, the RSA encryption method used in the Notes "International Edition" is less stringent than that used in the Notes "North American" edition. RSA stands for Rivest-Shamir-Adleman, the three developers of the encryption technique.

RS232C Established by the Electronics Industries Association (EIA) in 1969, RS232C is a standard defining the electrical signaling and cable connection characteristics of a serial port, the most common type of communications circuit used today.

RS/6000 RISC System/6000. IBM workstation systems employing RISC processors. Workstations, as opposed to PCs, are typically used for engineering or scientific work that requires intensive computation.

Scheduled Agents These agents are useful for running periodic workflow or maintenance tasks on selected Notes documents. Select the database and choose View Agents. Double-click the Agent. Select one of the On Schedule options. Click Schedule to change the default schedule for the interval you've chosen or to change the sever on which this agent runs. Specify the document selection and actions. Close and save the Agent.

Segment A segment is a single token-ring, FDDI ring, or Ethernet. Segments are interconnected via bridges or routers.

Server A system on a network that provides services to a requesting system, or client.

Server Farm An area reserved for a large number of server systems with a raised floor (to avoid wiring clutter), enhanced cooling, uninterruptible power, and centralized console-based management.

SHTTP Secure Hypertext Transfer Protocol. This is the proposed Internet protocol used for creating secure HTTP connections using Public Key technology.

SIG Site Interface Gateway. Part of the local dial access provided by the IBM Global Network. The SIG is located at a customer site, while the LIG is located outside the customer site, but within the local area.

SLIP Serial Line Interface Protocol. A protocol for IP data transmission over serial lines (RFC 1055), such as telephone circuits or RS-232 cables interconnecting two systems. SLIP is now being replaced by PPP. See *PPP*.

Smarticons The name given to the icons for the Lotus Suite of products (including Notes) that perform macro functions.

SMP Symmetric MultiProcessing. A computer architecture in which tasks are distributed among two or more processors.

SMTP Simple Mail Transport Protocol. This protocol utilizes TCP/IP and is a widely used e-mail protocol developed for the Internet (RFCs 821, 822).

SNA Systems Network Architecture. IBM's mainframe-based network protocol.

Sniffer A product of the Network General Company. The Sniffer is a powerful, user friendly local area network tracing and monitoring device.

SNMP Simple Network Management Protocol. SMNP includes a small but powerful set of facilities for monitoring and controlling network elements.

Socket A service interface to the Internet; a process opens a socket, identifies the network service required, binds the socket to a destination, and then transmits/receives data.

SONET Synchronous Optical Network. A standard for high-speed, 51.84 Mbps (OC-1) to 2.488 Gbps (OC-48), data transmission over a fiber-optic network. It is being used for telephone networks today. It is planned to be the platform technology for ATM data transmission services over a WAN.

SP2 This is IBM's Power Parallel computer, made of 4 to 512 RISC processors. SP refers to Scalable POWER parallel.

SPX Novell's Sequential Packet Exchange network protocol.

SSL Secure Sockets Layer. This proposed Internet security protocol secures the communications channel between client and server, in a layer above TCP/IP and below HTTP and other application protocols.

The protocol offers bidirectional authentication, message encryption, and message integrity.

Stub (or replica stub) A new database replica that has not yet been filled with documents. The database is no longer a stub once the first replication takes place.

Subnet A set of networking nodes where each node can communicate directly to every other (e.g., a LAN). A set of subnets form a network (e.g., a routed internet of LANs).

Systems View An IBM product consisting of an integrated set of applications for managing multiplatform, multivendor systems and resources.

T1 A digital data transmission service that can transport up to 1.544 Mbps. Digital communication is full-duplex. The bit stream can be viewed as 24 channels of 64 Kbps that are multiplexed on the aggregate of 1.544 Mbps stream. 64 Kbps is the bandwidth used for digital voice transmission.

T3 A digital data transmission service that can transport up to 44.746 Mbps, incorporating 28 T1 circuits.

TCP Transmission Control Protocol. A transport-level protocol for connection-oriented data transmission. It is the major transport protocol in the Internet suite of protocols providing reliable, connection-oriented, full-duplex streams. It uses IP for delivery.

TCP/IP The set of applications and transport protocols that uses IP to transmit data over a network. TCP/IP was developed by the Department of Defense to provide telecommunications for internetworking.

Telnet Telecommunications network protocol. The virtual terminal protocol in the Internet suite of protocols. Telnet allows users of one host to log into a remote host and interact as normal terminal users of that host.

Template A Notes database design that you can use as a starting point for a new database.

Token Ring LAN A Local Area Network technology that allows users to share network resources by sharing a "token." Token Ring LANs typically run at 4 Mbps or 16 Mbps speeds.

UAA Universally Administered Address. Each LAN adapter has a UAA that is a unique 12-digit hexadecimal address that is permanently encoded in the microcode on the adapter. Blocks of these addresses are assigned to each manufacturer by the IEEE, which ensures uniqueness. The UAA is also called the *burned-in address*. Certain products allow the UAA to be overridden by a Locally Administered Address (UAA), which can prove to be more convenient.

UNIX   The operating system originally designed by AT&T and enhanced by the University of California at Berkeley and others. Because it was powerful and essentially available for free, it became very popular at universities. Many vendors made their own versions of UNIX available; for example, IBM's AIX, based on OSF/1. The UNIX trademark and definition has since come under the control of X/Open, who will issue a unifying specification.

UPS   Uninterruptible Power Supply. A device that is used to provide temporary backup power to a computer system.

URL   Universal Resource Locator. This is the World Wide Web name for a document, file, or other resource. It describes the protocol required to access the resource, the host where it can be found, and a path to the resource on that host.

Usenet   This is a worldwide group of computers that store and exchange news among other computers. The Internet connects most of the Usenet computers together.

Usenet Newsgroups   These are collections of discussion groups that reside on various News Server computers on the Internet. Newsgroups resemble electronic bulletin boards where readers read, post, and reply to articles posted by other newsgroup readers.

User ID   A file (user.id) assigned to each user that uniquely identifies that user to Notes. All Notes users and servers must have a user ID. The user ID (or user.id) contains the "private" section of the public/private key Notes security system. The user ID must be kept in a secure location and should include a password to protect it from unauthorized use.

UUCP   UNIX-to-UNIX Copy Program. A standard protocol for exchanging newsgroup articles using asynchronous dial-up lines.

UUdecode and UUencode   Programs that convert files from text to binary. You use these programs to send file attachments across the Internet. Many e-mail applications automatically UUencode files when you send them and UUdecode files when you receive them. Both of these programs were developed as part of UUCP.

View   A view is a list of documents in a database that are sorted or categorized to make finding documents easier. A database can have any number of views. To change views, the user simply chooses another view from the View menu.

VM   Virtual Machine. IBM's operating system for mainframe computers.

VMSMail   The electronic mail system used on DEC computers running the VMS operating system.

WAN   Wide Area Network. A long-distance network for the efficient transfer of voice, data, and/or video between local, metropolitan, campus, and site networks. WANs typically use lower transfer rates (64Kbps) or high-speed services such as T1, which operates at 1.544Mbps. WANs also typically use common-carrier services (communications services available to the general public) or private networking through satellite and microwave facilities.

WebExplorer   WebExplorer is the name given to the World Wide Web Browser that comes with the IBM Internet connection packages (e.g., part of the BONUS PACK in OS/2 WARP).

Weblicator   The Lotus name for their product that provides Lotus replication technology to Web browsers.

Web browser   An application that provides an interface to the World Wide Web.

Web Navigator Database   The Notes database that resides on the InterNotes server and allows users access to the Web.

Web Navigator Template   The Notes template that contains all the design elements needed to create the Web Navigator database.

Web Publisher   Generic term for a tool or software package that converts documents to HTML so that they can be viewed on the World Wide Web.

Web Server   A server on the Internet that stores the information accessible from a Web browser.

Web Site   The set of pages that comprise a person or organization's presence on the Web.

Webmaster   A person who manages a Web site; similar to the network administrator.

Wide Area Network   See *WAN*.

Workflow Application   An application that facilitates the flow of work among users on a network.

Workspace   The Notes "desktop," which includes Notes windows, menus, SmartIcons, and the six tabbed pages where you display database icons. You can organize your Workspace by naming the tabbed pages; by adding, moving, and removing icons; and by customizing your SmartIcon palettes.

World Wide Web   The World Wide Web, or WWW, is a wide-area hypermedia information retrieval initiative aiming to give universal access to a large universe of documents. To access the Web, you use a Web browser that provides hypertext links to jump to information on many different Internet Web servers.

X.25   CCITT standard for data transmission over a public data network. It was designed originally for connection of terminals to host computers. X.25 transports packets from point-to-point (via virtual circuits) over a WAN.

X.400   CCITT standard for message-handling services. To conform to X.400, client e-mail applications maintain their user interfaces but would change the file format of each e-mail message produced to conform to the X.400 standard. Correspondingly, they would each accept the X.400 format for incoming e-mail messages. The X.400 server would then handle the messages from any number of unique e-mail applications transparently.

X.500   CCITT standard that defines a file organization and interface to distributed directory data for network users and resources.

X.PC   See *XPC*.

XPC   XPC is the dial protocol that comes with every copy of the Lotus Notes software (client or server). This is the Lotus Notes name for its version of the X.PC protocol from which it was derived. X.PC was developed by Tymshare in 1983 from the X.25 protocol.

X Recommendations   The CCITT documents that describe data communication network standards. Well-known ones include X.25 Packet Switching Standard, X.400 Message Handling System, and X.500 Directory Services.

BIBLIOGRAPHY

The following books and articles give additional information relating to the Lotus Notes, Domino, and Lotus Notes and Domino Network Design.

Aterwola, T., et. al. "SP2 System Architecture", *IBM Systems Journal*, 34, No. 2., pp. 152-184, 1995. Describes design of IBM SP2 processor, which is the strategic Notes server platform for IBM's internal and commercial service offerings.

Bolin, Barbara A. and Ordonez, R. Benjamin. *Lotus Notes: The Complete Reference*, Osborne McGraw-Hill, Berkeley, CA, 1996. As the name implies, this book covers areas for Notes administrators, applications developers, and end users.

Brown, Kenyon; Brown, Kyle; Koutchouk Francois; and Brown, Kevin. *Mastering Lotus Notes 4*, SYBEX Inc., San Francisco, CA, 1996. All you ever wanted to know about Lotus Notes as an end user. Over 1000 pages.

Calabria, Jane and Plumley, Sue. *Lotus Notes 4.5 and the Internet, 6 in 1*, Que Corporation, Indianapolis, IN, 1997. This book is organized into six parts: Lotus Notes 4.5, Domino 4.5, Application Development, InterNotes Web Publisher, Lotus Notes 4.5 Web Navigator, and Domino.Action, hence its title. However, most of the book seems to be aimed towards the end user, rather than the Notes administrator or designer.

Comer, Douglas E. *Internetworking with TCP/IP*, vol. 1, 2nd ed., Prentice-Hall, Inc., Englewood Cliffs, NJ, 547 pp. Classic text covering the TCP/IP protocol suite in technical detail at the bits and bytes level. Covers Class A, B, and C network addresses; subnetting; the 7 OSI protocol layers; ARP; and routing. This is a book for the serious networking student.

Costales, Bryan; Allman, Eric; and Rickert, Neil. *sendmail*, O'Reilly & Associates, Inc., Sebastopol CA, 1993, 792 pp. Understanding sendmail is essential for anybody involved with e-mail. Sooner or later, you will need to send SMTP mail to somebody on your internal network or on the Internet. This book is over 1-1/2" thick, but you must read it. To paraphrase an old Chinese saying, "A journey of 1000 pages begins with the first page."

Covey, Steven. *The Seven Habits of Highly Effective People*, Simon & Schuster, New York, 1989. 358 pp. This book fosters "working together," which is the heart of Lotus Notes.

Falkner, Mike. *How to Plan, Develop, and Implement Lotus Notes in Your Organization*, John Wiley & Sons, Inc., New York, 1996. This book gives a broad overview on the obstacles often met during the implementation of Lotus Notes and how to overcome them.

Falkner, Mike. *Using Lotus Notes as an Intranet*, John Wiley & Sons, Inc., New York, 1997. This book describes the tools that Lotus provides for Domino and includes 25 Domino/intranet databases on a CD.

Flanagan, David. *Java in a Nutshell*, O'Reilly & Associates, Inc., Sebastopol CA, 1996. This is one of the most popular books on Java. Because it's compact and yet comprehensive, it's the kind of book you'll take with you in your knapsack so that you can look up that Java command no matter where you are!

Frisch, Aeleen. *Essential System Administration*, O'Reilly & Associates, Inc., Sebastopol CA, 1991. UNIX systems require administration of items such as user accounts and passwords, file system and disk space, permission to access directories and files, and permission to run programs. These are some the items covered in this book. Familiarity of UNIX system administration is necessary for a full understanding of UNIX security.

Garfinkel, Simpson. *PGP: Pretty Good Privacy*, O'Reilly & Associates, Inc., Sebastopol CA, 1995. 393 pp. How to use PGP to encrypt mail sent over the Internet. Includes a colorful history of RSA cryptography and the controversies surrounding its use.

Garfinkel, Simpson and Spafford, Gene. *Practical UNIX Security*, O'Reilly & Associates, Inc., Sebastopol CA, 1991. 482 pp. This is a UNIX security book, but many principals and practices apply to any operating system, such as password selection, modems, data backups, firewalls, and encryption.

Gillmor, Steve. "Microsoft Exchange Server Needs Work: Yours," *Byte*, pp. 70-71, July, 1996. This short article describes the July, 1996 state of Microsoft Exchange and how it compares to Lotus Notes and Web Technology. With the Windows 95/NT workstation platforms, Microsoft Exchange will be a formidable Notes competitor.

Hawker, Marion, et. al. *Lotus Notes Release 4.5: A Developer's Handbook*, IBM Corporation, International Technical Support Organization, Research Triangle Park, NC, 1996. This book is a member of IBM's "Redbook" group and was written by IBM and Lotus personnel from different parts of the world. It provides detailed application development information for both Lotus Notes and the integrated Notes HTTP server (Domino). The book includes chapters on LotusScript, Agents, Components, Security, NotesPump, and MQSeries.

IBM. *Enterprise Deployment of Lotus Notes*, IBM Technical White Paper, 1997. This white paper describes the Lotus Notes architecture deployed within IBM. It also describes the reasons for the design decisions made for this internal deployment of Lotus Notes. A copy of this paper can be obtained by calling IBM Global Services at 1-800-IBM-4YOU.

IBM Global Services. *ISSC Lotus Notes Operations Cookbook*, IBM Internal publication, 1996. This IBM internal publication details on how ISSC (now called IBM Global Services) operates IBM's Internal Lotus Notes servers.

Krantz, Steve. *Real World Client/Server*, Maximum Press, Gulf Breeze, FL, 1995. Interesting book that tells the story of IBM Boca Raton's migration to Client/Server, including Lotus Notes.

Krantz, Steve. *Building Intranets with Lotus Notes and Domino*, Maximum Press, Gulf Breeze, FL, 1997. This book includes information on IBM's internal rollout of Lotus Notes and Domino.

Kreisle, William and Schultz, Dan. *Lotus Notes in the Enterprise*, M&T Books, 1996. This book covers Notes 4.0, the API(s), and the "Internet and Lotus Notes."

Kreisle, Bill. *Lotus Notes 4.5*, MIS Press, New York, 1997, 568 pp. This was the first book published on Lotus Notes 4.5. It is written in a very "friendly" style.

Lamb, John P. and Lew, Peter W. *Lotus Notes Network Design*, McGraw-Hill, New York, 1996. 264 pp. This book is the predecessor to the book you're reading. This original book on Lotus Notes network design covered Release 3 and 4 of the product, with an emphasis on Release 3.

Lamb, John P. and Papritz, Ronald E. "Enterprise LAN Standards and Guidelines, A Case Study", *Proceedings of the 18th Conference on Local Computer Networks*, IEEE Computer Society, Minneapolis, Minnesota, pp. 214-221, September, 1993. Gives summary of IBM LAN Council standards.

Lamb, John and Cusato, Tony. "LAN-Based Office for the Enterprise, a Case Study", *Proceedings of the 19th Conference on Local Computer Networks*, IEEE Computer Society, Minneapolis, Minnesota, pp. 440-447, October, 1994. Describes architecture for LAN-based office applications developed for the IBM Corporation. Includes Lotus Notes as one of the applications.

Layland, Robin. "Is Your Network Ready for Notes?", *Data Communications*, April, 1995, pp. 83-89. Tips on using Notes over your network.

Lewis, Jamie. "LDAP Unites the Internet", *Byte*, December, 1996, pp. 121-126. This article discusses the state of the Lightweight Directory Access Protocol (LDAP), the Internet standard for directory services. LDAP, in turn, is a distillation of the X.500 directory protocol standard for directory and resources management. Because X.500 has proven itself too complex for most users tastes, LDAP is fast becoming the directory standard protocol for many platforms, including Lotus Notes. This article looks at LDAP in a very positive light.

Liu, Cricket and Albitz, Paul. *DNS and BIND*, O'Reilly & Associates, Inc., Sebastopol CA, 1992. Understanding Domain Names System (DNS) is essential for anybody involved with UNIX system administration or TCP/IP networking. It is absolutely necessary for anybody involved with SMTP mail because sendmail is reliant on DNS. The first few chapters provide such a clear overview of DNS that every end user of a TCP/IP network (including the Internet) would benefit from reading this book. The mystery of how computers find each other on a TCP/IP network is revealed!

Lotus Development Corporation. *Lotus Notes and Domino Release 4.5 Administrator's Guide*, Second Revision, 1996. The "manual" that goes with Lotus Notes and Domino.

Lotus Development Corporation. *Working with Lotus Notes and the Internet*, First Edition, 1996. Details on using Lotus Notes and Domino with the Internet.

Lotus Development Corporation. *Lotus Notes Knowledge Base*, online system from Lotus Notes Network (LNN), 1997. This online database was the source of much of the information on Lotus Notes products described in this book. It was also the source of much information for chapter 13 of this book on "Migration to Lotus Notes from Existing Mail Systems."

Lotus Development Corporation. *Domino 4.5 Server Doc Pack*, 1996. This "pack" from Lotus contains 10 documents: The Lotus Notes/Domino 4.5 Administrator's Guide, Working with Lotus Notes and the Internet, Deployment Guide, Migration Guide, Install Guide for Servers, Install Guide for Workstations, SMTP/MIME MTA, Database Manager's Guide, Network Configuration Guide, and Release Notes 4.5. This documentation is a bargain because the whole pack is only $30 plus taxes and shipping. Call "Lotus Selects" at 1-800-635-6887 and request the "Domino 4.5 Server Doc Pack", part # 029355, and they'll ship it to your home or office.

Marshak, David S. *Mission-Critical Lotus Notes*, Prentice Hall, New York, August, 1996. Marshak explores two vital questions in this book: What is Lotus Notes?; and What problems can Lotus Notes best solve?

Mathers, Barbara and Newbold, Dave. *Lotus Notes Internet Cookbook*, New Notes Knowledge Base (online system from Lotus Notes Network), 1996. This "online" document is in the form of a FAQ (Frequently Asked Questions) document and is updated frequently. The Cookbook provides key information about how to take advantage of Notes' existing, built-in Internet support and a directory of Lotus and third-party products designed to extend Notes' Internet support. The Cookbook is included on the CD-ROM for the book you're reading.

Roberts, Bill. "Groupwar Strategies", *Byte*, pp. 68-78, July, 1996. The "war" in this article is between Lotus Notes and Web Technology. The conclusion is that, at the time of the article, Lotus Notes is still far ahead in groupware, but the gap is closing.

Russell, Deborah and Gangemi Sr., G.T., *Computer Security Basics*, O'Reilly & Associates, Inc., Sebastopol CA, 1991. 448 pp. As the title suggests, the basics about computer security including selecting passwords, public key vs. private key cryptography, DES, RSA, and physical security. Emphasis on U.S. government requirements, such as C2 level security specifications.

Schulman, Mark. *Using Lotus Notes*, Que Corporation, Indianapolis, IN, 1994. Good end-user book in the Que tradition, but now out-of date because it discusses R3 of Notes.

Stallings, William. *Data and Computer Communications*, Second Edition, Macmillan Publishing Company, New York, 1988. Good reference book for detailed information on different types of computer networks

Tannenbaum, Andrew. *Computer Networks*, Second Edition, Prentice Hall, Englewood Cliffs, NJ, 1988. This book has become the classic textbook on computer network design based around the OSI seven-layer reference model.

Technology Investments, Inc. *Installation and User's Guide: Adaptive Replication Engine for Lotus Notes, Version 1.0*, 1995. Describes use of MQSeries as transport layer for Lotus Notes replication using a message queuing system.

Wainwright, Andrew, et. al. *Secrets to Running Lotus Notes: The Decisions No One Tells You How to Make*, IBM Corporation, International Technical Support Organization, Research Triangle Park, NC, 1996. This book is a member of IBM's "Redbook" group and was written by IBM and Lotus personnel from different parts of the world. As the name implies, it provides a lot of "hints and tips" on how to run Lotus Notes.

Walsh, Tony. *Lotus Notes Release 3 in the OS/2 Environment*, Van Nostrand Reinhold, New York, NY, 1995. Useful information on OS/2 Notes servers and the use of OS/2 with computer networks.

Weider, Chris and Reynolds, Joyce. "Executive Introduction to Directory Services Using the X.500 Protocol", *Request for Comments: RFC 1308*, Network Information Center, Menlo Park, CA, March, 1992. Gives a brief overview of the X.500 Protocol by authors from the Internet's "Network Working Group."

Weider, Chris; Reynolds, Joyce, and Heker, Sergio. "Technical Overview of Directory Services Using the X.500 Protocol", *Request for Comments: RFC 1309*, Network Information Center, Menlo Mark, CA, March, 1992. This RFC gives a technical overview of the X.500 Protocol and also compares and contrasts Directory Services based on X.500 with several of the other Directory services currently in use in the Internet.

Yavin, David. "Replication's Fast Track", *Byte*, pp. 88A-90, August, 1995. Information on what to watch out for with Notes replication. Includes Release 4 vs. Release 3 replication comparisons.

Summary of IBM Redbooks on Lotus Notes and Domino

IBM Redbooks, named for their red covers, are "how to" books, written by very experienced IBM professionals from all over the world. They are written at the International Technical Support Organization (ITSO), which has centers close to IBM's development divisions. The following are IBM Redbooks related to Lotus Notes and Domino. IBM Redbooks can be ordered on the Web at http://www.redbooks.ibm.com using the Redbook number indicated here.

SG24-4875, *Secrets to Running Lotus Notes: The Decisions No One Tells You How to Make.* An excellent guide to planning and deploying a Lotus Notes and Domino implementation. Covers everything from the choice of hardware to backup and recovery to training and support. You can use this as a step-by-step project plan.

SG24-4848, *The Domino Defense: Security in Lotus Notes and the Internet.* As the name describes, this Redbook covers the details of the security features of Lotus Notes and Domino. I like the diagrams that show how challenge/authentication works, how SSL works, and how firewalls use different TCP/IP ports.

SG24-4694, *Lotus Domino Server Release 4.5 on AIX Systems.* Lotus Notes and Domino look and function the same regardless of operating system from the end-user viewpoint. Of course, there installation and administration differences unique to the particular operating system. This Redbook has many hints and tips for installing and managing the Domino/Notes server on AIX, IBM's version of UNIX. Useful scripts will help automate administrator tasks.

SG24-4876, *Lotus Notes Release 4.5: A Developer's Guide.* This Redbook uses examples and case studies to illustrate how to use LotusScript,

LotusScript Extensions, Lotus Components, and program for the integrated Notes/Web server. Of special interest is how to access relational databases, how to use NotesPump, MQSeries, and the Notes C++ API.

SG24-4856 *LotusScript for VisualBasic Programmers*

SG24-4837 *Lotus Notes: An Enterprise Application Platform*

SG24-4779 *Using Lotus Notes on the IBM Integrated PC Server for AS/400*

SG24-4977 *Mail Integration for Lotus Notes 4.1 on the IBM Integrated PC Server for AS/400*

SG24-4649 *Lotus Notes Release 4 in Multiplatform Environment*

SG24-4756 *Local Area Network Concepts & Products: LAN Operation*

SG24-4512 *Using the IBM CICS Gateway for Lotus Notes*

SG24-4851 *Image and Workflow Library: Integrating IBM FlowMark with Lotus Notes*

SG24-4741 *Implementing LAN Server for MVS in a Lotus Notes Environment*

SG24-4811 *Enterprise Calendaring and Scheduling with Lotus Notes: The Notes to OfficeVision Connection*

SG24-4657 *IBM PC Server and Lotus Notes Integration Guide*

SG24-2102 *IBM PC Server and Lotus Domino Integration Guide*

SG24-4335 *Using ADSM to Back Up Databases*

SG24-4808 *NetFinity V5.0 Database Support (pending release)*

SG24-4865 *An Approach to ODBC: Lotus Approach to DB2*

INDEX

SOFTWARE AND INFORMATION LICENSE

The software and information on this diskette (collectively referred to as the "Product") are the property of The McGraw-Hill Companies, Inc. ("McGraw-Hill") and are protected by both United States copyright law and international copyright treaty provision. You must treat this Product just like a book, except that you may copy it into a computer to be used and you may make archival copies of the Products for the sole purpose of backing up our software and protecting your investment from loss.

By saying "just like a book," McGraw-Hill means, for example, that the Product may be used by any number of people and may be freely moved from one computer location to another, so long as there is no possibility of the Product (or any part of the Product) being used at one location or on one computer while it is being used at another. Just as a book cannot be read by two different people in two different places at the same time, neither can the Product be used by two different people in two different places at the same time (unless, of course, McGraw-Hill's rights are being violated).

McGraw-Hill reserves the right to alter or modify the contents of the Product at any time.

This agreement is effective until terminated. The Agreement will terminate automatically without notice if you fail to comply with any provisions of this Agreement. In the event of termination by reason of your breach, you will destroy or erase all copies of the Product installed on any computer system or made for backup purposes and shall expunge the Product from your data storage facilities.

LIMITED WARRANTY

McGraw-Hill warrants the physical diskette(s) enclosed herein to be free of defects in materials and workmanship for a period of sixty days from the purchase date. If McGraw-Hill receives written notification within the warranty period of defects in materials or workmanship, and such notification is determined by McGraw-Hill to be correct, McGraw-Hill will replace the defective diskette(s). Send request to:

Customer Service
McGraw-Hill
Gahanna Industrial Park
860 Taylor Station Road
Blacklick, OH 43004-9615

The entire and exclusive liability and remedy for breach of this Limited Warranty shall be limited to replacement of defective diskette(s) and shall not include or extend to any claim for or right to cover any other damages, including but not limited to, loss of profit, data, or use of the software, or special, incidental, or consequential damages or other similar claims, even if McGraw-Hill has been specifically advised as to the possibility of such damages. In no event will McGraw-Hill's liability for any damages to you or any other person ever exceed the lower of suggested list price or actual price paid for the license to use the Product, regardless of any form of the claim.

THE McGRAW-HILL COMPANIES, INC. SPECIFICALLY DISCLAIMS ALL OTHER WARRANTIES, EXPRESS OR IMPLIED, INCLUDING BUT NOT LIMITED TO, ANY IMPLIED WARRANTY OF MERCHANTABILITY OR FITNESS FOR A PARTICULAR PURPOSE. Specifically, McGraw-Hill makes no representation or warranty that the Product is fit for any particular purpose and any implied warranty of merchantability is limited to the sixty day duration of the Limited Warranty covering the physical diskette(s) only (and not the software or information) and is otherwise expressly and specifically disclaimed.

This Limited Warranty gives you specific legal rights; you may have others which may vary from state to state. Some states do not allow the exclusion of incidental or consequential damages, or the limitation on how long an implied warranty lasts, so some of the above may not apply to you.

This Agreement constitutes the entire agreement between the parties relating to use of the Product. The terms of any purchase shall have no effect on the terms of this Agreement. Failure of McGraw-Hill to insist at any time on strict compliance with this Agreement shall not constitute a waiver of any rights under this Agreement. This Agreement shall be construed and governed in accordance with the laws of New York. If any provision of this Agreement is held to be contrary to law, that provision will be enforced to the maximum extent permissible and the remaining provisions will remain in force and effect.